# 3rd CCPS Global Summit on Process Safety 2016



Dammam, Saudi Arabia
4 – 5 December 2016

ISBN: 978-1-5108-4167-3

**Printed from e-media with permission by:**

Curran Associates, Inc.
57 Morehouse Lane
Red Hook, NY 12571

**Some format issues inherent in the e-media version may also appear in this print version.**



Printed by Curran Associates, Inc. (2017)

For permission requests, please contact AIChE
at the address below.

AIChE
120 Wall Street, FL 23
New York, NY 10005-4020

Phone: (800) 242-4363
Fax: (203) 775-5177

www.aiche.org

**Additional copies of this publication are available from:**

Curran Associates, Inc.
57 Morehouse Lane
Red Hook, NY 12571 USA
Phone: 845-758-0400
Fax: 845-758-2633
Email: curran@proceedings.com
Web: www.proceedings.com

# TABLE OF CONTENTS

# 3rd Global Summit Welcome

Shakeel H. Kadri
Executive Director, CCPS
4 December 2016

"The Global Community Committed to Process Safety"

# Chemical Engineering & Process Safety

- What good we bring to the many?
- What harm we bring to some?
- The harm all remember.....

"The Global Community Committed to Process Safety"

2

## CCPS 30th Anniversary Film

"The Global Community Committed to Process Safety"

## The Harm All Remember…

# The Bhopal disaster

"The Global Community Committed to Process Safety"

## The Bhopal Disaster...

"The Global Community Committed to Process Safety"

## Bhopal -- Lessons Learned

- Bhopal is the worst process safety incident that's ever occurred in the chemical industry
- It served as a bellwether event for the industry and a catalyst for a safety reform
- Bhopal has lead to improved process safety practices worldwide.
- Much more focus on consequence analysis, off-site risk and risk tolerance criteria
- AIChE was asked to create a Safety Center --- Center of Chemical Process Safety [CCPS] – to lead a collaborative effort to eliminate catastrophic process incidents

"The Global Community Committed to Process Safety"

## About CCPS

- CCPS was formed on 23 March 1985
- Not for profit organization supported by Corporate Members globally
- It is an Industrial Technology Alliance that was founded by--and is part of--AICHE.
- It's mission is the elimination of catastrophic process safety incidents.
- It is a global organization has 196 Corporate members, 40% of the members are outside Americas
- It's headquarter is in New York City, with offices in Frankfurt, Mumbai, Singapore, Ningbo [China] and Houston.

"The Global Community Committed to Process Safety"

## CCPS Vision

An AIChE Technology Alliance

CCPS

Center for Chemical Process Safety

**"To protect people, property and the environment by bringing the best process safety knowledge and practices to industry, academia, the governments and the public around the world through collective wisdom, tools, training and expertise."**

"The Global Community Committed to Process Safety"

# CCPS Mission

Eliminate catastrophic process incidents globally by:

- Advancing global PS technologies, culture, and management practices
- Establishing Process Safety as foundation for responsible operation
- Serving as premier worldwide resource of Process Safety
- Fostering knowledge and understanding of Process Safety
- Promoting Process Safety as key societal value and expectation

# Leading Process Safety since 1985

CCPS Certified
**CCPSC**

Creating Books and Publications

Creating Industry-wide Tools, Programs and Guidelines

Sharing Best Practices

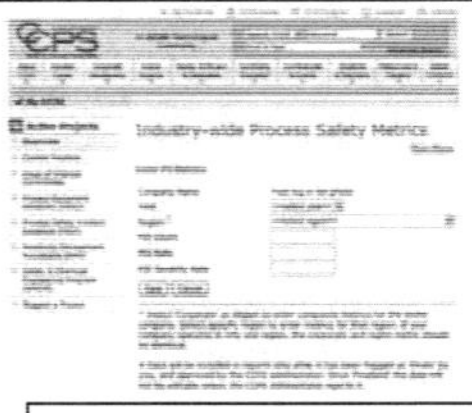


Educating Educators

Process Safety Beacon

Conducting Global Conferences and Training

"The Global Community Committed to Process Safety"

CCPS Center for Chemical Process Safety

## Center of Chemical Process Safety

### Responsible Collaboration

Chemical Safety Board [CSB]
US Environmental Protection Agency [EPA]
US Occupational Safety and Health Administration [OSHA]
Contra Costa Health Service [CCHC]
Energy Institute (UK)
Society of Petroleum Engineers (SPE)
Synthetic Organic Chemical Manufacturers Assoc. (SOCMA)
American Chemical Council (ACC)
API [American Petroleum Institute]
AFPM [American Fuel and Petroleum Manufacturing]
ABQUIM [The Brazilian Association of Chemical Industry
OISD [Oil Industry Safety Directorate] - India
SCIC [Singapore Chemical Industry Council]
JSSE [Japan Society for Safety Engineering]
European Process Safety Center [EPSC]
IChemE Safety Center [ISC]
Mary Kay O'Conner Process Safety Center [MKO]

11

CCPS Center for Chemical Process Safety

## CCPS Global Membership Today

### 2016 Geographical distribution of 196 member companies [Oct 2016]

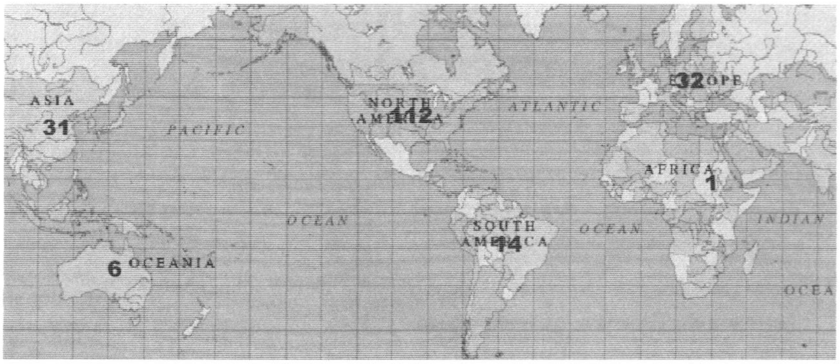


"The Global Community Committed to Process Safety"

## Increasing Global Engagement
## CCPS Regional TSCs Formed in 2016

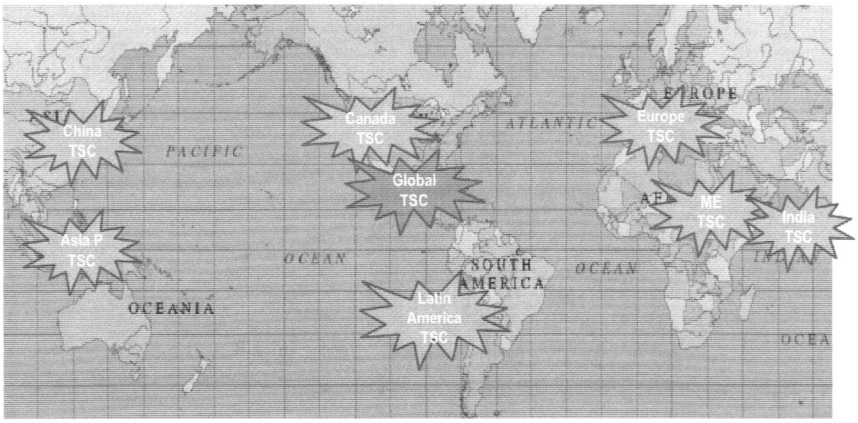


**Seven [7] CCPS Regional TSCs Formed**

## CCPS Global / Regional Conferences "Plan"`

| North America GCPS | Latin America | Europe | Asia-Pacific-Oceania | Middle East & Africa |
|---|---|---|---|---|
| 12th | Peru 2016 | Germany 2016 | China 2016 | Saudi Arabia 2016 |
| 13th | Trinidad & Tobago 2017 | WCCE Spain 2017 | China, Japan 2017 | UAE 2017 |
| 15th | Brazil / Mexico 2018 | Czech Republic 2017 | China, Thailand 2018 | Nigeria 2018 |
| 15th | Columbia 2019 | Europe 2018/19 | China, S. Korea 2019 | UAE 2019 |

*"The Global Community Committed to Process Safety"* 14

## Credentialing Launched
## CCPS Certified [CCPSC]

| CCPSC Metric | October 2016 |
|---|---|
| Applications open | ~ 700 |
| CCPS Certified todate | ~ 100 |
| Exams offered in a year | 3 times |

## Undergraduate Process Safety Education

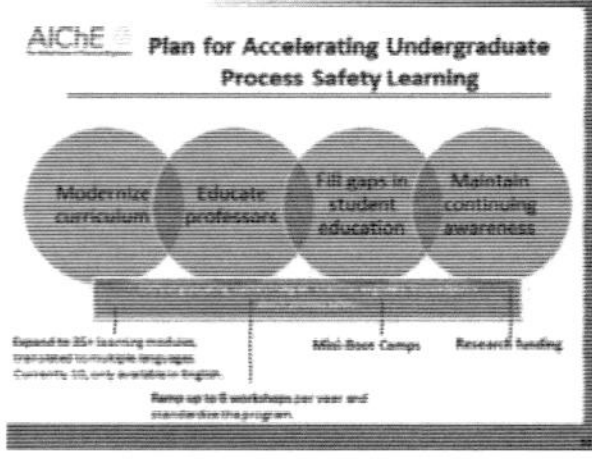

**Undergraduate Process Safety Education Program is managed effectively by the AIChE cross-functional Project Team**

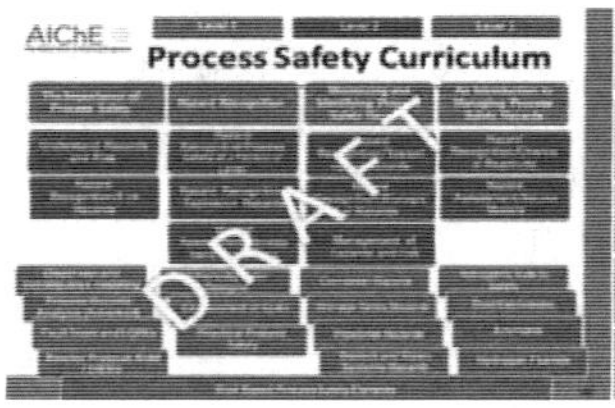


| 2016 Faculty Workshops | Planned attendees As June 10, 2016 |
|---|---|
| Dow | 27 |
| ADM | 23+ |
| Cargill | 17+ |
| Chevron | 30 |

# GSPS Program at a Glance

| 4th December | | 5th December | |
|---|---|---|---|
| 08:00-08:20 | CCPS Welcome | 08:00-08:20 | SABIC Welcome |
| 08:20-08:30 | Conference Opening | 08:20-08:30 | Leadership Forum Introduction |
| 08:30-09:00 | Inauguration by Chief Guest- CEO, SABIC | 08:30-09:00 | Summary of Leadership Forum on Process Safety |
| 09:00-09:30 | **Keynote – 1** | 09:00-09:30 | **Keynote – 3** |
| 09:30-10:00 | **Keynote – 2** | 09:30-10:00 | **Keynote – 4** |
| 10:00-10:30 | Break / Poster Session | 10:00-10:30 | Break / Poster Session |
| 10:30 - Noon | **Leadership Panel** | 10:30 - Noon | **Academia Panel** |
| Noon-13:00 | Lunch Break / Poster Session | Noon-13:00 | Lunch Break / Poster Session |
| 13:00-17:00 | **Leadership Forum [Closed Door]** | | |
| 13:00 – 14:30 | Technical Session 1A, 1B and 1C | 13:00 – 14:30 | Technical Session 3A, 3B and 3C |
| 14:30- 15:00 | Break / Poster Session | 14:30- 15:00 | Break / Poster Session |
| 15:00 – 16:30 | Technical Session 2A, 2B and 2C | 15:00 – 16:30 | Technical Session 4A, 4B and 4C |
| 16:30 | Day 1 End | 16:30 | Concluding Remarks CCPS and SABIC |
| 17:00-19:00 | Evening Reception | 17:00 | Conference End |

*The Global Community Committed to Process Safety*

17

# Key Program Changes

1. Keynote by Mr. Shaikh [1st Day]
   – Title change to: "Process Safety Top Down Approach".
2. Keynote by Ms. Vanessa Sutherland [2nd Day]
   – Keynote address by WebEx [Ms. Vanessa could not get KSA visa]
3. Leadership Panel [1st Day]
   – Replacing Ms. Vanessa by Mr. VV Surya Rau of Reliance
4. Academia Panel [2nd Day]
   – Replacing Ms. Kristen [who also did not receive the visa] by Shakeel Kadri

❖ Our apologies for a few last minute unavoidable change

3RD CCPS GLOBAL SUMMIT ON PROCESS SAFETY

*The Global Community Committed to Process Safety*

18

3RD CCPS GLOBAL SUMMIT ON PROCESS SAFETY

**MAIN SPONSOR**

سابك
sabic

**PLATINUM SPONSOR**

صدارة
Sadara

**GOLD SPONSORS**

AcuTech
PROCESS RISK MANAGEMENT

PETROTECHNICS
*Excellence is now operational*

**SILVER SPONSOR**

TÜV SÜD

Kienbaum
*Inspiring. Different.*

**BRONZE SPONSORS**

ABS Group
Saudi Arabia Company Ltd.

DUPONT

ioMosaic
*Minimizing risk. Maximizing potential.*

CCPS
An AIChE Technology Alliance
Center for Chemical Process Safety

CCPS
An AIChE Technology Alliance
Center for Chemical Process Safety

3RD CCPS GLOBAL SUMMIT ON PROCESS SAFETY

سابك
sabic

# 3rd Global Summit

## THANKS!

"The Global Community Committed to Process Safety"

# PROCESS SAFETY TOP DOWN APPROACH

December, 4th 2016 – Dammam

CHEMISTRY THAT MATTERS™

## HAZARD AWARNESS

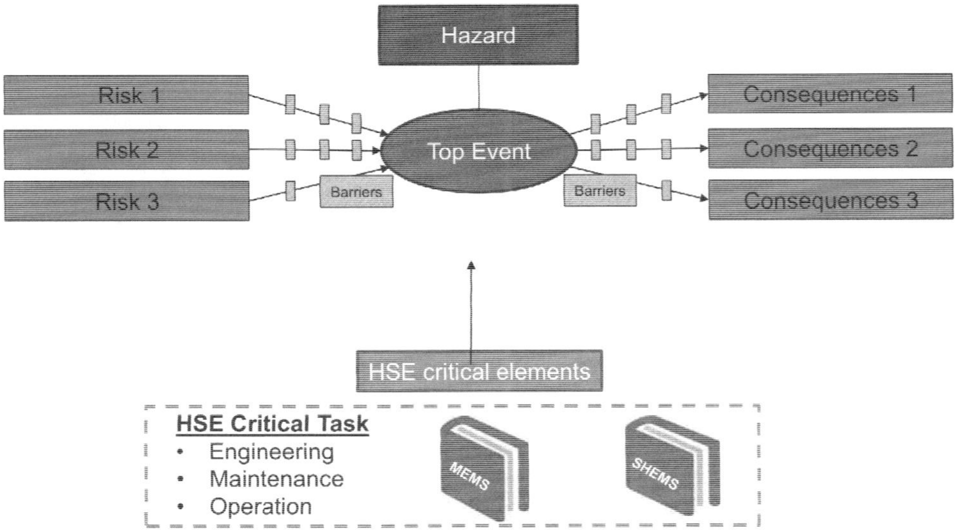


No. 1

# ASSET INTEGRITY – SUMMARIZING KEY ELEMENTS

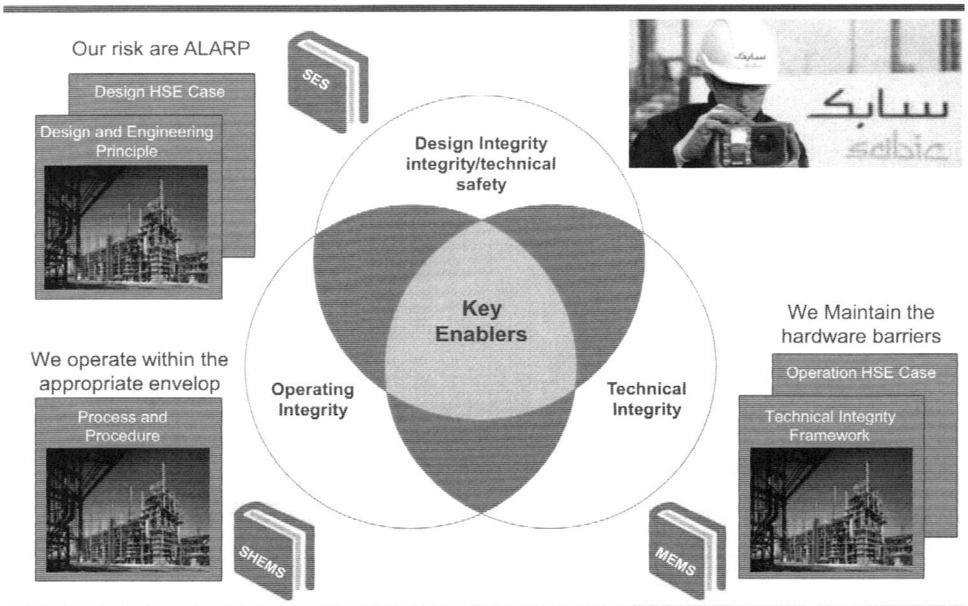


No. 2

## A Case Study in Improving Process Safety Culture Using the CCPS PSM Culture Guidelines

CCPS 3rd Global Summit on Process Safety
December 4-5 2016
Dammam, Kingdom of Saudi Arabia

David A. Moore, PE, CSP, CPSA
President and CEO
AcuTech Consulting Group
Washington DC | Houston | Philadelphia
Dubai | Shanghai | Mumbai
www.acutech-consulting.com



## Outline

- Need for improved PSM culture to improve overall PSM performance
- CCPS PSM Culture Guidelines Overview
  - Definitions
  - Models
- Example of the use of the CCPS PSM Culture Principles for analyzing PSM culture
- Conclusions and Q&A

## Process Safety Management (PSM) Performance

- Process Safety Management (PSM) is a "blend of engineering and management skills focused on preventing catastrophic accidents, particularly explosions, fire and toxic releases associated with the use of chemicals and petroleum products" (AIChE, CCPS, 2010)
- The complexity of PSM makes for a difficult assignment to maintain excellence in PSM performance
- Unfortunately we remain challenged by a number of significant Process Safety Events
- Up to this point most companies have focused on the technical side of PSM.
- We believe enhancing PSM culture is key to a potential breakthrough in more effective program results.

## Process Safety Evolution

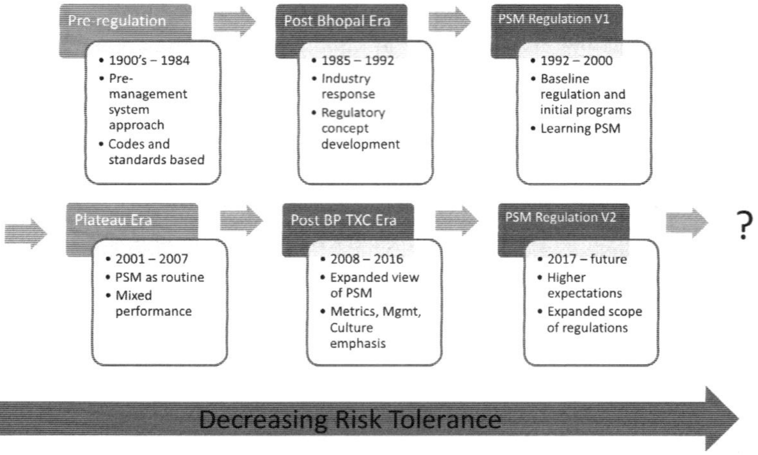

## Challenge - Improvement of Performance in the Management of Risk

AcuTech PROCESS RISK MANAGEMENT

- PSM Culture is facilitated by:
  - collaborative effort to learn, share and address issues
  - an operating discipline
  - a mindset of improvement and operational excellence
  - Committed leadership
  - Guidance on how to accomplish

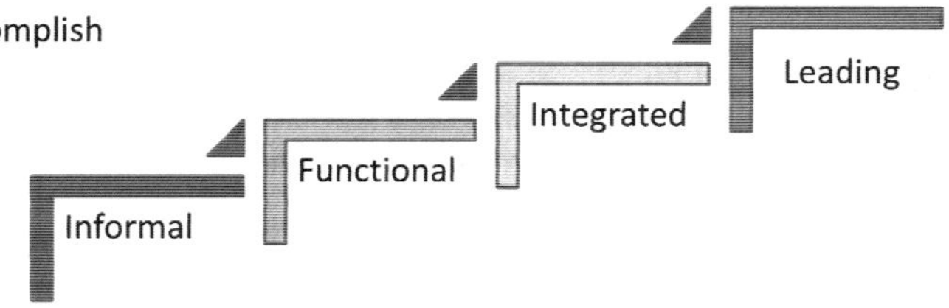




5

AcuTech PROCESS RISK MANAGEMENT

## Barriers to Successful Performance

| Barrier | Effect |
|---|---|
| Without dedicated management commitment | PSM will not be effective |
| Without the right PSM culture | There will be barriers to excellence |
| Without specific advice and guidance on how to facilitate an effective PSM culture | Even committed managers will not be effective to make a difference |
| Without a strategic plan to improve PSM culture | The PSM culture will not improve |
| Without recognizing success in occupational safety and health culture is not a guarantee of PSM culture | PSM culture will not improve since they involve different subjects and emphasize different necessary behaviors |



6

## Benchmarking and Goal Setting to to Successful Performance

- PSM culture should be analyzed and then monitored for status, or it will be difficult to improve – structure the issue to strategizeimprovement.
- AcuTech uses a maturity scale as a way to define relative performance of the entire system and to set goals for improvement, including for PSM Culture

| Element Score | Description of Element Implementation and Functionality | Maturity Level |
|---|---|---|
| Above 90% | The element demonstrates best practice performance | Leading |
| 80% to 90% | The element is implemented and functional | Integrated |
| 60% to 80% | The element is implemented and partly functional | Functional |
| Below 60% | The element is not implemented and not functional | Informal |





## Lessons Learned From PSM Audits

From many PSM audits we have conducted, it is clear that:

1. Management commitment is a clear indicator of success and is often obtuse or missing
2. Poor implementation of PSM is always a failure of leadership
3. A good documented management system does not guarantee good PSM performance
4. PSM excellence = business excellence
5. Culture is one of the least understood and receives the least focus and lowest scores in many cases of all the elements
6. Yet, the underlying PSM culture of the organization determines the effectiveness of the program

| Element – (element weight) | % |
|---|---|
| Applicability (2) | 50.0% |
| Process Safety Culture (2) | 38.9% |
| Compliance with Standards (1) | 25.0% |
| Process Safety Competence (1) | 60.4% |
| Employee Participation/Trade Secrets (2) | 17.2% |
| PKM/Process Safety Information (3) | 70.7% |
| HIRA/Process Hazard Analysis (3) | 52.6% |
| Operating Procedures (3) | 64.3% |
| Safe Work Practices (2) | 90.3% |
| Asset Integrity/MI (3) | 95.7% |
| Contractors (2) | 62.7% |
| Training (2) | 62.8% |
| Management of Change (3) | 71.3% |
| Operational Readiness/PSSR (2) | 35.7% |
| Conduct of Operations (1) | 50.0% |
| Emergency Planning & Response (2) | 86.5% |
| Incident Investigation (1) | 61.0% |
| Metrics and Measurement (1) | 40.2% |
| Audits (1) | 64.0% |
| Management Review (2) | 50.0% |
| **Overall Weighted Score** | 70.4% |

## CCPS 20/20 Vision

- CCPS, in its 20/20 Vision project document, stated that a committed culture consists of three key aspects (McCavit):
  - **Felt leadership from senior executives** - Felt leadership means more than a periodic mention of process safety in speeches and town hall meetings. It means that the executives are personally involved in process safety activities.
  - **Operational discipline** - the performance of all tasks correctly every time.
  - **Maintaining a sense of vulnerability** – essential to motivate and to set priorities.

Acuity is a registered trademark of AcuTech
Patent Pending



## CCPS PSM Culture Guidelines

- Process safety management (PSM) has been focused primarily on technical and management elements that must be implemented as essential to managing the risks of catastrophic releases of highly hazardous chemicals.
- Many companies have not focused on the underlying PSM culture necessary to facilitate the management system for it to be effective.
- CCPS recognized this and formed a committee to understand the issues and recommend an approach.
- This culminated in 2014 in the start of a new CCPS guideline project "Essential Practices for Developing, Strengthening and Implementing Process Safety Culture – Project # 249".
- AcuTech is preparing the guidelines to be released in 2017.

## CCPS PSM Culture Guidelines

- A significant amount of literature addresses working culture in industry, however, little literature exists which is solely dedicated to the process safety culture in the processing and allied industries.
- This book itemizes *"essential practices,"* to assess and strengthen process safety culture, but it is not expected to provide the comprehensive guidance for defining detailed management systems, as is often provided in a CCPS "guidelines" book.





## Organization of the Book

- The book covers the evolution of process safety culture by examining key incidents that have shaped the process safety culture discussion and examining definitions that have been proposed over the years.
- A key aspect of the book is the development of a model of ten Core Principles.
- These principles are described in detail, including means to ascertain the status of the principles. Practical applications will then be presented for each principle. The guideline will be organized into the following chapters:
- Introduction – including a discussion of Process Safety Culture definitions and a review of key incidents that have impacted the culture discussion.
  - Process Safety Core Principles
  - Leadership for Process Safety Culture within the Organizational Structure
  - Process Safety Culture Applications – organized by the Core Principles
  - Continuum - this chapter will discuss sustaining process safety culture improvements and maintaining a good culture. Warning signs of cultural degradation and remedies will be discussed

## CCPS Guidelines Definition of PSM Culture

- ***The pattern of shared written and unwritten attitudes and behavioral norms that positively influence how a facility or company collectively supports the development of and successful execution of the management systems that comprise its process safety management program, resulting in the prevention of process safety incidents.***







## Core Principles and Relation to RBPS Framework

- AcuTech mapped key PSM culture references and developed a new CCPS model for essential elements of a PSM culture initiative. These included the prior publications of the CCPS Core PSM Culture Principles based on the Columbia Space Shuttle Investigation (Kadri/Jones), the PSM culture goals of the CCPS Vision CCPS 2020 statement (CCPS Vision 2020), and the 10 key principles from the Culture Chapter of the RBPS Book (Chapter 3, 2007).
- These formed the basis of a holistic model based on ten Core Principles:
  1. Establish an Imperative for Safety
  2. Provide Strong Leadership
  3. Maintain a Sense of Vulnerability
  4. Understand and Act Upon Hazards/Risks
  5. Empower Individuals to Successfully Fulfill their Safety Responsibilities
  6. Defer to Expertise
  7. Ensure Open and Frank Communications
  8. Foster Mutual Trust
  9. Combat the Normalization of Deviance
  10. Learn to Assess and Advance the Culture

## Typical Approach to Process Safety Excellence

- Ordinarily follows a classic Deming Plan-Do-Check-Act approach for the development, implementation and management of multiple tasks and disciplines to manage risk:
  - **Plan**
    - Documenting the management system
    - Defining requirements and objectives
    - Regulatory, Internal, Industry Best Practices
  - **Do**
    - Implementation of the system
  - **Check**
    - Audits, risk assessments, internal review
    - Checklists
    - Key performance indicators review
  - **Act**
    - Management review
    - Setting of improvement objectives
    - Monitoring progress on implementation or action items

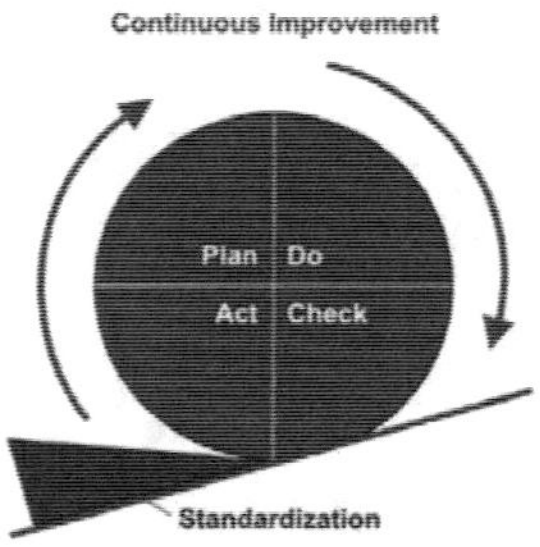






## Typical Approach to PSM Culture Assessment and Improvement

**Task 1**
***Awareness*** *- Perform a Culture Survey Prior to PSM Mgmt Systems Assessment*

→

**Task 2**
***Assessment*** *- PSM and Culture Assessment (Interviews, Observations, and Data Review)*

→

**Task 3**
***Analysis*** *– PSM and Culture Assessments Analysis and PSM Program Improvement Plan*

→

**Task 4**
***Enhancement*** *– Support the Ongoing Improvement of PSM Culture*

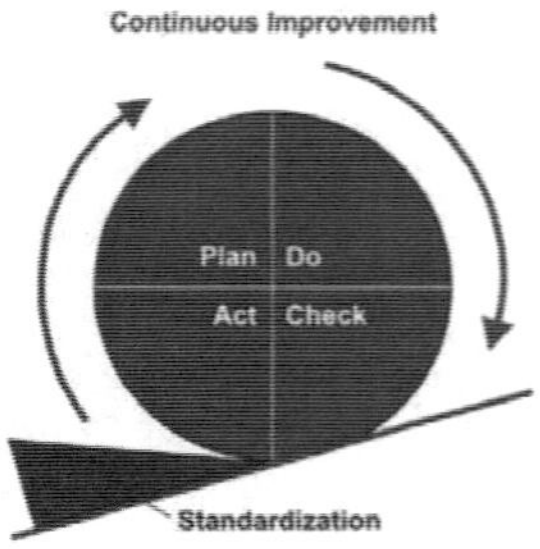

## Implementation of Process Safety Improvement

| Steps | Key Points |
|---|---|
| *Develop a safety strategy* | Goals, aspirations, and metrics together within a defined framework. |
| *Perform a baseline assessment* | Know where culture is starting |
| *Create clarity of purpose* | The members of the organization should be trained and organized |
| *Create the right safety climate* | Commitment, caring, cooperation, and coaching |
| *Create the right safety chemistry* | Passion, focus, expectations, proactive accountability, reinforcement, vulnerability, communication, measurement, and trust |
| *Control the PSM program* | Plan modifications to improve PSM Culture and prioritize. |
| *Continuously improve the PSM Culture* | Sustainability of the process |





## Case Study in PSM Culture Improvement - Refinery

- AcuTech consulted on PSM culture improvement to a refinery using the CCPS methodology.
- Using the 10 CCPS Principles as a basis, both positive and negative findings were made that will be important to understand and improve the PSM culture:
  - It was important to survey first to provide a baseline understanding and to give direction to the efforts.
  - The interviews and field observations validated the survey results and in some cases were necessary to clarify divergent opinions.
  - The identified areas of improvement may be more complex to solve than typical technical PSM issues.
  - The survey needed to differentiate between the facility, group, and individual opinions since there isn't only one culture in a company, but a number of subcultures working under a common group culture.

## Management/Operational Discipline Questions

37. Operators are not asked to operate a process that is in an unsafe condition?

| Strongly Agree (5) | Agree (4) | Neutral (3) | Disagree (2) | Strongly Disagree (1) | N/A |
|---|---|---|---|---|---|
| O | O | O | O | O | O |

38. Disabled or failed safety/protective devices are restored to service as soon as possible?

| Strongly Agree (5) | Agree (4) | Neutral (3) | Disagree (2) | Strongly Disagree (1) | N/A |
|---|---|---|---|---|---|
| O | O | O | O | O | O |

39. The process equipment is properly maintained and safe to operate, e.g., the number of overdue inspection, testing, and preventive maintenance tasks is minimized?

| Strongly Agree (5) | Agree (4) | Neutral (3) | Disagree (2) | Strongly Disagree (1) | N/A |
|---|---|---|---|---|---|
| O | O | O | O | O | O |





## Team Level Subculture Behavioral Questions

34. My team and I are encouraged to ask challenging questions about process safety concerns without fear of reprisal? *(That is, we trust the motivations and behaviors of each other; and have confidence that a just system exists where honest errors can be reported without fear of reprisal.)*

| Strongly Agree (5) | Agree (4) | Neutral (3) | Disagree (2) | Strongly Disagree (1) | N/A |
|---|---|---|---|---|---|
| O | O | O | O | O | O |

35. My team and I follow our procedures and treat them as mandatory requirements? *(For example, procedures needed to do my job well, like SOPs, Permit to Work, LOTO, and other Safe Work Practices are understood and consistently followed)*

| Strongly Agree (5) | Agree (4) | Neutral (3) | Disagree (2) | Strongly Disagree (1) | N/A |
|---|---|---|---|---|---|
| O | O | O | O | O | O |

36. We do not allow exceedance of established safe operating limits?

| Strongly Agree (5) | Agree (4) | Neutral (3) | Disagree (2) | Strongly Disagree (1) | N/A |
|---|---|---|---|---|---|
| O | O | O | O | O | O |

## Individual Behavioral Questions

46. I believe that when reporting a personal safety incident or near miss, the focus is to understand what happened and to learn from it, rather than assign blame & punishment?

| Strongly Agree (5) | Agree (4) | Neutral (3) | Disagree (2) | Strongly Disagree (1) | N/A |
|---|---|---|---|---|---|
| O | O | O | O | O | O |

47. If I were to have a workplace injury, even a first aid case, I would report it as there are no barriers to reporting injuries?

| Strongly Agree (5) | Agree (4) | Neutral (3) | Disagree (2) | Strongly Disagree (1) | N/A |
|---|---|---|---|---|---|
| O | O | O | O | O | O |

48. I believe that personal safety & health concerns are listened to and addressed in a timely fashion?

| Strongly Agree (5) | Agree (4) | Neutral (3) | Disagree (2) | Strongly Disagree (1) | N/A |
|---|---|---|---|---|---|
| O | O | O | O | O | O |





## PSM Culture Survey Results

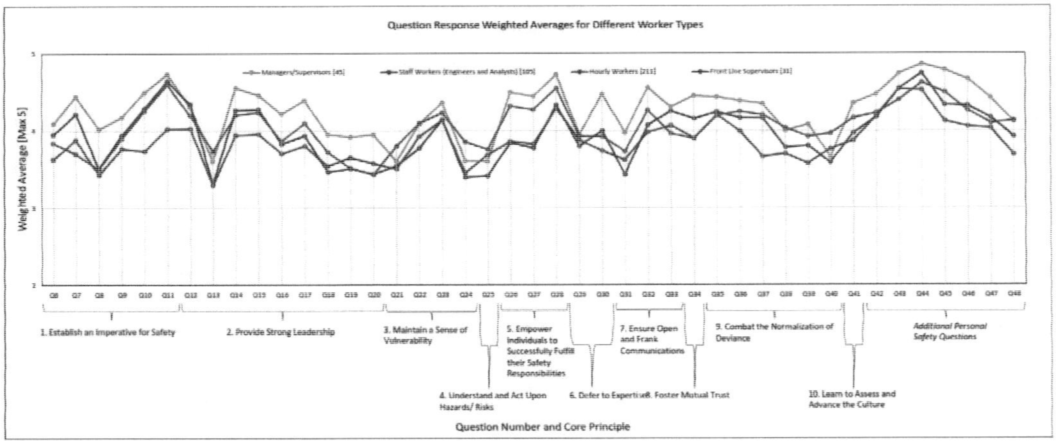

## Case Study in PSM Culture Improvement - Refinery

- AcuTech consulted on PSM culture improvement to a refinery using the CCPS methodology
- Using the 10 CCPS Principles as a basis, both positive and negative findings were made that will be important to understand and improve the PSM culture





## Case Study in PSM Culture Improvement - Refinery

- Provide Strong Leadership – Issues identified included:
  - Rate of Personnel Movement without adequate assurance of process safety competency:
    - Engineers may not have adequate experience in their positions to handle their responsibilities
    - As managers change, there is a reprioritization of safety program directions and loss of momentum of past initiatives.
    - The rapid advancement of operators resulting in some less experienced operators working in positions of authority (lead outside operator, responsibilities for safe work permitting, etc.)

## Case Study in PSM Culture Improvement - Refinery

- Combat the Normalization of Deviation: A number of examples of Normalization of Deviation were identified by various interview groups. Among these were:
  - Field issues such has tolerance of minor leaks.
  - Toxic detection system alarms that were in need of repair.
  - Numerous nuisance alarms and standing alarms on the process control system.
  - Operating procedures not being finalized as a practice that has become normalized.
  - Lack of completeness of safe work permits.



25

## Benefits to the Organization

- The reward for global culture leadership is clear. CEOs and executives who can define their culture and manage its strengths and weaknesses create better places to work, and they make more money.
- Kotter and Heskett at Harvard produced four major findings in *Corporate Culture and Performance* (Free Press, 1992). These four corporate culture tenets still hold true today:
  - Corporate culture can have significant impact on long-term financial performance.
  - Culture will become an even more important factor in determining corporate success or failure in the future.
  - Cultures that inhibit long-term financial strength are common and develop easily, even in companies full of reasonable and bright people.
  - Corporate culture can be managed and changed.
- Healthy culture is essential to effectiveness and business sustainability.

26

Reliance
Industries Limited
Growth is Life

# Process Safety Management & Leadership Challenges

Vivek Bichave
RELIANCE INDUSTRIES LIMITED.

# Contents

- Reliance Overview
- Introduction to PSM
- Leadership & Challenges
- Process Safety leadership 2020
- Safety Culture excellence Model

1

# Group Overview

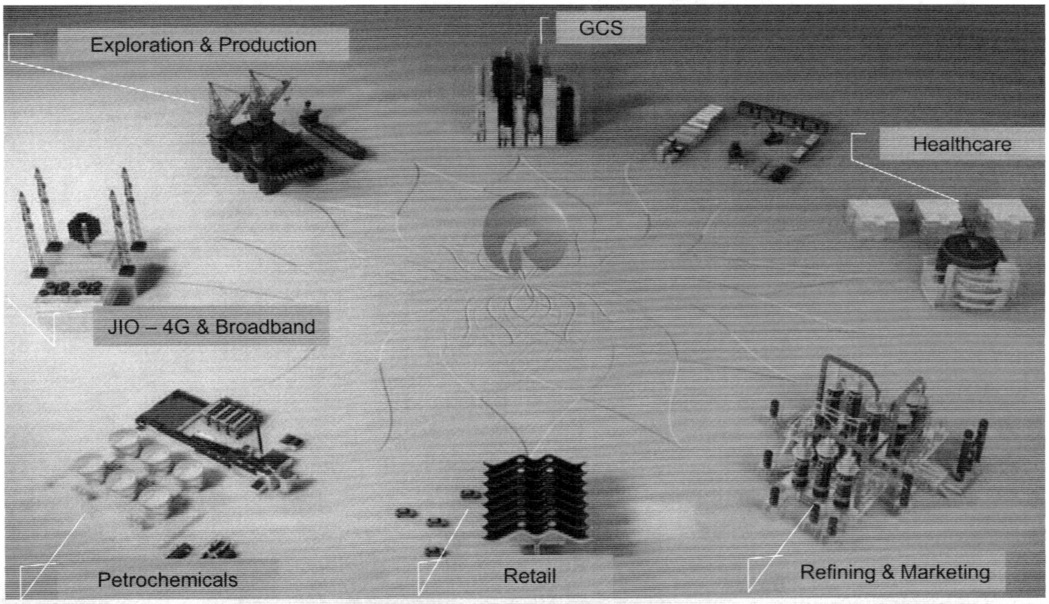


2/15

# Group Overview

- Fortune 500 company since last 10 Years
- India's Largest Private sector company Contributes 4% to GDP
- World's Largest Refining & Petrochemical Complex
- World's Largest Polyester yarn and Fiber producer
- India's Largest Retailer -1700+ stores
- Bound to Shape Future of Digital India through Jio

3/15

# PSM : Journey towards excellence

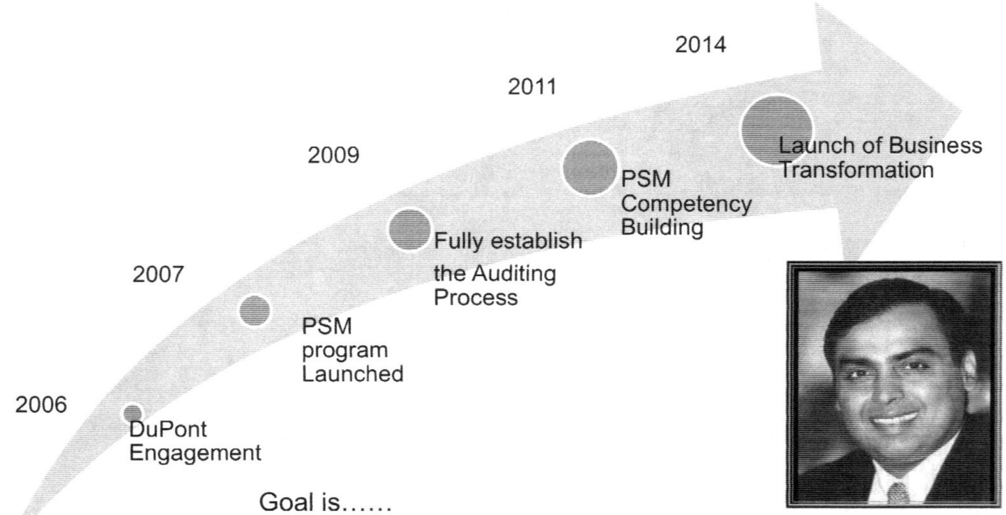


Goal is......
"Radical Complete Transformation (of Safety Culture) to achieve World Class" ------------October 2007

4/15

# What is Process Safety

Not all hazards are the same or can cause equal consequences. Personal or occupational safety hazards, such as slips, falls, cuts, and vehicle accidents usually affect one individual worker.

Occupational Safety Hazard: May Affect One or two persons

Bus accident: May Affect 50 – 60 persons

Plane crash: May affect 300 -400 persons

**But if it is process incident?**

5/15

## What is Process Safety

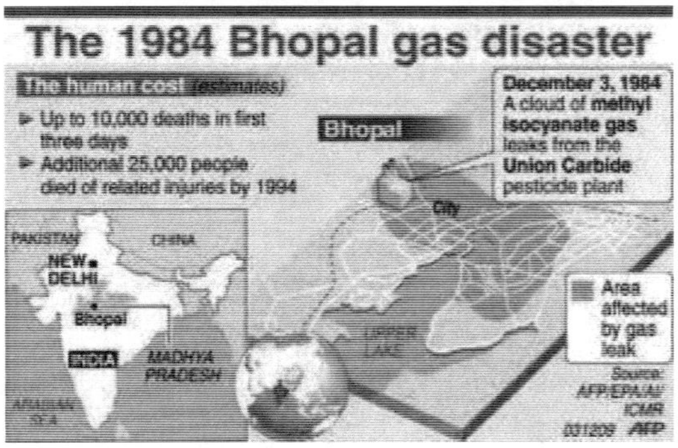


The majority of large scale losses are due to process incidents.

More than 80% of the largest losses on record have been experienced in the past 25 years.

6/15

## Process Safety at RIL

**Process Safety Management** is a Key Management system adopted at RIL, Intended to prevent an incident like the 1984 Bhopal Disaster

And

To prevent release of,

- Toxic
- Reactive
- Flammable, or
- Explosive Chemicals

7/15

# Barrier Management: Importance

**Process industry disasters were due to barrier failures:**

- Flixburough 1974
- Bhopal 1984
- Piper alpha 1988
- Buncefield 2005
- Texas City, March 23, 2005
- BP Oil Spill
- Jaipur 2009
- West Texas, 2014
- Tinajin, China 2015
- VCM Explosion Mexico, 2016
- Dombivli, 2016

*"who always claims `zero risk´, has no credibility in his ability to control risks!"*

8/15

# Health of Barriers

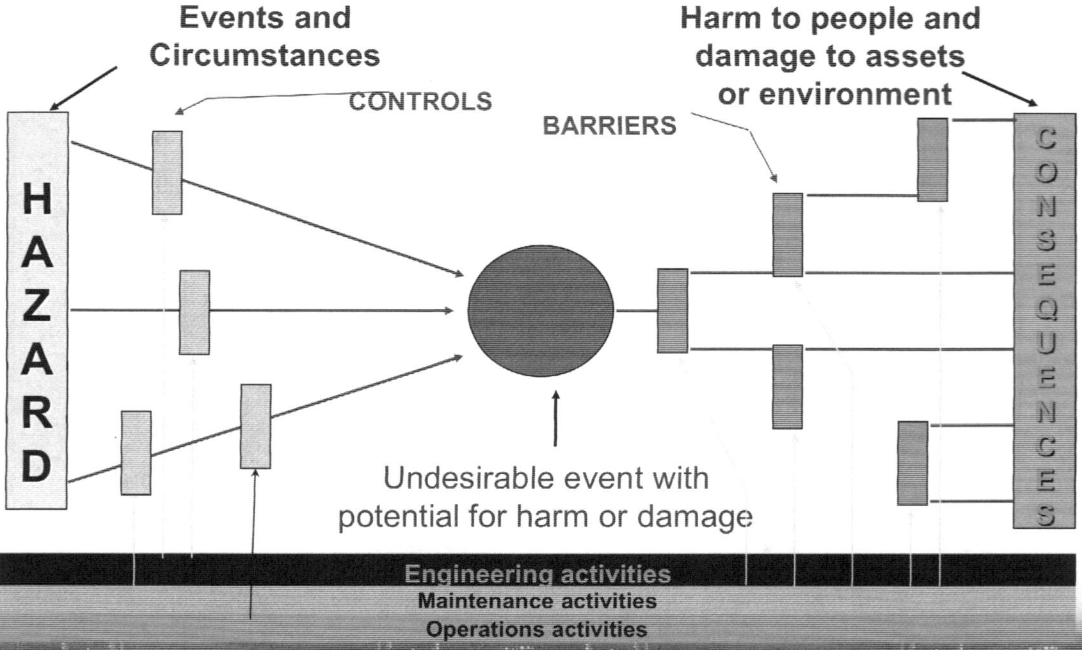


9

## What is Leadership (Process Safety Perspective)

Who are the leaders ?

- Corporate to political
- Every member of workforce to the persons sitting in board rooms
- Last person in the row is also having skill and will to understand what is process safety risk and what he / she must or must not do for managing the process safety
- Board room articulate the vision, plant leaders execute the vision

10/15

## Leadership Challenges

- Risk Posed by High Hazards industry on the society
- Poor Knowledge transfer
- Complacency : Chronic Sense of unease vs False reassurance (Tick the box)
- Emerging risks : Population around Chemical industry , threats from small industries, man made threats like terrorism
- Society is becoming less and less tolerant of mistakes
- Internalizing lessons learned
- Global competition

11/15

# How to overcome the challenges

- Ensure right feed (talent) to Chemical industry : Skill India
- Improved Process Safety competency among industry and regulators
  - Robust Process for Managing Process Safety risk : (Organisation and process)
  - Monitoring Right KPIs (Leading)
  - Vital Internal audit process
- Improved collaboration between Education institutes, Industry, Regulators and organizations like CCPS
- Improved stakeholders knowledge with the support from regulators
- Provide access for training to small entrepreneurs on risk based process safety

12/15

# Leadership 2020

- Process safety becomes a part of the DNA of running a MAH industry
- Leaders accepts what they don't know and are willing to learn
- Leaders lives with a chronic sense of unease and spot the warning signs early
- Take risked based Decisions
- Leaders giving the same priority to process safety wherever they are in the world
- Pipeline of Process safety experts laid and commissioned
- Eradicate "Tick the Box" , No Place for "Chalta Hai"
- Improved engagement and communications with the workforce

13/15

## Process safety excellence model

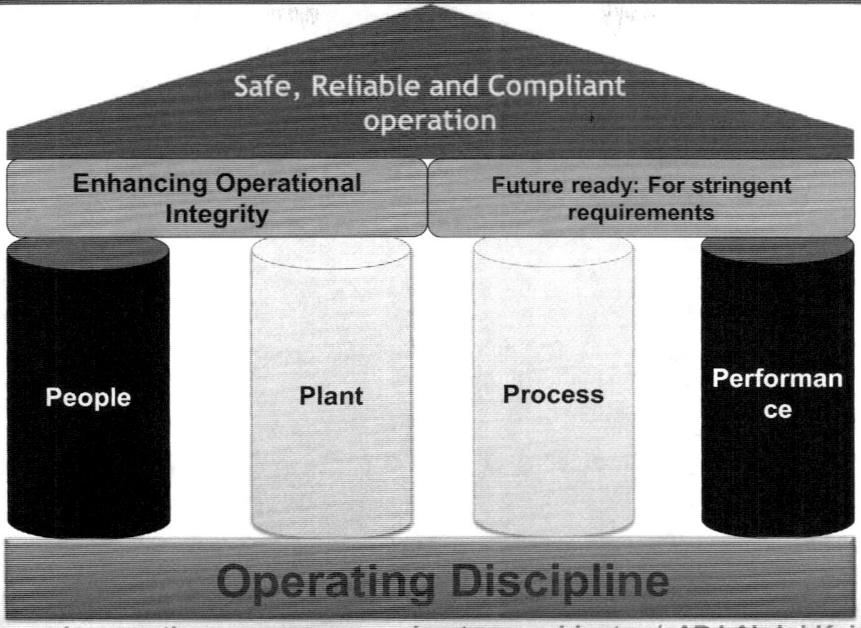


Excellence is a continuous process and not an accident .../. APJ Abdul Kalam



## Every One Goes Home safely

To be Globally Competitive,
We Must be
Globally Competent

THANKS !

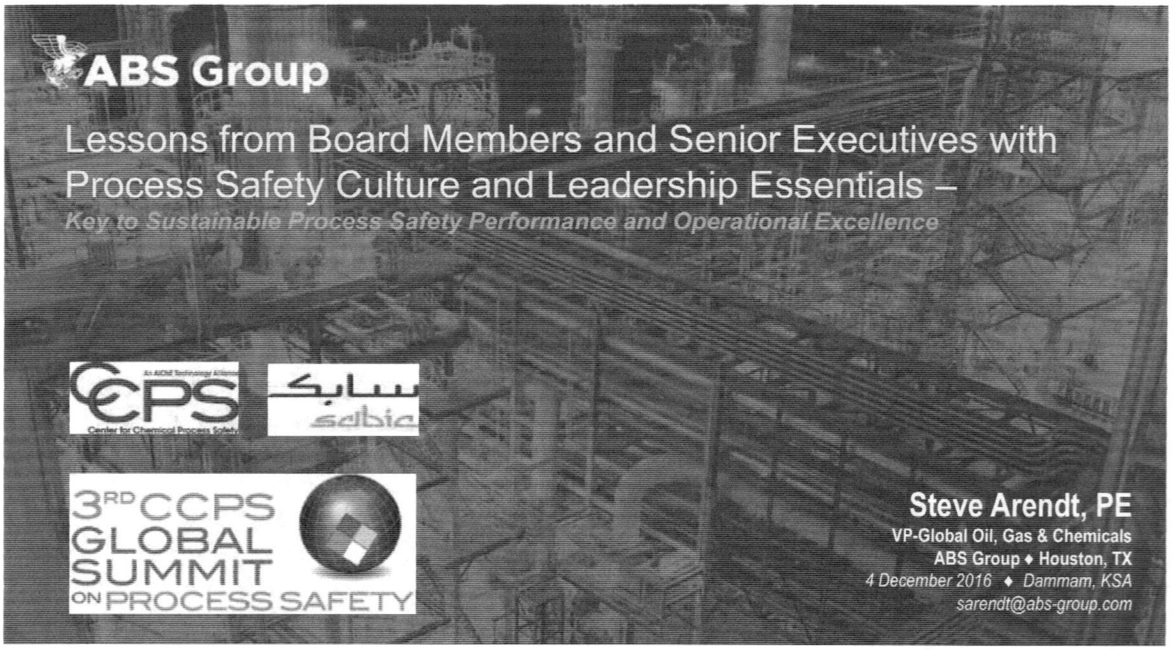


## Presentation Outline

Relationship between process safety culture, leadership and operational discipline

Key attributes of process safety culture/leadership

Preventing process safety culture disease

Promoting proper ownership of process safety

Developing process safety leaders

Challenges for Senior Executives and Board members

Lessons for sustaining process safety performance

ABS Group

## Improvements in Safety/HSE

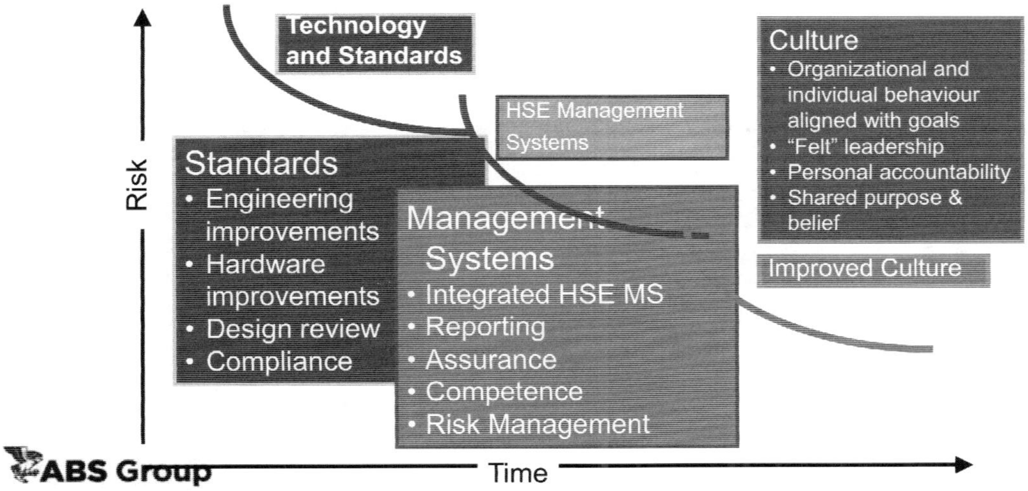


## ABS Group Safety Culture Credentials and Experience

Developed Center for Chemical Process Safety Culture Awareness Tool

Developed Risk Based Process Safety management system structure – Process Safety Culture element – 1st time in an "official" SMS framework

BP Texas City/Baker Panel evaluation of PSM performance and culture

CCPS Culture Committee

Conducted 35+ safety culture evaluations

- Variety of industries-oil and gas, offshore, refineries, terminal and pipelines, drilling, chemical, petrochemical, Pharma, biogen, consumer, mining
- Single plant, up to 80 representative facilities, normally 8-12 globally distributed NA, LA, EU, ME, AP
- Surveys and interviews done Eng, Span, Port, Danish, French, Arabic, Tagalog, Mandarin
- Always anonymous and includes contractors. Goes up to SVP or President, wherever safety and money decisions flow up to. 50% of the time get to the CEO
- Most studies are motivated by external influence, some by owners/Boards; rarely organic

ABS Group

## Culture, Leadership and Operating Discipline

- **Culture** is the tendency in all of us – and our organization - to want to do the **right thing** in the **right way** at the **right time**, **ALL the time** – even when/if no one is looking – ***ABSG definition***
  - Culture is the result of all the actions - **and inactions** - in institutional/workforce memory
  - Organizational culture is a complex combination of individual cultures + environment = **Safety climate**
  - Culture is hard to measure and more difficult to change; the "root cause of the next decade"
- **Leadership** is an essential feature of a good culture – ***CCPS RBPS characterization***
  - CCPS's 20 elements for success, Vision 2020 and RBPS **Culture** and **Conduct of Operations** elements
  - UK **Process Safety Leadership** Group's report following Buncefield
  - AFPM/API have been focusing on **Process Safety Leadership** as one of its assessment areas
- **Operational discipline** (or the lack thereof) is a behavioral result of your culture and leadership
- **Conduct of Operations** is a system designed to create and manage **operational discipline**

**ABS Group**

## Center for Chemical Process Safety Made Culture and Operational Discipline Official PSM Elements

- Evaluated major organizational accidents and prepared Safety Culture Awareness tool
- ABSC included Process Safety Culture as an element in CCPS ***Guidelines for Risk Based Process Safety***
- Defined the twelve essential features of a good culture
- Defined a management practice for ***Conduct of Operations*** to help create operational discipline and excellence
- Created structure for a culture management practice

**ABS Group**

## Senior Management's Role in Nurturing Culture

Senior management

- Sets the standards
- Sets the tone
- Provides the resources
- Provides reinforcement

Senior management can **enable** and nurture a sound safety culture, but cannot mandate it

"...one could argue that the only thing of real importance that leaders do is to create and manage culture..."

- MIT Professor Dr. Edgar Schein

ABS Group

7

## Safety Culture Aspects

Definitions

Scales/descriptions

Frameworks for understanding

Measurement/analysis methods

Data sources

Ways to improve culture

Ways to monitor culture

ABS Group

## Example of a Safety Culture Ladder

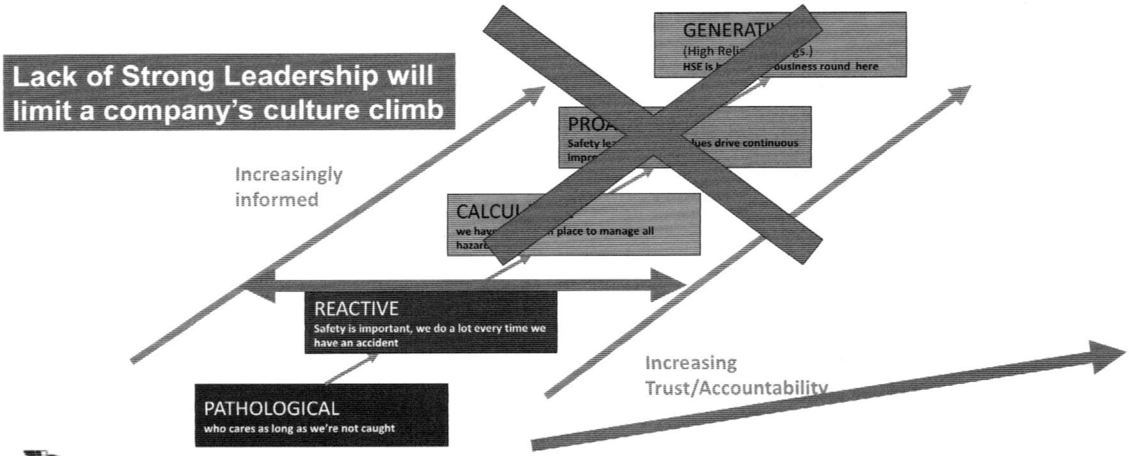


ABS Group

## HSE Culture Essential Features

1. Establish HSE as a core value
2. Provide strong leadership
3. Establish and enforce high standards of performance
4. Formalize the safety culture emphasis/approach
5. Maintain a sense of vulnerability
6. Empower individuals to successfully fulfill their safety responsibilities
7. Defer to expertise
8. Ensure open and effective communications
9. Establish a questioning/learning environment
10. Foster mutual trust
11. Provide timely response to safety issues and concerns
12. Provide continuous monitoring of performance

ABS Group

## Process Safety Performance Assurance Review (PAR)© Strategy

*How to Evaluate Leadership and Culture*

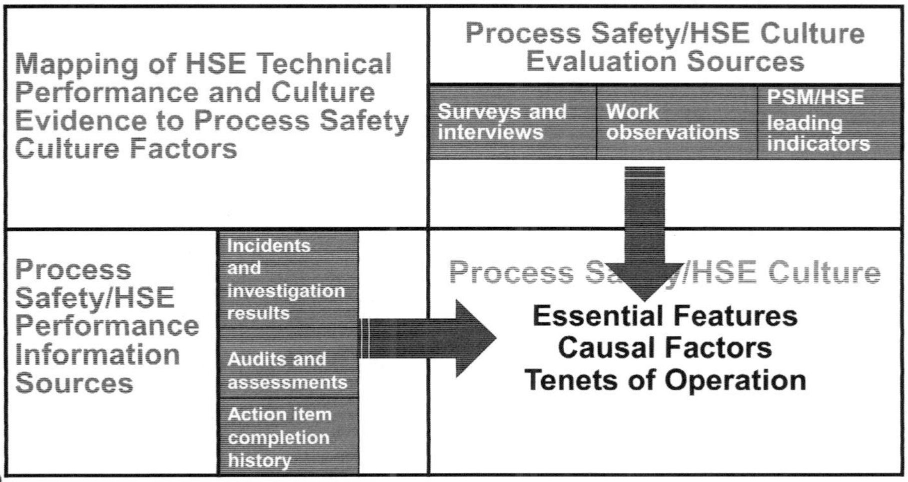


ABS Group

## Process Safety Leadership Attributes

1. Ethical
2. Committed
3. Vision
4. Competent
5. Courage
6. Emotional intelligence
7. Sense of vulnerability
8. Influence
9. Visibility
10. Communication
11. Accountability
12. Connectivity
13. Develop others
14. Empowering
15. Change navigator
16. Responsive
17. Consistent

ABS Group

12

## Example 1 - Culture Disease Pathology

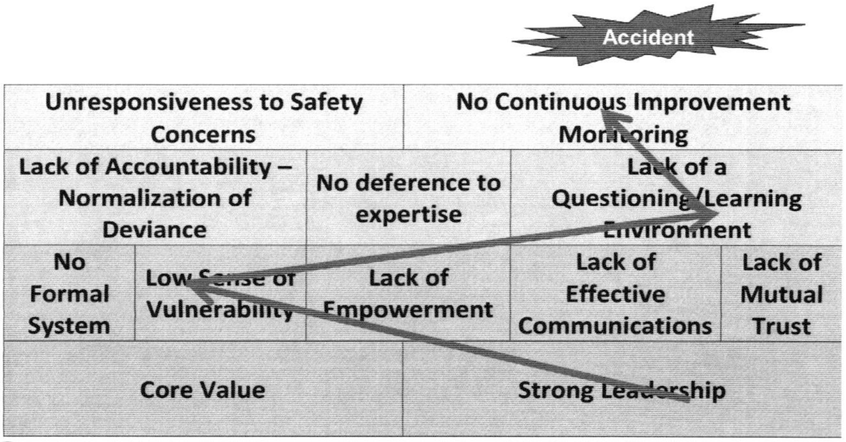


ABS Group

## Culture Disease Vaccine

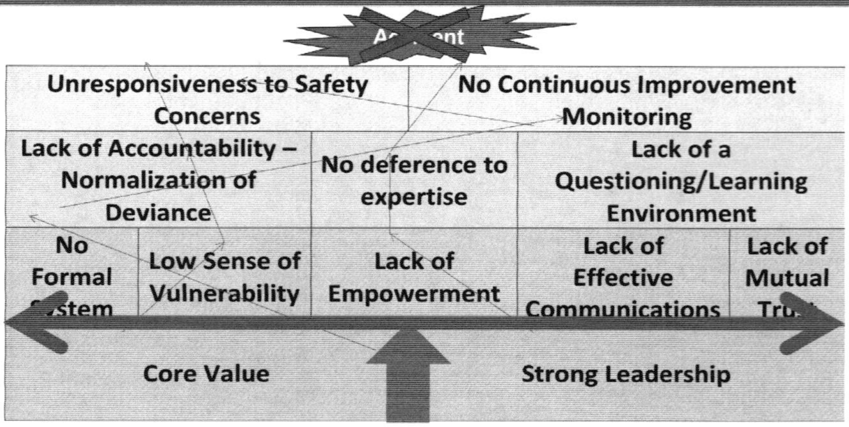


ABS Group

**Leadership excellence throughout the organization is the best process safety culture disease vaccination there is**

## Leadership/Culture Improvement Thought Process

- Personalize to each person's job
- Connect to their "ownership basis"
- Create belief in results
- Brainstorm how they got to where they are
- List current and future conditions that could affect culture
- Select one or two leadership/culture weaknesses/strengths
- Discuss how they are relevant to daily jobs
- Provide culture improvement idea menu
- Ask them to create/commit personal daily/weekly plan
- Solicit ideas for department/facility/company plan
- Monitor culture improvement – plan, personal growth, metrics

**Top leaders first, then Supervisors, then workers + contractors, then up the organization**

**They "plug the cheese holes"**

ABS Group

ABS Group

## Today's Challenges in Process Safety

## Organizational Challenges for Process Safety Leadership

- Where in the organization to establish the process safety function
- At what level to place the head of the process safety organization
- Balancing leadership with ownership
- Promoting leadership at every level of the organization
- Creating and renewing the process safety vision
- Keeping process safety as an imperative - preventing mixed messages
- Maintaining trust, openness and accountability
- Providing sufficient process safety resources
- Developing a sustainable learning organization

ABS Group

## Learning Organizations

*"A learning organization is an organization skilled at creating, acquiring, interpreting, transferring, and retaining knowledge, and at purposefully modifying its behavior to reflect new knowledge and insights."*

Garvin, Learning in Action, Harvard Business School Press, 2000

**Litmus tests**

- Defined learning agenda
- Open to discordant information
- Avoid repeated mistakes
- Maintain critical knowledge when people leave
- Act on what it knows

ABS Group

## Characteristics of a Learning-Disabled Company
*(Steve Arendt, Armchair Process Safety Psychologist)*

Dysfunctional safety culture

- **People hide things and kill messengers**
- **Fail to question; procedures not followed without accountability**
- **Mixed/improper safety/production messages**
- **Complacency, low trust, silo mentality**
- **Misplaced safety ownership, invisible/ineffective leadership**

Superficial causal analysis of problems

Things don't get fixed

No company memory

Add in your own…

ABS Group

## Learning from Experience

Outside your industry

Inside your industry

Inside your company

One BIG problem in industry is that we get taught the same lesson over and over, but don't truly LEARN so that the problems don't repeat

A critical skill is to find something relevant to learn and improve on out of EVERY significant incident

ABS Group

20

ABS Group

# What an Executive Needs to Know

## What an Executive Needs to Know

Board oversight - role of Board in PSM and safety in general

Questions Executives/Boards should be asking - Process safety assurance

- **Do you know your major risks?**
- **Do you know what you do to manage those hazards/risks?**
- **What information do you have to be confident that it is happening?**

Process safety stewardship and governance

Management of change for organizational changes

Key issues to track

Challenges and potential conflicts - financial, operational, organizational – watch for sending “mixed signals”

ABS Group

## Process Safety Stewardship and Governance

Corporate-level management review

- **Quarterly to annually**

Site management review

- **Monthly to quarterly**

Similar issues are tracked as at Board level, but more detail – Tier 3 and 4 metrics – leading indicators

Challenges and potential conflicts - financial, operational, organizational – watch for sending "mixed signals"

**ABS Group**

## What an Executive Needs to Know

Board oversight - role of Board in PSM and safety in general

Questions Executives/Boards should be asking - Process safety assurance

- **Do you know your major risks?**
- **Do you know what you do to manage those hazards/risk?**
- **What information do you have to be confident that it is happening?**

Process safety stewardship and governance

Management of change for organizational changes

Challenges and potential conflicts - financial, operational, organizational – watch for sending "mixed signals"

**ABS Group**

## Ongoing Challenges in Process Safety

Learning and REMEMBERING from experience

Sustaining focus in times of business stress

- *Good leaders understand and have control over their "inner game,"* **Art Petty**
- That means understanding the confusing and chaotic nature of organizations and making it their job to slice through the noise, embrace ambiguity and create spaces in which they and others can think before taking action.
- *"Their command of the inner-game helps them fight through the paralyzing potential of self-doubt and the debilitating effects of hubris and arrogance,"*

Developing an ability to sort and act on "weak signals"

**ABS Group**

## Sense, Learn, and Fix at Every Level

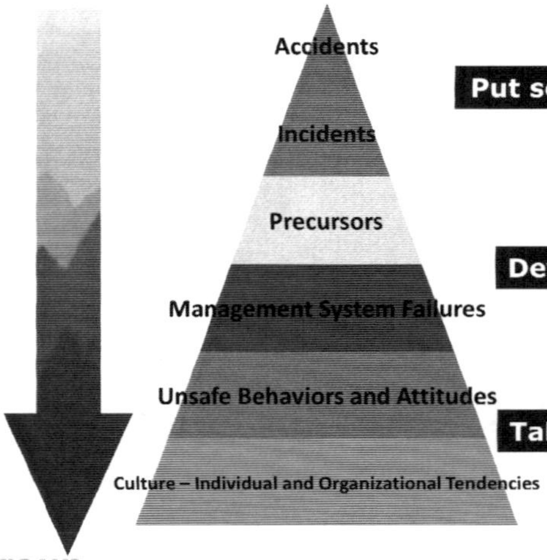


**Put sensors, not *censors*, at every level**

**Develop learnings at every level**

**Take corrective action at every level**

**ABS Group**

## Continuous, Sustainable Improvement in Process Safety Performance Demands…

**Effective RCA and corrective action creates improvement**

**Use of leading indicators to be continuous**

**Addressing culture and behaviors to be sustainable**

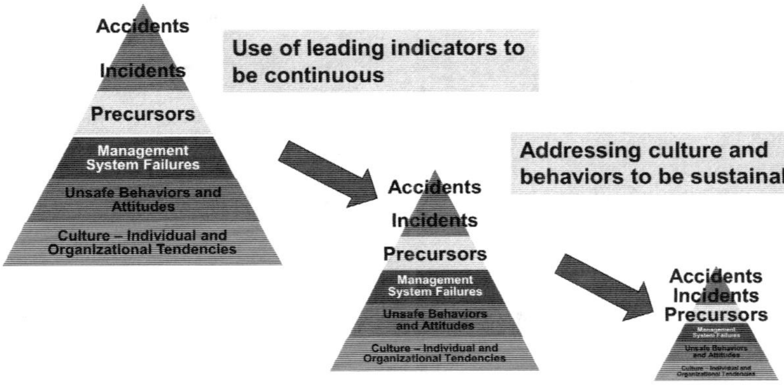


ABS Group

**Improving the Integrity Assurance of Bolted Joints**

David Brown

hydratight

## My Themes..........

- Bolted Joints do not need to be Weak Points
- Bolted Joint Integrity Programs can be Developed
- Focus on Joint Integrity improves Safety & Plant Performance

12/19/2016

## Some Thoughts.....

- Simple Structures
- Critical Function

- Treatment **HAS TO** Reflect the Function
- Every Intervention Introduces Risk

12/19/2016

## Challenger Space Shuttle 1986

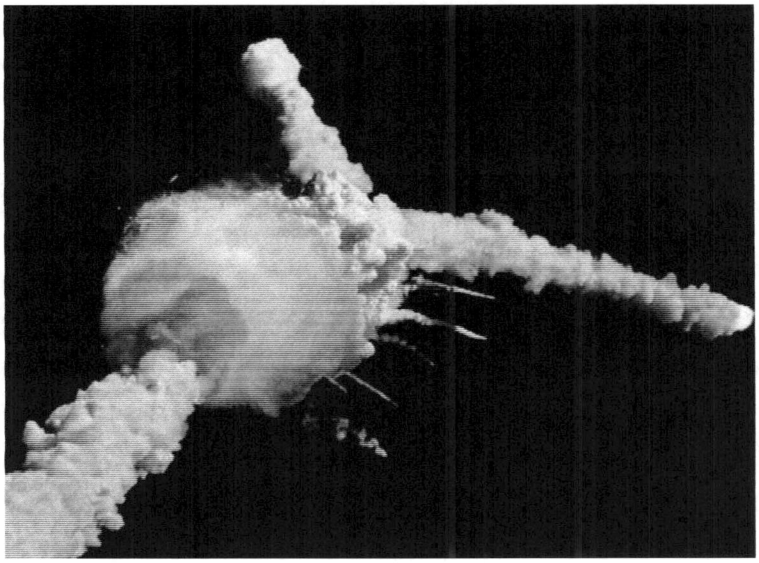

12/19/2016

## Joint Failures Cost Money.......

This Hub Connector passed a Nitrogen Helium Leak test but leaked 3 years later causing an explosion. The cause -fitting a Carbon Steel Seal Ring instead of a Stainless one. ($30m in damage)

12/19/2016

## "Our Joints Don't Leak.........."

12/19/2016

## Leaks Cause Fires – Fires Are Expensive

*Fire presumed to start from a safety valve; the bolting was believed to be incorrectly tightened.*

*Because of joint leakage, the process was altered leading to fires in four columns*

**Examples of Hydrocarbon asset losses (2003):**

- *USA - $80m*
- *Mexico - $136m*
- *Italy - $65m*
- *Saudi - $60m*

**Source :-Marsh Report**

12/19/2016

## 2013 – A Great Year for the Bolted Joint........

**CEN EN 1591 – 4**: Flanges and their joints – Part 4: Qualification of personnel competency in the assembly of the bolted connections of critical service pressurized systems.

Sets a performance standard for technician competency.

**ASME PCC-1 2013**: Guidelines for Pressure Boundary Bolted Flange Joint Assembly.

Defines a performance standard for bolting assemblers & provides comprehensive guidance on the execution & assurance of joint integrity.

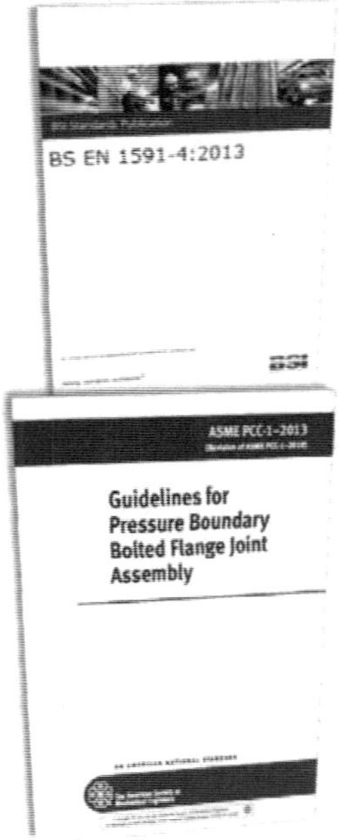


12/19/2016

## Competence

- Training ≠ Competence
  - Safe and Correct Practice
  - Assessment
  - Over Time

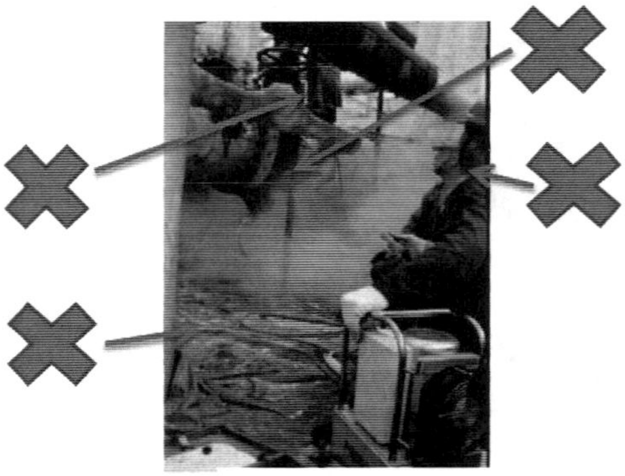

12/19/2016

hydratight

## The ASME PCC1 2013 stated Guiding Principle.....

ASME stated principle:
Treat every bolted joint as though it were a weld

12/19/2016



hydratight

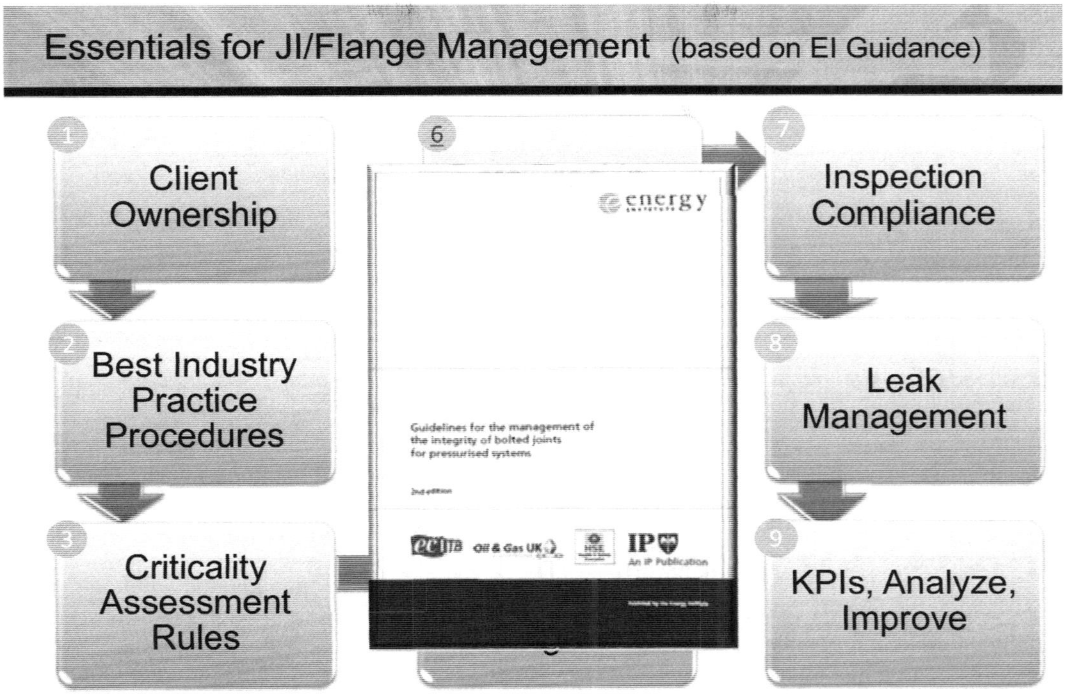
Essentials for JI/Flange Management (based on EI Guidance)
1
Client Ownership
2
Best Industry Practice Procedures
3
Criticality Assessment Rules
6
energy
Guidelines for the management of the integrity of bolted joints for pressurised systems
2nd edition
Oil & Gas UK
IP
An IP Publication
7
Inspection Compliance
8
Leak Management
9
KPIs, Analyze, Improve

12/19/2016

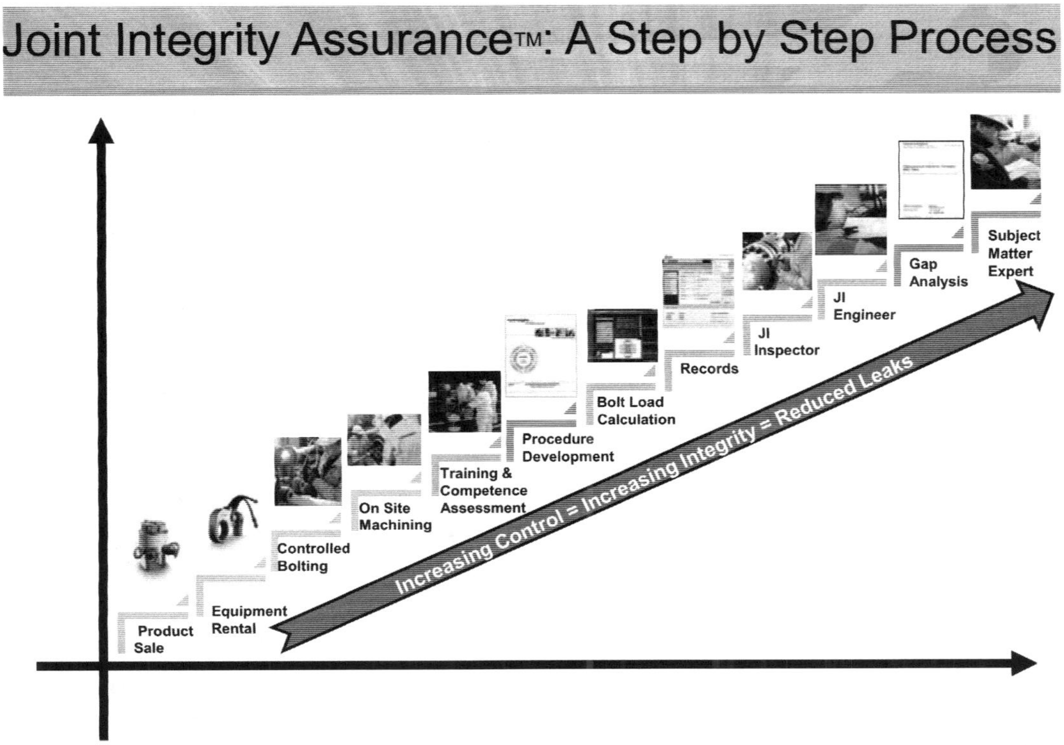
Joint Integrity Assurance™: A Step by Step Process
Product Sale
Equipment Rental
Controlled Bolting
On Site Machining
Training & Competence Assessment
Procedure Development
Bolt Load Calculation
Records
JI Inspector
JI Engineer
Gap Analysis
Subject Matter Expert
Increasing Control = Increasing Integrity = Reduced Leaks

12

## Joint Integrity Assurance: Benefits

- Fewer Leaks during Testing ahead of Start-Up
- Lower Repair Costs and Testing Costs
- Safer Start-Up & Operations
- Faster Start-Up
- Fewer Unplanned Outages

13

## Joint Integrity Assurance: Real Results

- Review of All Procedures
- Adoption of Updated Bolt Loads
- Pre-Shutdown Awareness Campaigns
- Mandatory Training of **ALL** Mechanical Contractors
- Close Monitoring/Supervision/Inspection

14

# Joint Integrity Assurance: Real Results

- Pre-Campaign – up to 5% Joints Leaked under Test
- Pre-Campaign – Average of 1.4 Tests per System
- Examinations – Need to Understand Why?
- Implement Improvements

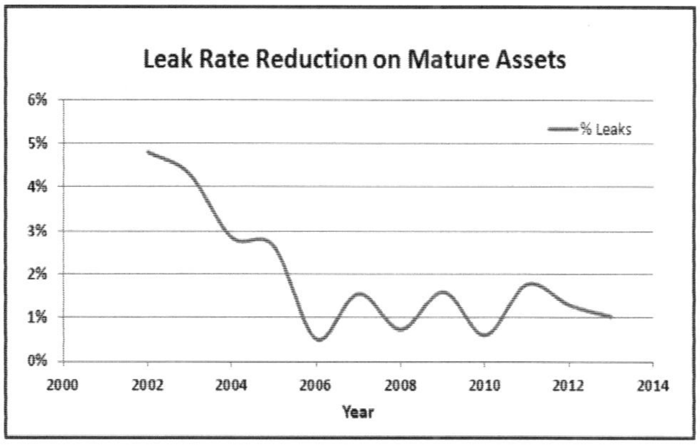


- Post-Campaign – Less than 1% Joints Leaked under Test
- Post-Campaign - Average of 1.1 Tests per System

15

# Joint Integrity Assurance: Summary

- Shutdowns Cause Leaks: Leaks Cause Shutdowns
- Increasing Global Focus on Bolted Joints
- Be Aware of the Latest Information Available
- Understand Your JI Process – Introduce/Validate/Seek Help
- Leak Free Start-Up is Achievable

16

## Joint Integrity Assurance: CCPS Global Summit

- A Committed Culture and Process Safety Leadership
- Human Factors
- Contractor Management
- Disciplined Adherence to Standards
- Enhanced Applications of Lessons Learned
- Asset Integrity Management, Aging Facilities & Facility Siting
- Intentional Competency Development
- Metrics & Process Safety Performance Improvement
- Operational Discipline / Excellence
- Enhanced Stakeholder Knowledge
- Responsible Collaboration
- Meticulous Verification
- Vibrant Management Systems

17

## Questions?

18

## Presentation Outline

بابكو Bapco

1. Introduction to Bapco
2. The OE Business Case
3. OE Culture
4. Our Journey
5. Reaping the Benefits
6. Key Takeaways

2

Bapco

# Introduction to Bapco



بابكو
Bapco

## The Bahrain Petroleum Company (Bapco)

Bapco is 100% owned by the Government of the Kingdom of Bahrain, an is an oil and gas company:

- Natural gas distribution
- Active oil and gas exploration
- Refining of crude to produce valuable petroleum products
- Storage and export of the refined products
- Marketing of the products locally and internationally

4

Bapco

## About Bapco

- Company was formed in 1929: in Canada
- Oil discovered in 1932
- Refinery first day of operation July 1936
- Today annual sales of refined products is > $6.0 billion
- Contributes about 70% of Bahrain's economy
- Currently employs over 2500 people and is one of the largest employers in the Kingdom

5

Bapco

## The OE Business Case

Bapco

## Major Incidents with Global Impact

- 1974 Flixborough
- 1976 Seveso
- 1984 Mexico City
- 1984 Bhopal
- 1988 Piper Alpha
- 2005 Texas City Incident
- 2005 Buncefield Fire
- 2010 Macondo Deepwater Horizon

8

Bapco

## Key Lessons Learned from Major Incidents

- Failure to equip employees with appropriate knowledge
- Failure to make available the necessary up to date information
- Normalization of deviation
- No clear focus or visibility on process safety
- Many incidents not reported or investigated

**Our Solution : Operational Excellence**

9

Bapco

## The Compelling Business Case

- Operational Excellence is how we implement the Company's Mission.
- Getting this right in a low margin environment is imperative.
- We will show you what has worked for Bapco and hopefully you can gain some insight to help your business.

10

Bapco

## How do we define OEMS?

- Operational Excellence is the systematic management of process safety, personal safety and health, environment, reliability and efficiency to achieve world class performance

- Focus on effectiveness rather than (not just) certification

11

باپكو Bapco

## Our Philosophy

- Only acceptable safety performance is **incident and injury free**
- Create an environment that encourages all employees (and contractors) to be actively involved Health & Safety initiatives
- Shift mindset : **“Thinking we are OK”** to **“Knowing We are Safe”**

© Bapco 12

Bapco

## OE Culture

© Bapco 7 October 2014

بابكو
Bapco

## OE Culture

- What do we mean by Operational Excellence (OE) culture?
- Our people <u>wanting to do</u> the "right thing" the "right way" all the time (always).
- The basic principles constituting the OE culture:
  - Zero injuries and incidents is attainable
  - Safety is not to be compromised for production or cost
  - Compliance with safe operating procedures and policies is not optional
- To us in Bapco; Leadership is key.





بابكو
Bapco

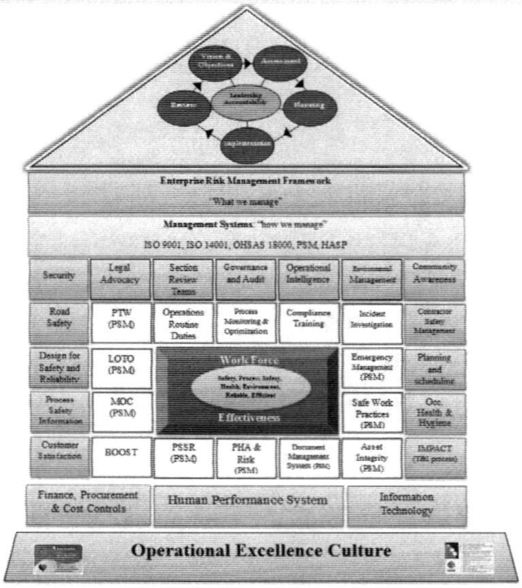


**OE**
**is being Operationally Focused**

## Management System Process - Annual Continual Improvement Cycle

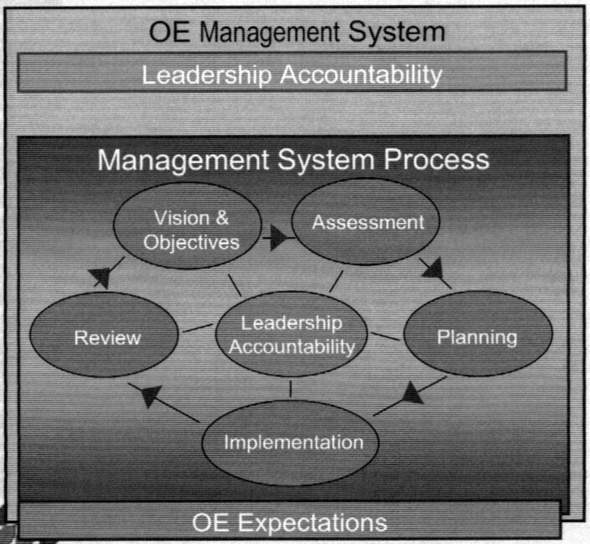


- Set an OE Vision
- Conduct gap Assessments
- Prioritize gaps
- Develop 3-year Plan
- Integrate into Business Plans
- Assign tasks and process roles in PMP
- Implement OE plans
- Build or improve OE Processes and Standards
- Regularly Review and measure progress.
- Build an OE culture

16

## The Three Legged Stool Analogy

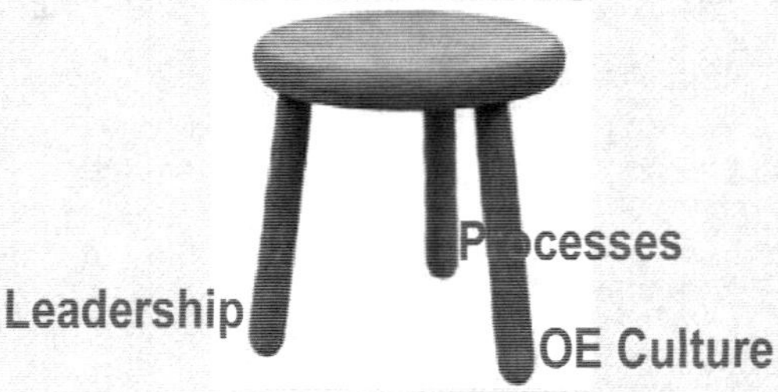


= Operational Excellence

17

Bapco

# Our Journey



بابكو
Bapco

## Map Bapco Processes to OEMS Expectations

- There are 46 OE Expectations detailed under 13 OE Elements.
- These 46 Expectations have been mapped to 29 Bapco OEMS processes.
- Each Process is given an Owner and Custodian.
- The roles are critical.

Bapco

## Culture : Fluency and Alignment

- Our Leaders set simple yet clear objectives in Osool Bapco (Codes of Conduct)
- The operative word in each principle is "ALWAYS"
- Good leaders seek and gain commitment from their employees to use Osool Bapco
- These elements properly actioned = An aligned Organization

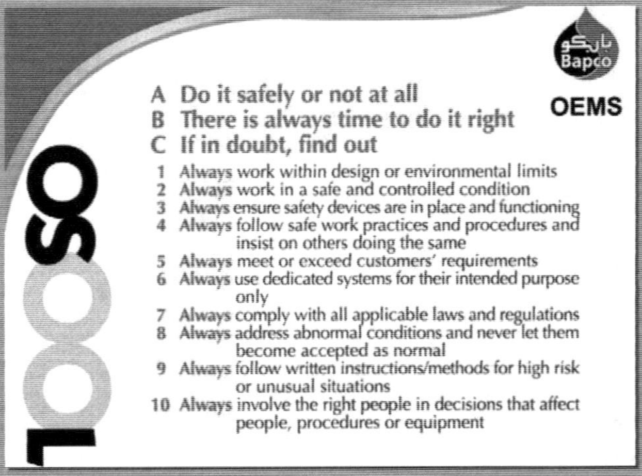





Bapco

## Enabling Leaders : The Bapco Dashboard : XHQ

- Dashboard connects to 17 different databases
- Provides real-time window into company performance
- Available to all staff

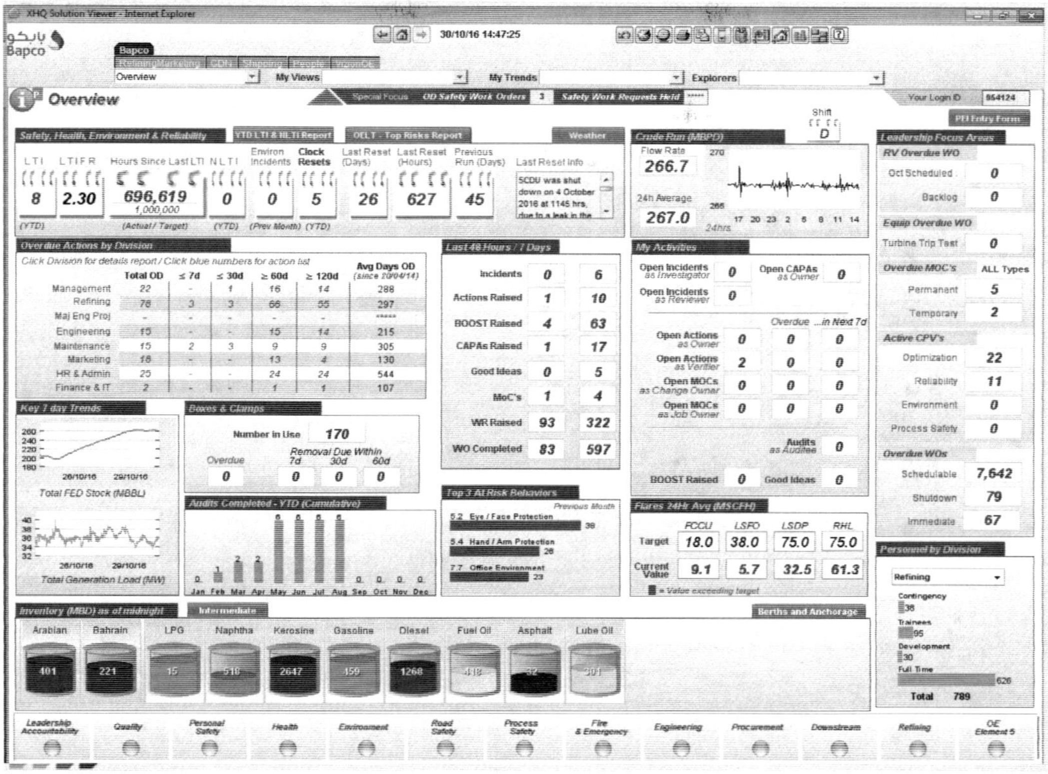

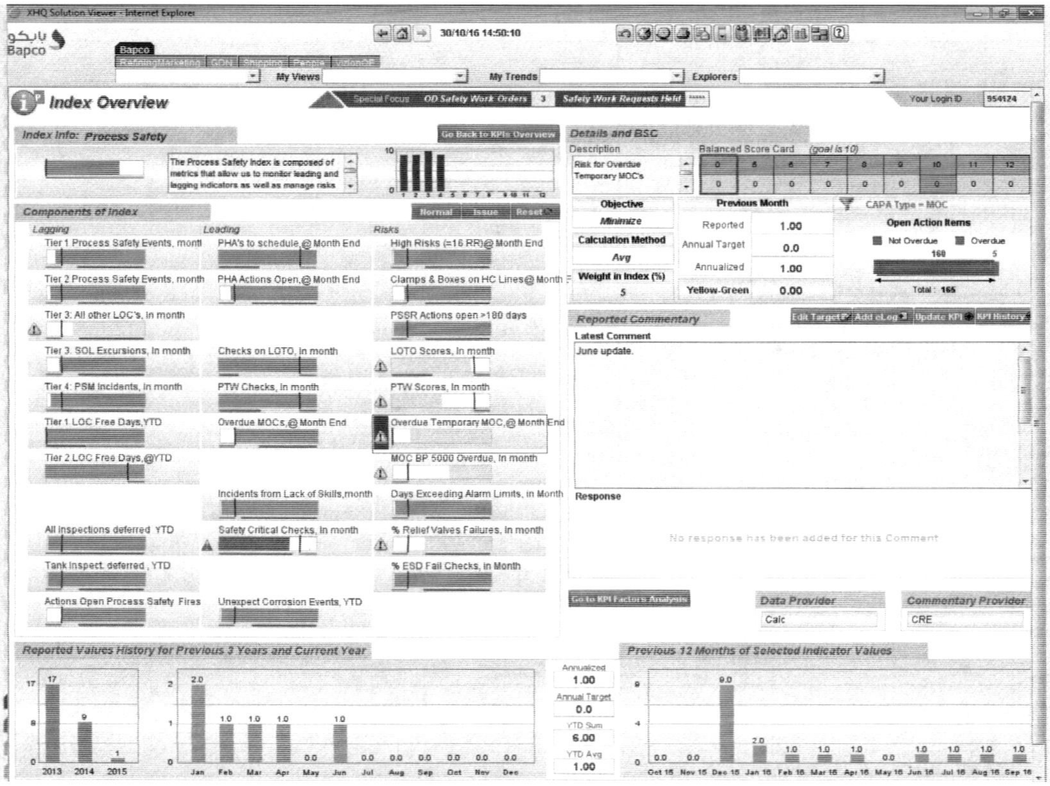

Bapco

## Sectional Governance

- Superintendent
  - Operations (6) *or*
  - Power & Utilities (2) *or*
  - Oil Storage & Export (3)
- Maintenance Superintendent
- Plant Reliability Specialist
- Central Reliability Engineer
- Technical Services Engineer
- Engineering Projects Representative
- OE Facilitator
- *Invited Guest*

- Leadership
  - Line OE Governance the "right level"
  - Functional OE Governance at top leadership
- Processes
  - Gained control and compliance across the whole OEMS
- Culture
  - A "doing" culture




Bapco

## Annual Gap Assessment

- We self assess our processes on Design and Effectiveness against a Matrix weighted to Effectiveness.
- We assess 5 Components: Purpose, Documentation, Roles, Metrics and Continuous Improvement in Performance.
- We give a rating 1-5 (good) for each.

| | | DESIGN | | | | |
|---|---|---|---|---|---|---|
| | | 1 | 2 | 3 | 4 | 5 |
| EFECTIVENESS | 5 | LTS | Satisfactory | Good | At or Near Best Practice | |
| | 4 | LTS | Satisfactory | Good | | |
| | 3 | LTS | Satisfactory | Satisfactory | Good | Good |
| | 2 | LTS | LTS | LTS | Satisfactory | Satisfactory |
| | 1 | LTS | LTS | LTS | LTS | LTS |

Bapco

## Annual Gap Assessment (cont'd)

Here is an example of how we rate Metrics (Design) and CI (Effectiveness)

| | Less Than Satisfactory (1) | Satisfactory (S1) (2) | Satisfactory (S2) Systems and controls are in place; reasonable assurance of compliance with OE requirements (3) | Good (4) | At or near Best Practice Only minor weaknesses; controls verify compliance with OE requirements (5) |
|---|---|---|---|---|---|
| Measurement and Verification | There is no authorized Leading and Lagging indicators available to verify and validate the process. | There are documented Leading and Lagging indicators and the verification and validation processes are addressed in the 5C Model. | The process in place to formally review and report the leading and lagging indicators in the 5C model. | The process in place to formally review and report and validated by the process owners/custodians the leading and lagging indicators in the 5C model. | The process in place to formally review the appropriateness and suitability of the metrics annually. |
| Continual Improvement | Significant incidents due to process failure | Not meeting all objectives (Lagging indicators) | Meeting all objectives (Lagging indicators) | Exceeding objectives (Lagging indicators) | Process has been identified as good by independent body |

26

Bapco

## Governance and Gap Closure

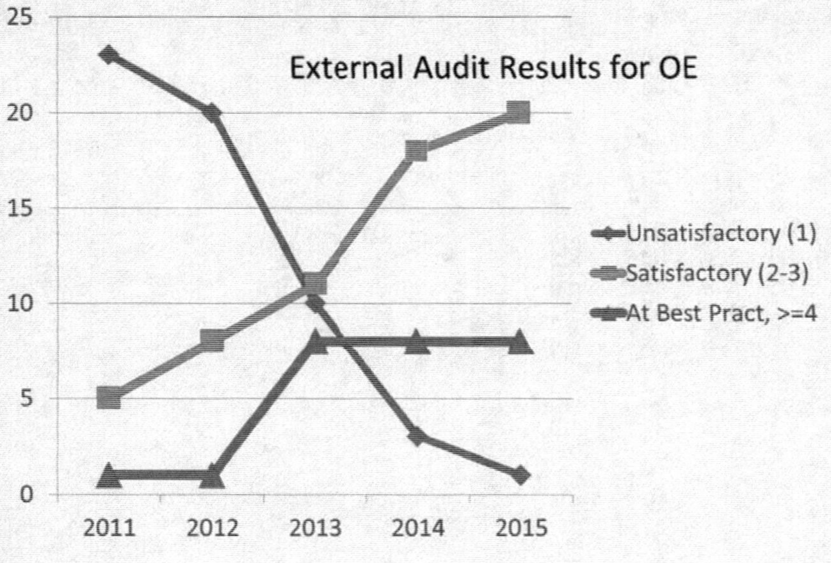


27

Bapco

## Our Journey : Review MOC Compliance

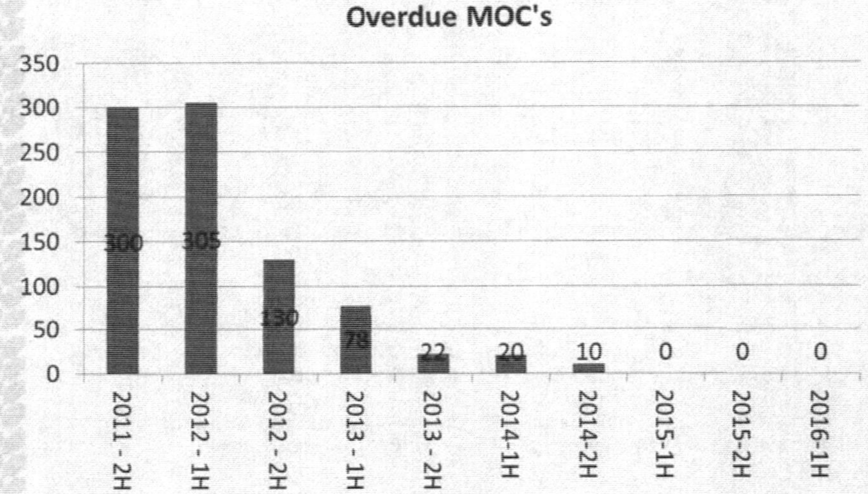


28

Bapco

## Our Journey: Review Safety Work Orders

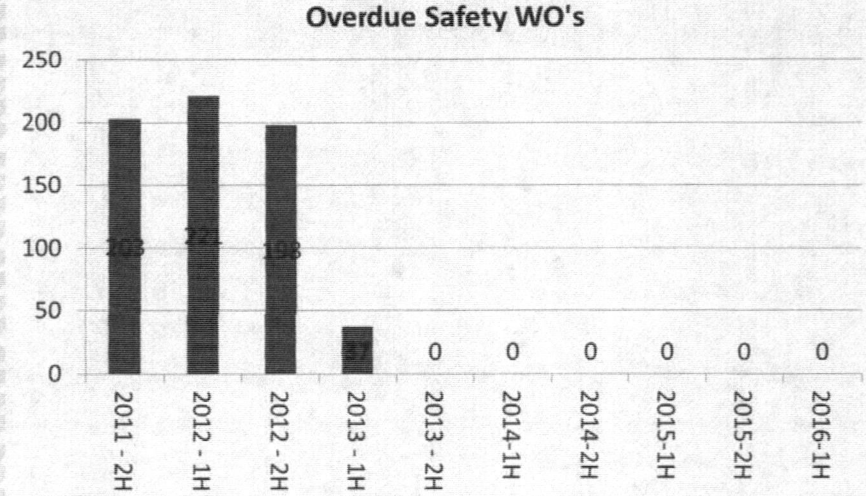


29

Bapco

# Reaping the Benefits



بابكو
Bapco

## Reaping the Benefits in 2015

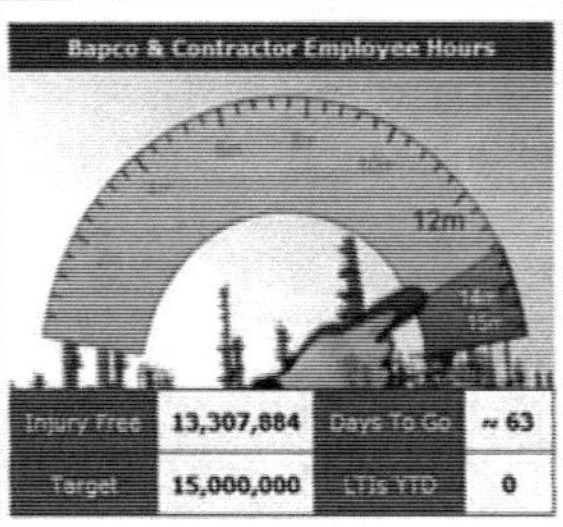


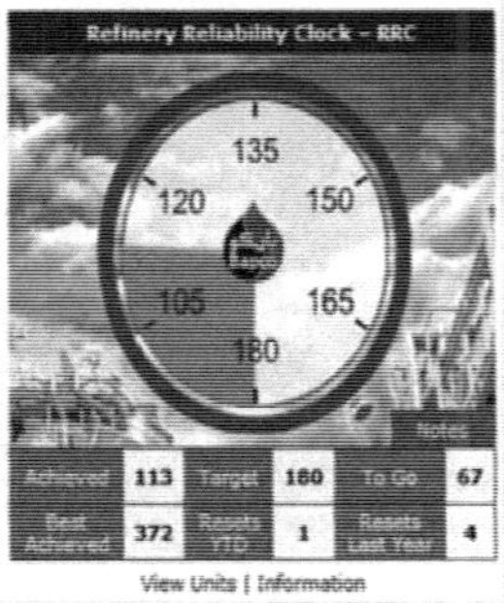


- •Leadership
  - Focus on Leading Indicators
  - Safety is not compromised for production or cost
- •Processes
  - Best Reliability Records
  - Best Process Safety Records
- •Culture
  - Best Loss Time Injury Record
  - Work force truly believes zero injuries and incidents is attainable

Bapco

## Key Takeaways



بابكو
Bapco

## Key Takeaways

- OE cannot be implemented and sustained successfully without having the right culture that supports it.
- OEMS Process improvements are not enough.
- Visible leadership engagement and commitment is key in nurturing the OE culture.
- OE rewards will include improved cost efficiency and cost effectiveness: Low oil price environment is the opportunity to prove this.
- It is a long journey, but it pays off.
- We are seeing the rewards but we cannot sit on our laurels!

Bapco
Thank You
© Bapco 7 October 2014

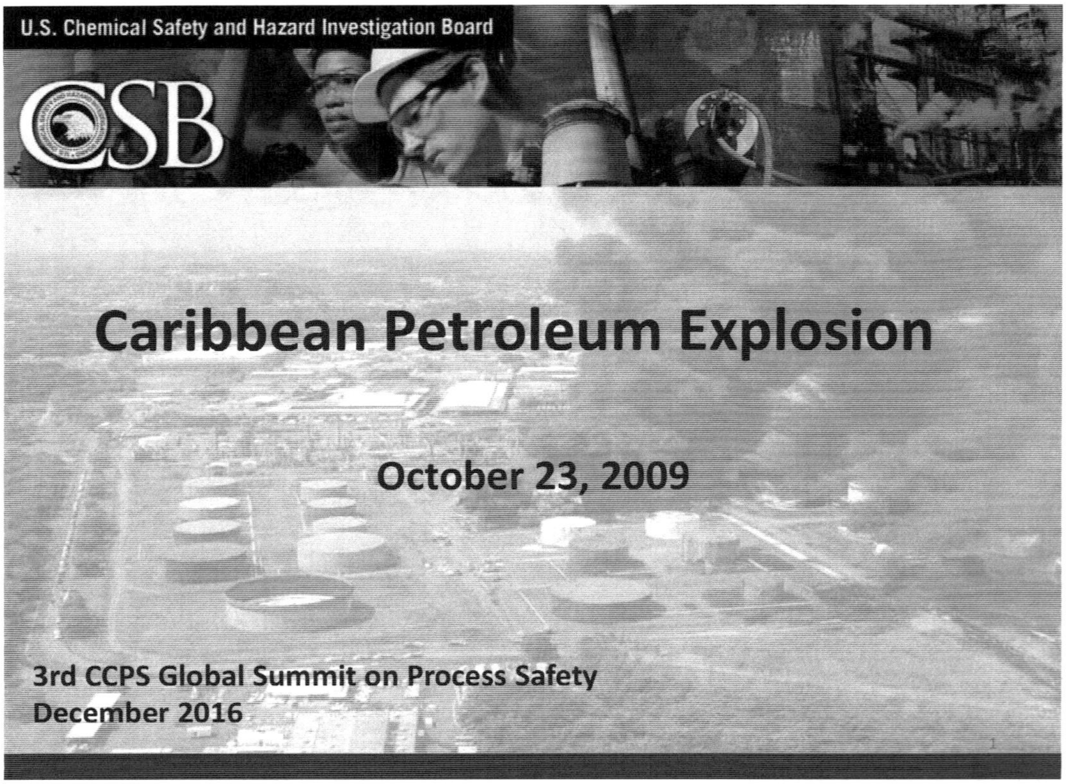


U.S. Chemical Safety and Hazard Investigation Board

# Number of Tank Terminals in US

**There are at least 2,900 bulk petroleum storage terminals within one mile of communities with >300,000 residents.**

**2012 Petroleum Bulk Storage Tank Terminals within 1 mile of Metropolitan Areas and Populations Over 300k**

2012 EPA Toxic Release Inventory database

2

U.S. Chemical Safety and Hazard Investigation Board

# Background: Caribbean Petroleum Corporation (CAPECO)

- Bayamón, PR
- 1955: Refining operations estab.
- 1955-87: Multiple owners
- 2000: Refining operations terminated
- 2001: Declared bankruptcy
- 2003: Reorganized to petroleum storage and distribution facility
- 2009: Incident occurred
- 2010: CAPECO declared bankruptcy

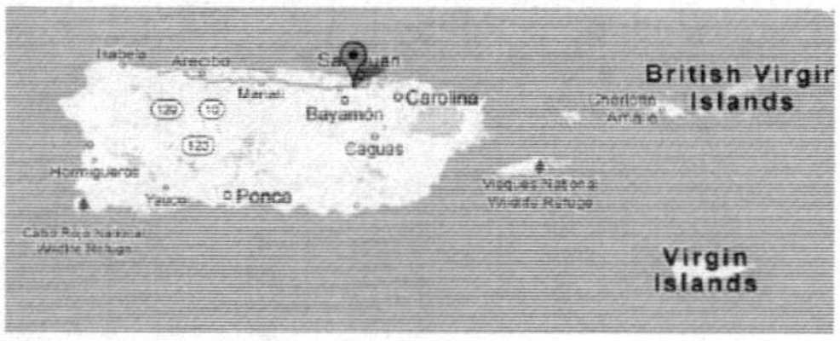


U.S. Chemical Safety and Hazard Investigation Board

CSB U.S. Chemical Safety and Hazard Investigation Board

## Aerial View of Terminal

Pumped to Puerto Rico Electric Power Authority and the airport

Loaded onto tanker trucks for distribution to gas stations across Puerto Rico

CSB U.S. Chemical Safety and Hazard Investigation Board

## Site Operations

- **Staffing**
  - **2 tank farm operators and 1 wastewater treatment operator per shift**
  - **3 shifts, rotating every 8 hours**
- **Tank-filling**
  - **Operators manually opened and closed valves to transfer and blend gasoline before pumping to various locations**
  - **During fuel transfer, operators recorded tank levels in the morning then hourly**

U.S. Chemical Safety and Hazard Investigation Board

## Communication During Fuel Transfer Operations

**Operators communicated via radio:**

- Manual nature of operations required operators to remain in constant contact
- Tank size varied and operators were often manually switching between multiple tanks

U.S. Chemical Safety and Hazard Investigation Board

## Tank Gauging System at CAPECO

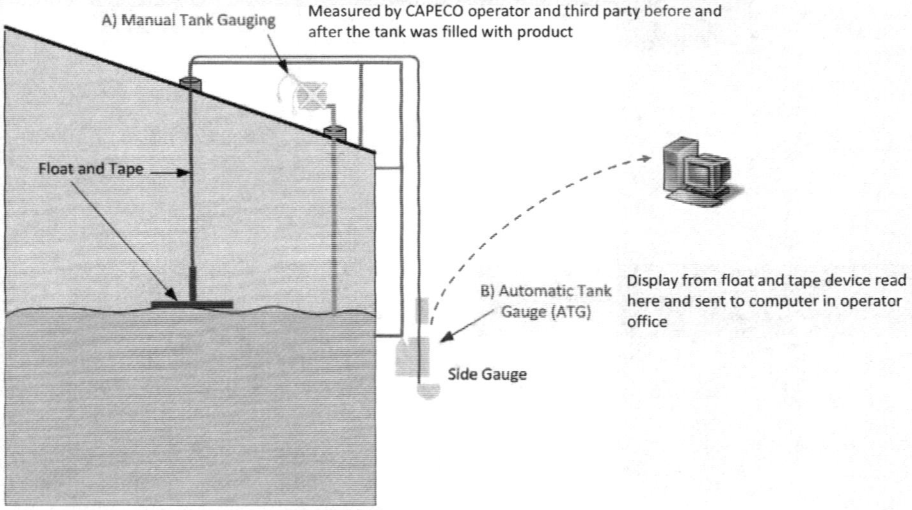

CSB U.S. Chemical Safety and Hazard Investigation Board

## CSB Animation of CAPECO Incident

Video Presentation in
Original e-media

9

CSB U.S. Chemical Safety and Hazard Investigation Board

## Surveillance Footage of Explosion

Video Presentation in
Original e-media

10

U.S. Chemical Safety and Hazard Investigation Board

U.S. Chemical Safety and Hazard Investigation Board

## Federal and Local Emergency Response

- 530 firefighters and 900 National Guard responded
- Federal Emergency Order declared by the President
- Federal Emergency Management Agency (FEMA) awarded $3.4 M to 27 municipalities and agencies

CSB U.S. Chemical Safety and Hazard Investigation Board

## Environmental Impact

- Contaminated runoff released into nearby creek, wetlands, storm water channels
  - 171,000 gallons of oil and 22 million gallons of contact water collected and shipped offsite
- CAPECO fined $8.2 M by the Environmental Protection Agency (EPA)
  - Declared bankruptcy August 2010
- EPA assumed responsibility for clean-up

CSB U.S. Chemical Safety and Hazard Investigation Board

U.S. Chemical Safety and Hazard Investigation Board

## Safety Management System Deficiencies

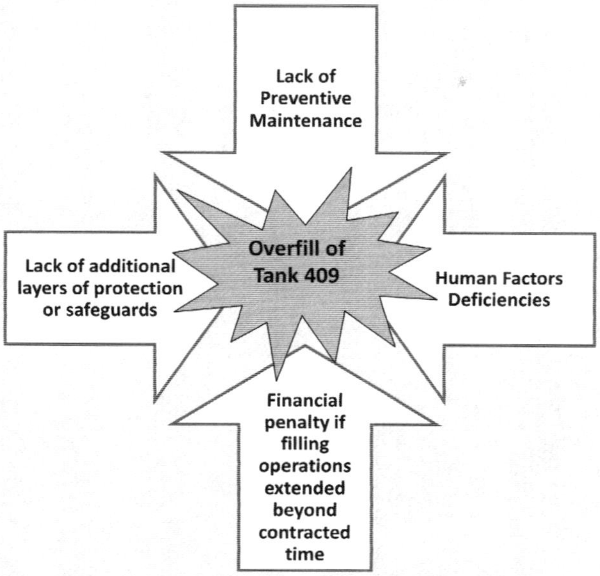


U.S. Chemical Safety and Hazard Investigation Board

## Breakdown of the Level Control and Monitoring System

Safety critical equipment prone to failure

- Float and tape
- Transmitter cards for Tank 409 out of service
- Liquid levels for Tank 409 not visible on operations computer

U.S. Chemical Safety and Hazard Investigation Board

# Tank Gauging System at CAPECO

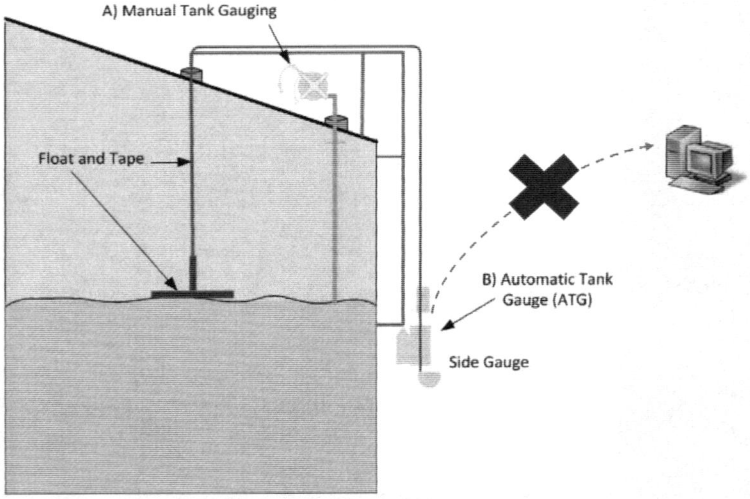


17

U.S. Chemical Safety and Hazard Investigation Board

# Poor Preventive Maintenance Program

**EPA inspection reports (1992-2004) found lack of investment in:**

- **Secondary containment**
- **Tank valves**
- **Appropriate level gauges**
- **Engineering controls**

CSB U.S. Chemical Safety and Hazard Investigation Board

## Human Factors Deficiencies

- Poor lighting at tank farm; difficult to observe overfill
- Lack of formal procedures to fill a tank
  - SOPs not updated since 1999 when refinery was in service
  - Had a two-page document listing activities without assigning responsibility

CSB U.S. Chemical Safety and Hazard Investigation Board

## Human Factors Deficiencies

Differing valve designs made it difficult to tell whether the secondary containment valves were open or closed

# Human Factors Deficiencies

- Lack of sufficient staffing for fuel transfer operations
- Offloading to multiple tanks required more than one operator to open valves
- Operations staff "cracked" valves open to reduce pressure



CSB U.S. Chemical Safety and Hazard Investigation Board

# Safety Management System Deficiencies

**Reliance on *one* layer of protection to prevent an overfill incident**

- No high-level alarm to measure tank levels
- Tanks not equipped with a redundant or independent level alarm

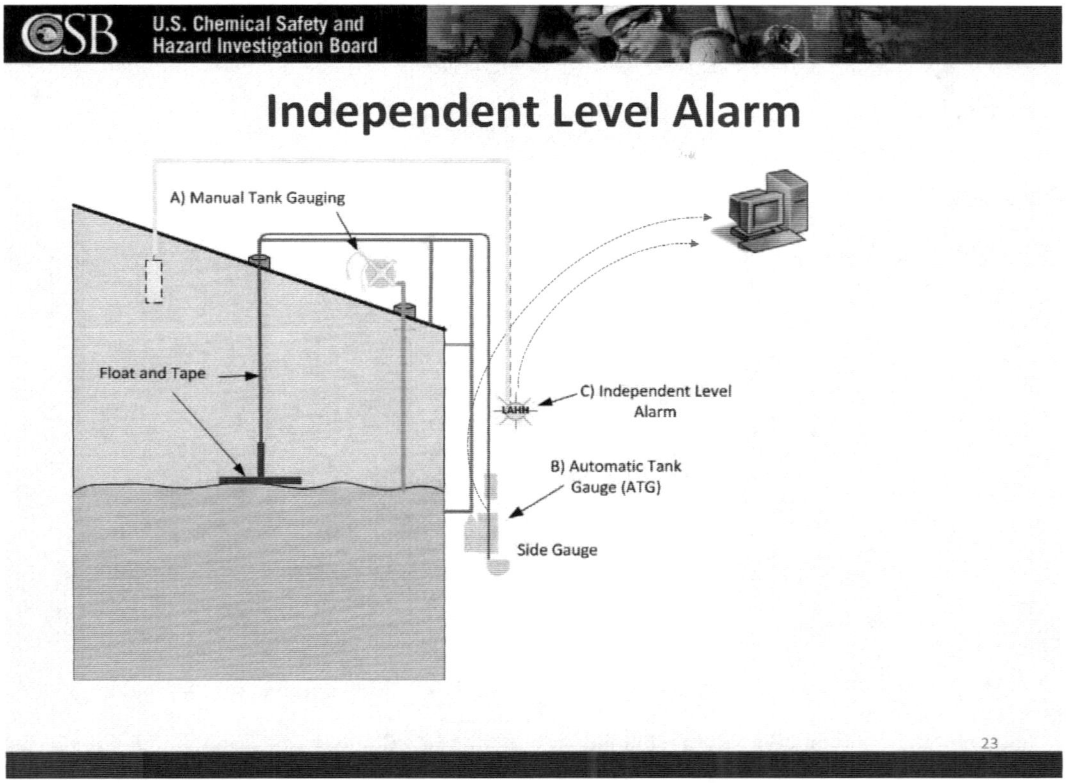
U.S. Chemical Safety and Hazard Investigation Board
Independent Level Alarm
A) Manual Tank Gauging
Float and Tape
C) Independent Level Alarm
LAHH
B) Automatic Tank Gauge (ATG)
Side Gauge
23

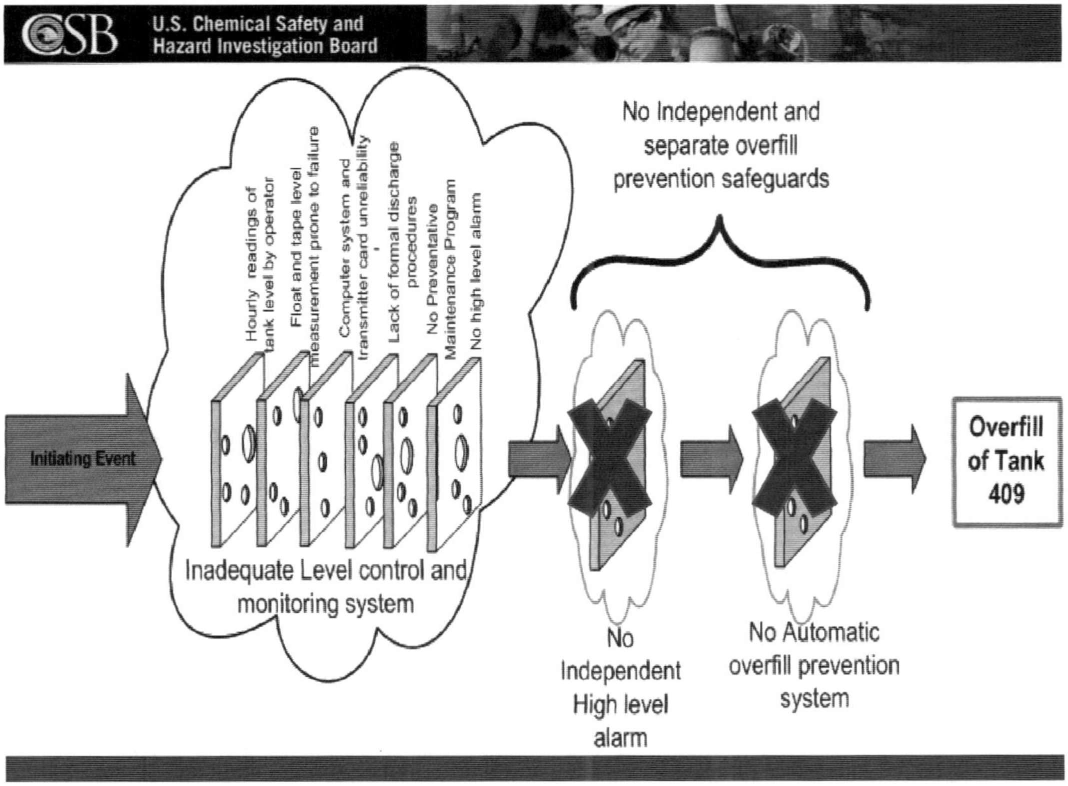
U.S. Chemical Safety and Hazard Investigation Board
Hourly readings of tank level by operator
Float and tape level measurement prone to failure
Computer system and transmitter card unreliability
Lack of formal discharge procedures
No Preventative Maintenance Program
No high level alarm
Initiating Event
Inadequate Level control and monitoring system
No Independent and separate overfill prevention safeguards
No Independent High level alarm
No Automatic overfill prevention system
Overfill of Tank 409

U.S. Chemical Safety and Hazard Investigation Board

## US Regulatory Environment: Overfill Prevention

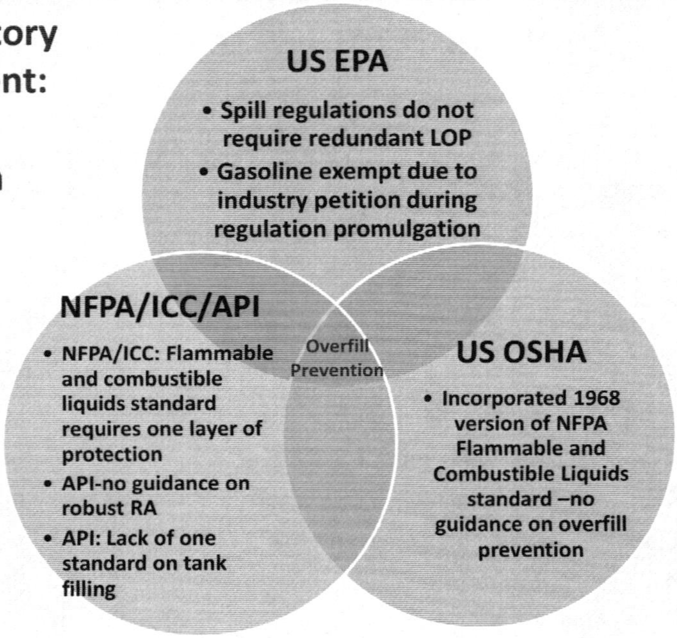


25

U.S. Chemical Safety and Hazard Investigation Board

## Summary of CSB Recommendations

CSB issued 9 recommendations to the US EPA (3), OSHA (1), International Code Council (1), National Fire Protection Association (1), American Petroleum Institute (3) that, if adopted, will:

- Strengthen rules governing above ground storage facilities
- Strengthen overfill prevention systems standards and regulations
- Require risk assessments that consider potential impact on community

26

U.S. Chemical Safety and Hazard Investigation Board

**Thank You.**

27

# Understanding the Risk of Unstable Pressure Relief Valve Operation

3rd Global Summit on Process Safety
Dammam, Saudi Arabia – 4 December 2016

 Dr. Abdulrehman Aldeeb

## Disclaimer

SIEMENS

The information contained in this presentation represents the current view of the authors at the time of publication.

Process safety management is a complex subject and this document cannot embody all possible scenarios or solutions related to compliance.

This document is for informational purposes only.

Siemens makes no warranties, express or implied, in this paper or presentation.

## Today's Agenda

SIEMENS

- Fundamentals of pressure relief valves
- Unstable modes of operations
- Risks of unstable pressure relief devices
- Research program to better understand instability
- Summary




## Fundamentals of pressure relief valves

SIEMENS

- Provide last mechanical line of defense against process operation upsets that may lead to hazardous loss of containment.
- Required proper sizing, selection, and installation.
- Designed to open and relieve excess pressure and to reclose and prevent the further flow of fluid after normal conditions have been restored.
- Are relatively simple and reliable (PFOD ~$10^{-4}$)

## Pressure relief valves are relatively simple & reliable

SIEMENS

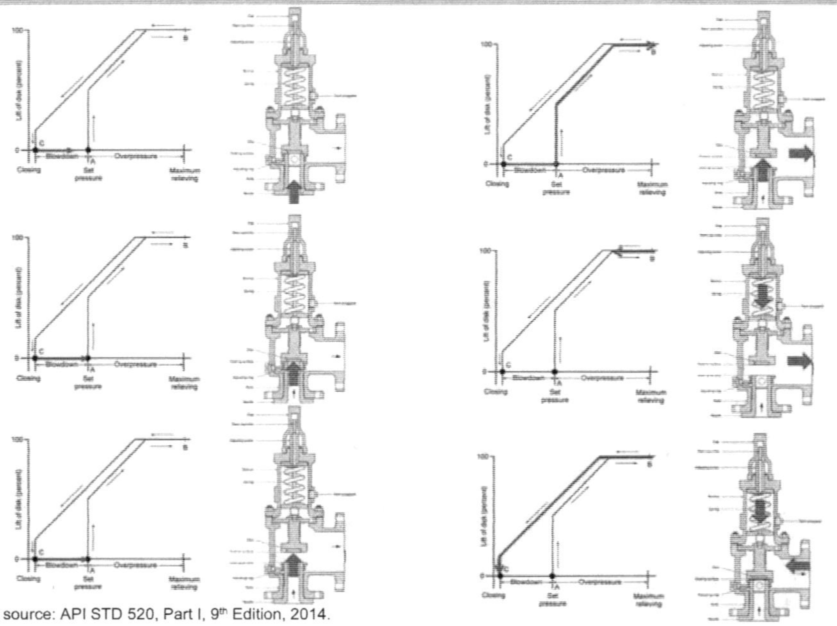


Figures source: API STD 520, Part I, 9th Edition, 2014.





## Installation Requirements

SIEMENS

API STD 520, Part II, 6th Edition, 2015, defines pressure relief devices installation requirements.

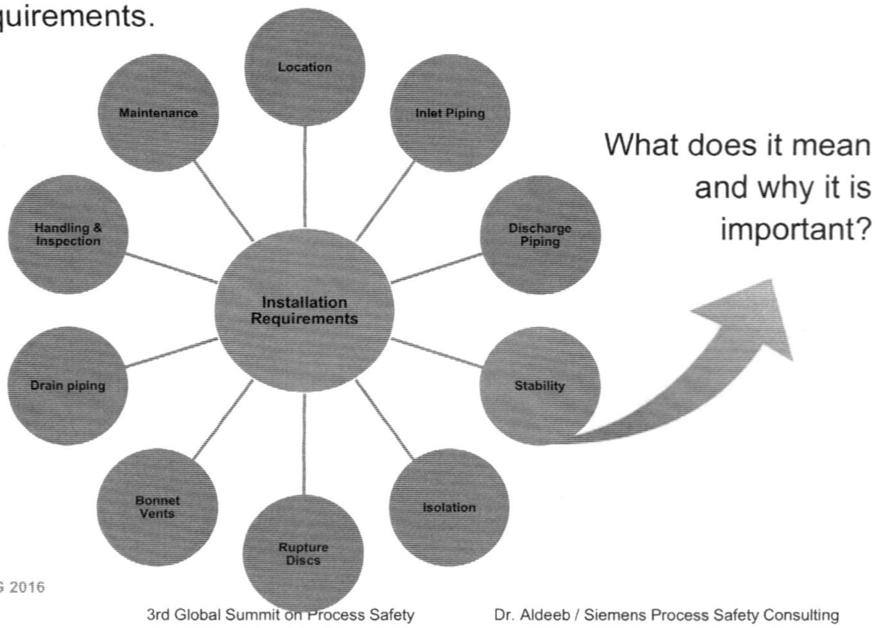

## PRV Dynamic Reponses upon Activation

SIEMENS

**PRV stability is a function of the disc lift magnitude and opening/closing cycles frequency**

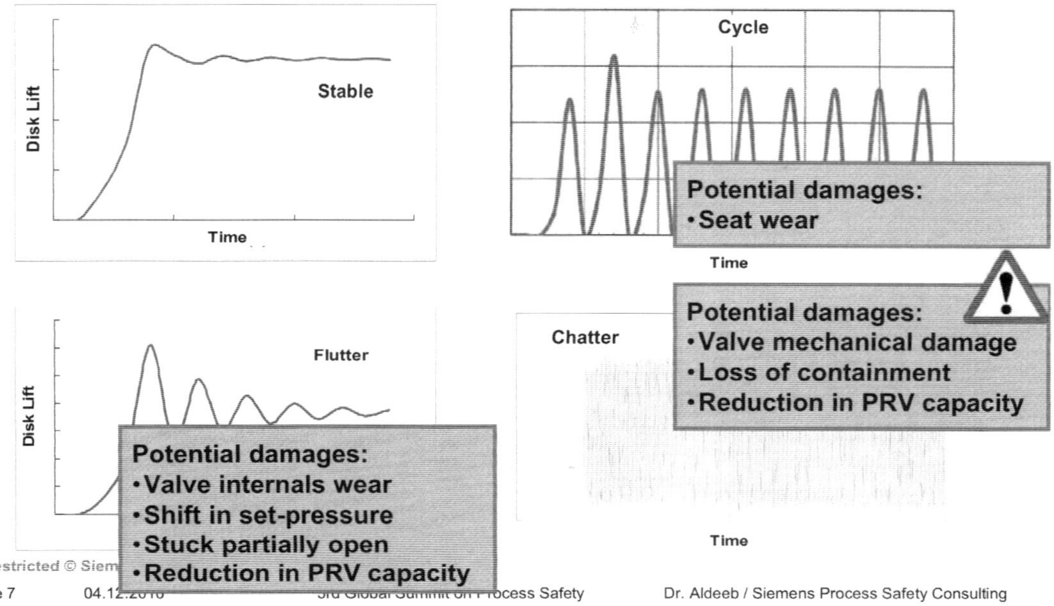




## Potential Causes of Instability

SIEMENS

- **Excessive inlet piping pressure losses**
  Valve will open and re-close rapidly if the nozzle pressure fall more than 3% of PRV set pressure.
- **Excessive build-up backpressure**
- **Acoustic interaction between the disc and fluid waves**

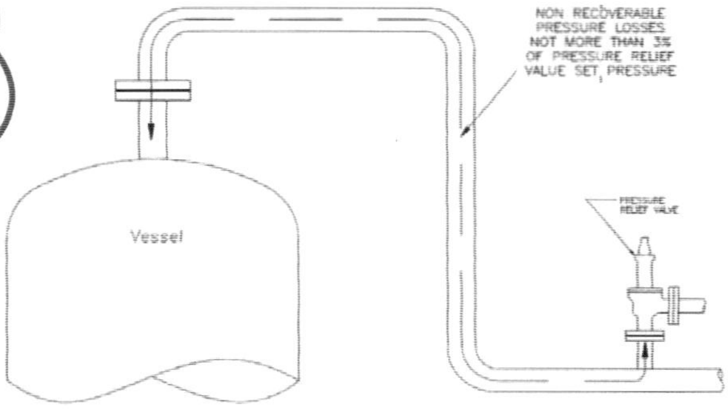


Figure source: API STD 520, Part I, 9th Edition, 2014.

## Industry Survey

SIEMENS

### Scope

- >3,500 pressure relief audits
- Span 2000 to 2014
- Sample pool:
  - >175,000 Equipment
  - >80,000 PRVs

### Installations Deficiencies Criteria

- Inlet pressure drop
- Build-up backpressure
- Installation

### Results

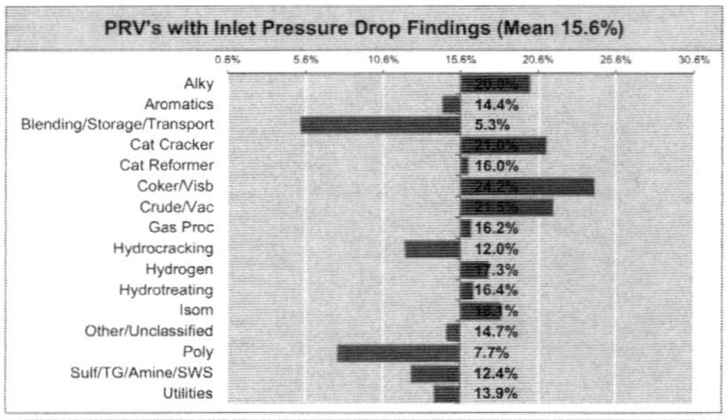


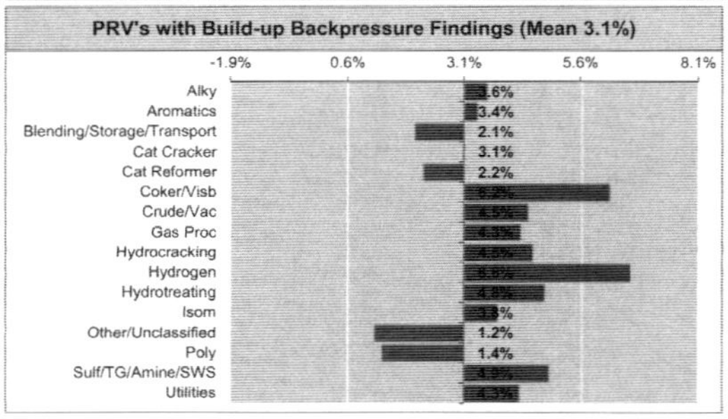






## Industry Survey

SIEMENS

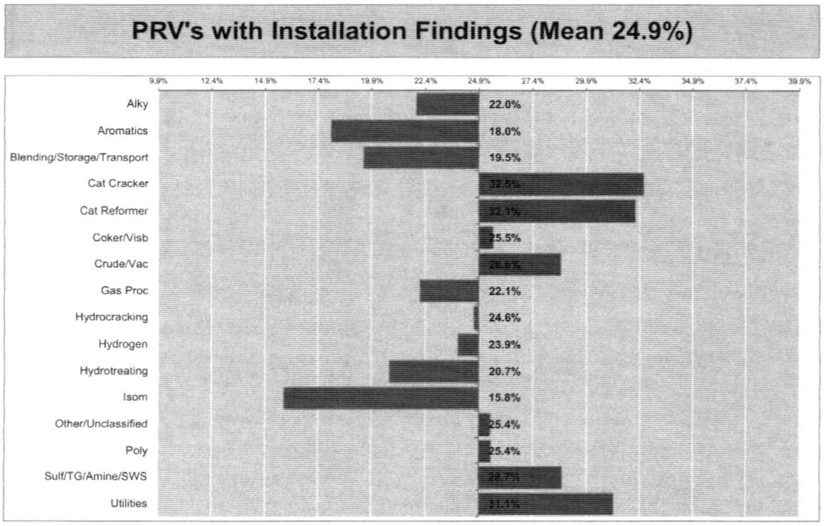

## Potential Causes of Instability

SIEMENS

- **Excessive inlet piping pressure losses**
  Valve will open and re-close rapidly if the nozzle pressure fall more than 3% of PRV set pressure.
- **Excessive build-up backpressure**
- **Acoustic interaction between the disc and fluid waves**

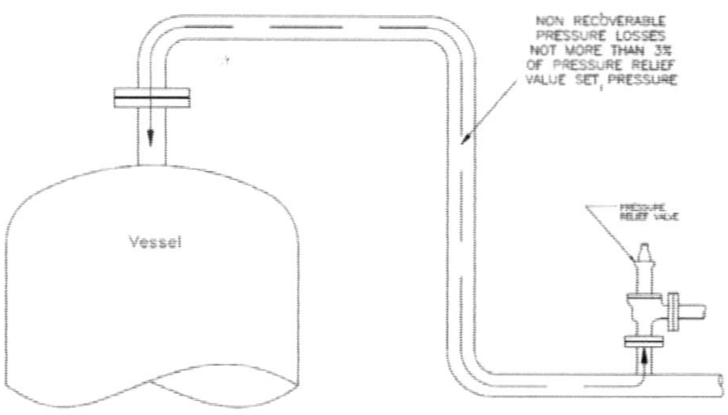


Figure source: API STD 520, Part I, 9th Edition, 2014.

**Steady-state criteria that rarely a reflection of field conditions.**




## Need Better Understanding

SIEMENS

**API PRV Stability Research Project**

Literature & Incident Database Search → Mathematical Modeling → PRV Testing → Model Validation → Findings

**5 coupled nonlinear equations for disk dynamic response:**

$$x = fn(F_x, t, \dot{x}, \ddot{x})$$
$$F_x = fn(P_N, P_B, \dot{m}_N)$$
$$P_N = fn(\dot{m}_o, \dot{m}_N, t)$$
$$P_B = fn(\dot{m}_o, \dot{m}_N, t)$$
$$\dot{m}_N = fn[x, P_N, P_B]$$

*$x$ = valve lift , $F_x$ = force on disk , $t$ = time*

*$P_N$, $P_B$ = pressure at nozzle , backpressure*

*$\dot{m}_o$ = relief mass rate , $\dot{m}_N$ = nozzle mass rate*

- Forces on disc:
  - Total fluid pressure on disc
  - Fluid momentum force on disc
  - Propagating acoustical pressure wave in inlet line
  - Blowdown and backpressure effects
- Dynamic response of disc vs. time:
  - Spring-mass impulse response
  - Capacitance of vessel and inlet line

SIEMENS

## Experimental Program

**18 valves in nitrogen service; >300 tests, inlet and outlet piping, various pressurization rates.**

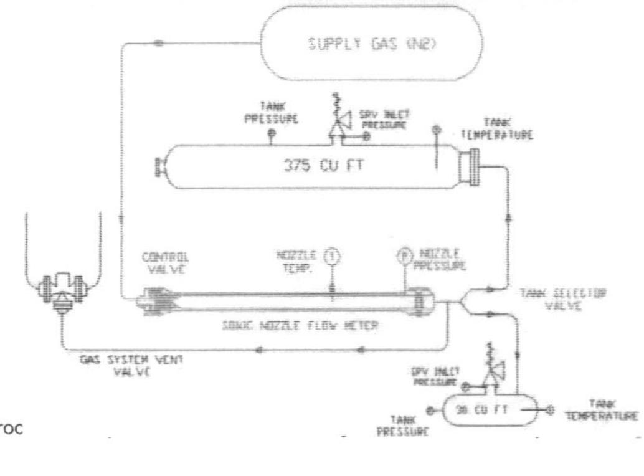





SIEMENS

## Research Program Main Finings

**1. PRV stability is affected by several parameters.**

Stability is affected by the following factors:

- damping of the moving parts
- stiffness of the spring
- opening time of the valve
- blowdown setting
- diameter of inlet piping
- diameter of discharge piping
- length of inlet piping
- length of outlet piping
- mass of moving parts
- geometry of valve/disc

## Research Program Main Finings

SIEMENS

1. PRV stability is affected by several parameters.
2. Mathematical modeling of valve lift response is possible. >80% success!
3. Challenges with modeling parameters availability.

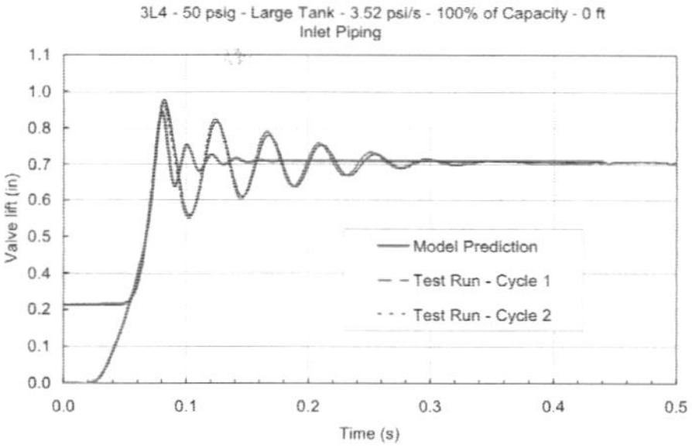


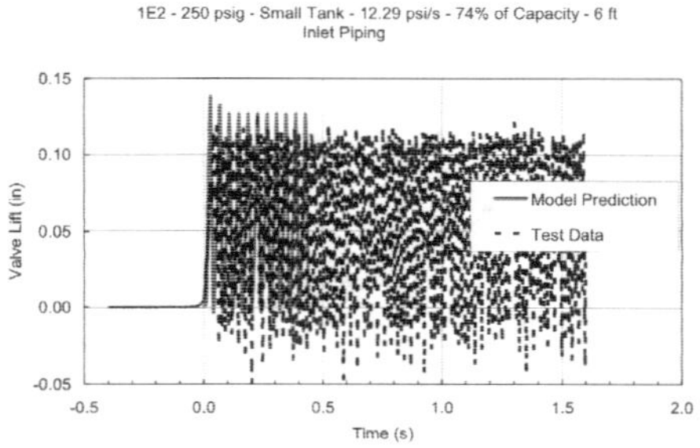





## Summary

SIEMENS

**Pressure relief valves...**

... critical mechanical layer of protection and its integrity is highly dependent on the correct selection, sizing, and installation

... have several operation modes, with some being unstable with potential high risk of loss of containment and mechanical damages. It important to understand the causes of such instable operation and mitigate.

... understanding and meeting stable installation requirements extends beyond simple inlet pressure loses rules. Mathematical modeling may be necessary to identify instability.

## Thank You. Questions?

SIEMENS

**USA:**
Siemens Energy, Inc.
4615 Southwest Freeway
Houston, TX 77027, USA
Phone: +1 (713) 570-2900
E-mail: support.energy@siemens.com

**KSA:**
Al-Raja Tower
Khobar-Dammam Expressway
Al-Khobar 31952, Saudi Arabia
Phone: +966 (13) 865-7748

**UAE:**
Siemens Building
Unit No B-05, Masdar City
P.O. Box 47015, Abu Dhabi, UAE
Phone: +971 (55) 200-3380
abdul.aldeeb@siemens.com

## Content

- Who is GEXCON?
- Why use 3D CFD FLACS software for process safety, quantitative consequence and risk analysis
- FLACS-Risk add-on to FLACS
- Example application: H2S analysis & zoning
- FLACS-VR for immersive 3D training environment and communication
- Closing remarks

GEXCON
SOFTWARE

## GEXCON today

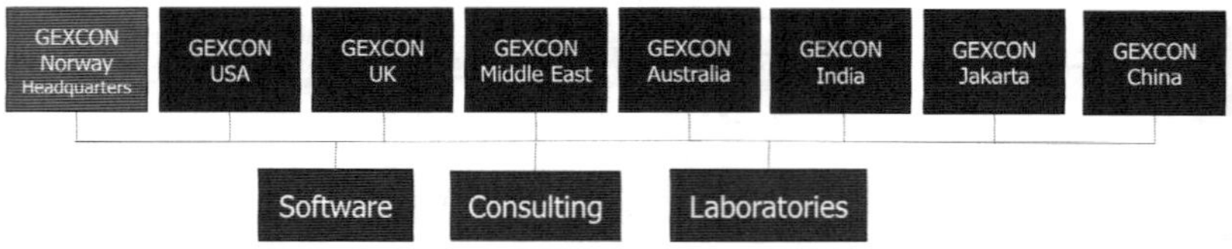

- 35 years of focused research and consultancy in the parent company Christian Michelsen Research (CMR) resulting in the industry standard explosion model FLACS
- GEXCON has over 100 employees worldwide
- Headquarters located in Bergen, Norway
- Other offices are located in Trondheim (Norway), Dubai (UAE), Pune (India), Ormskirk (UK), Bethesda and Houston (USA), Perth (Australia), Shanghai (China) and Jakarta (Indonesia).

## GEXCON Software

- Commercial software house of GEXCON
- Independent separate business unit
- 4 departments:
  - Software Research
  - Software Development
  - Software Support
  - Software Sales
- Flagship product: **FLACS®**
- Recent acquisition of Stormfjord added Scope XP and Scope Planner

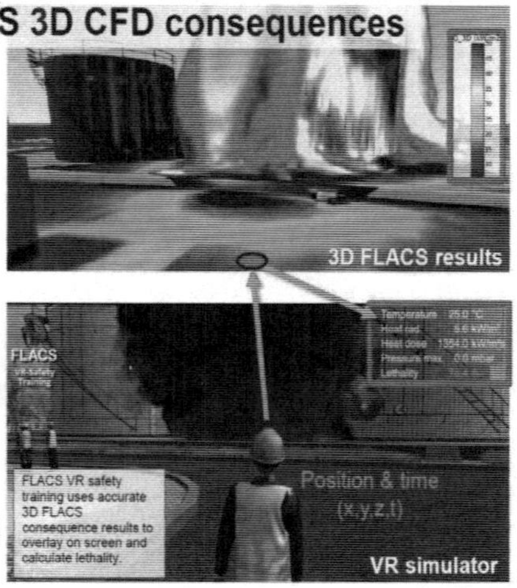

## FLACS
## CFD Dispersion, Fire and Explosions simulator

### Brief history:

**1980s Gas Safety Programs (GSP) =>**
**FLACS development started**

*Major Sponsors: ExxonMobil, BP, Total, Statoil, GDF, Philips, Hydro, Norwegian Power Producers*

**1986 FLACS-86 (First version of FLACS)**
**1990s Various Gas Safety Programs**
(GSP 90-92, 93-96, 97-99, 2000-2002)
**1996 FLACS-software is commercialized**
**2011 Formal US DOT approval of FLACS for LNG studies**
**2014 FLACS-Fire released**
**2016 FLACS-Risk**

FLACS- 95
(1995)

FLACS v10.4
(2015)

GEXCON
SOFTWARE

Selected FLACS customers
Oil & Gas
ExxonMobil
Statoil
TOTAL
TOKYO GAS
GDF SUEZ
PDVSA
bp
PETROBRAS
PETRONAS
PERTAMINA
ADNOC
Saudi Aramco
SINOPEC
中国石油天然气集团公司
CHINA NATIONAL PETROLEUM CORPORATION
Universities
NEWCASTLE
ATM
Indian Institute of Technology Roorkee
China Association for Science and Technology
MEMORIAL UNIVERSITY
The University Of Sheffield.
부산대학교
PUSAN NATIONAL UNIVERSITY
NTNU
Engineering & Consultancies
Technip
AkerSolutio
WorleyParsons
resources & energy
eni
MES
saipem
CBI
HYUNDAI
HEAVY INDUSTRIES CO.,LTD.
中国寰球工程公司
DSME
ARCADIS
EC HARRIS
SAMSUNG
SAMSUNG HEAVY INDUSTRIES
arc
Risktec
TÜV Rheinland Group
Hyder
An ARCADIS Company
BMT Fluid Mechanics
"Where will our knowledge take you?"
GENESIS
JGC CORPORATION
ERM
MMI
Classification
ABS
DNV·GL
CCS CHINA CLASSIFICATION SOCIETY
Lloyd's Register
BUREAU VERITAS
KR KOREAN REGISTER
GEXCON
SOFTWARE

FLACS
Why use 3D CFD FLACS
GEXCON
SOFTWARE

## Typical consequence and risk results from analytical methods

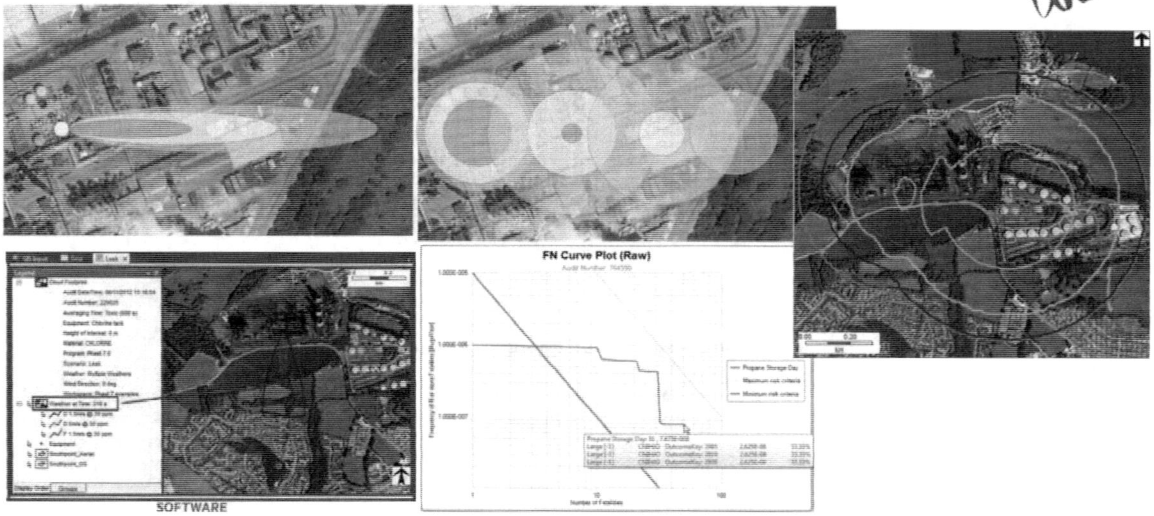


## Unique advantages of FLACS

- FLACS can quantify accurately the impact & consequence of significant releases and thus highlight the real threats
- The software application can assess the impact of mitigation measures
- FLACS can do multiple analyses from one true model e.g. facility sitting, building risk assessment, fire and gas mapping, etc.
- It is possible to visualize the consequence and risk in an easy-to- understand manner

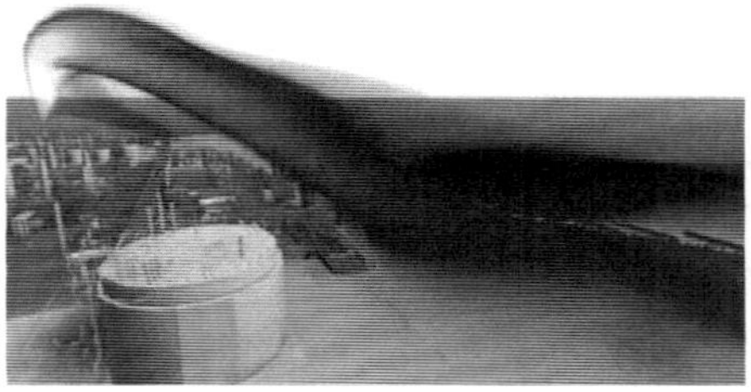

## FLACS Overview Modules

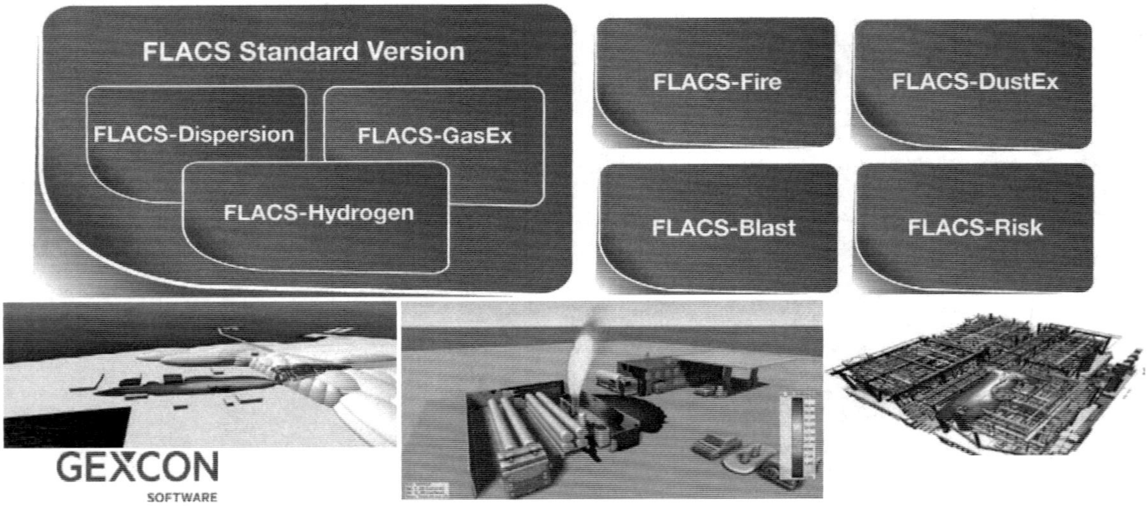


## 3D Geometry does matter...

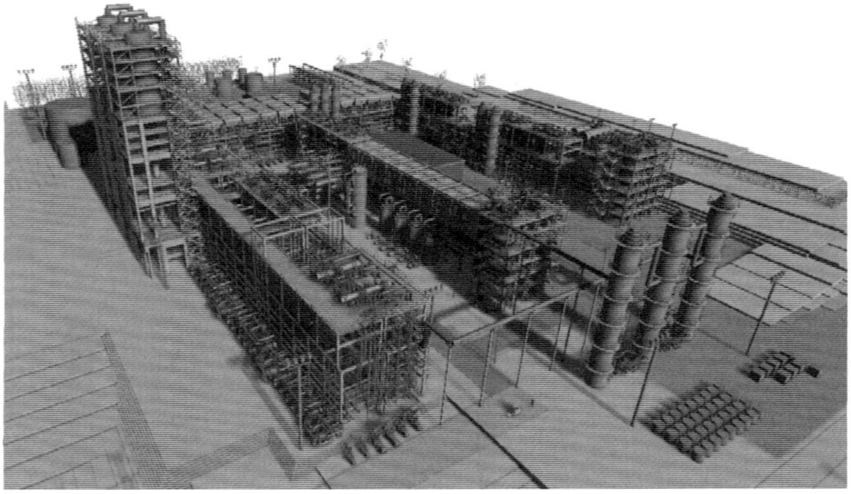

## Typical toxic dispersion results from FLACS

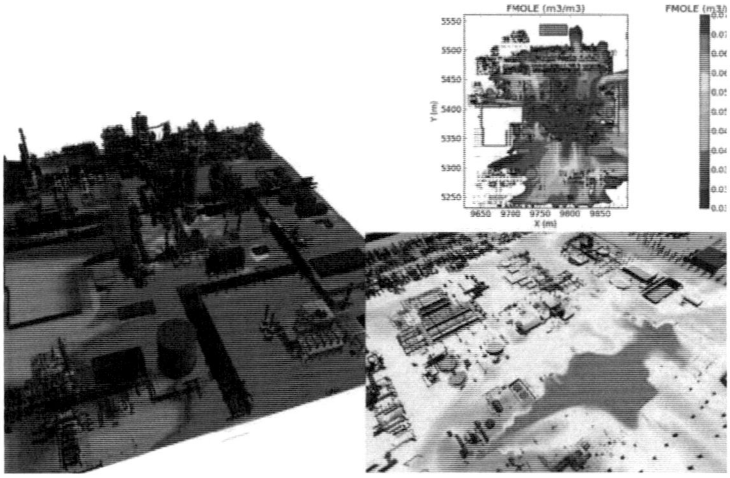


GEXCON
SOFTWARE

## Explosion Results from FLACS

## Fire and Radiation Results from FLACS

## FLACS explosions results for FLACS-Risk

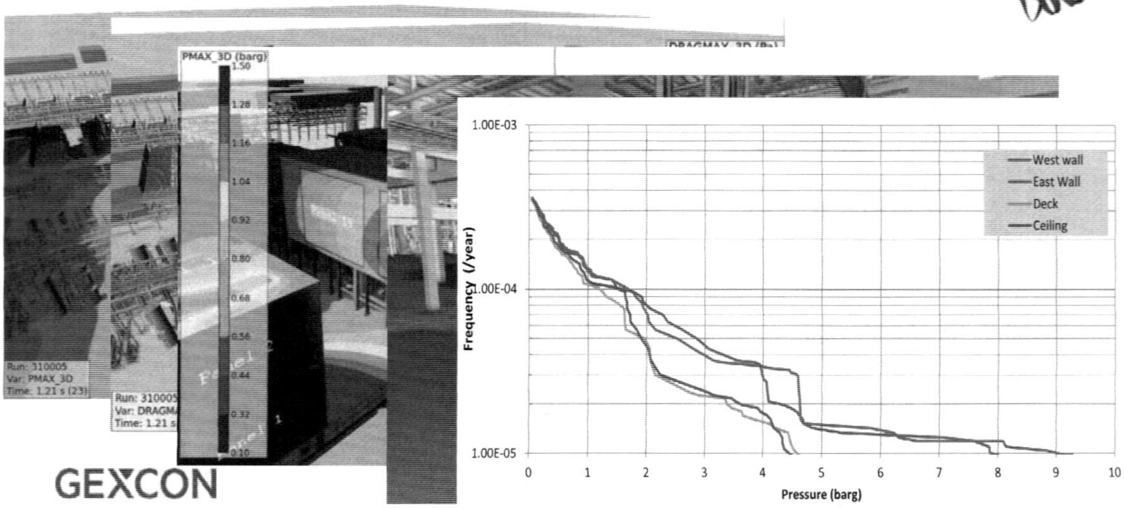


GEXCON
SOFTWARE

FLACS

GEXCON
SOFTWARE

## H2S Toxic Model Assumptions

- Onshore facility
- Different releases 7.5barg pressure, propane (two different isolatable sections, one with 5% H2S and one with 3% H2S)
- 16 Wind condition combinations modelled
- 7 Leak locations / 15 Leak hole size/location combinations modelled (22mm, 70mm and 150mm) – steady state releases
- Assumed steady state release and simulation time 100s
- Total combinations 1,440 (e.g. FLACS simulations)
- Lethality calculated using one of the default TNO Probit function in FLACS-Risk
- A = -10.8338
- B = 1
- N = 1.9

$$PROBIT(-) = a + b \cdot \ln(TDOSE)$$

$$TDOSE(mg/m^3 \cdot minute) = \int_0^t C^n dt$$

$$PDEATH(-) = 0.5 \cdot \left[1 + erf\left(\frac{PROBIT - 5}{\sqrt{2}}\right)\right]$$

$$erf(x) = \frac{2}{\sqrt{\pi}} \int_0^x e^{-t^2} dt$$

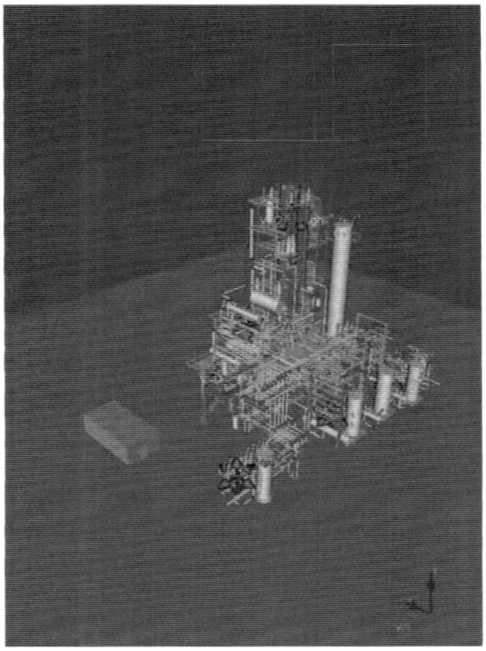

GEXCON
SOFTWARE

## Analysis setup in FLACS-Risk

- Define Project & analysis type
- Load geometry or build one from scratch
- Define gas composition for the various streams and define isolatable sections
- Specify weather conditions and incoming probabilities
- Define leak locations, conditions and frequencies
- In this case there is no need to define the ignition probabilities as this is a toxic only analysis
- Adjust the grid for the analysis and simulations parameters
- Launch the calculations
- Ready for viewing the results

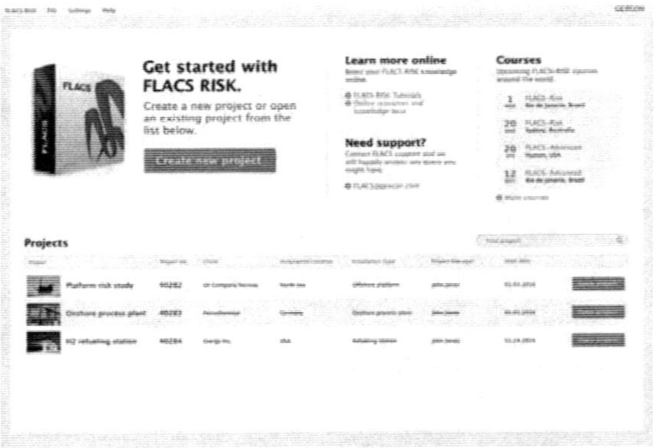


FLACS-Risk Workflow

GEXCON
SOFTWARE

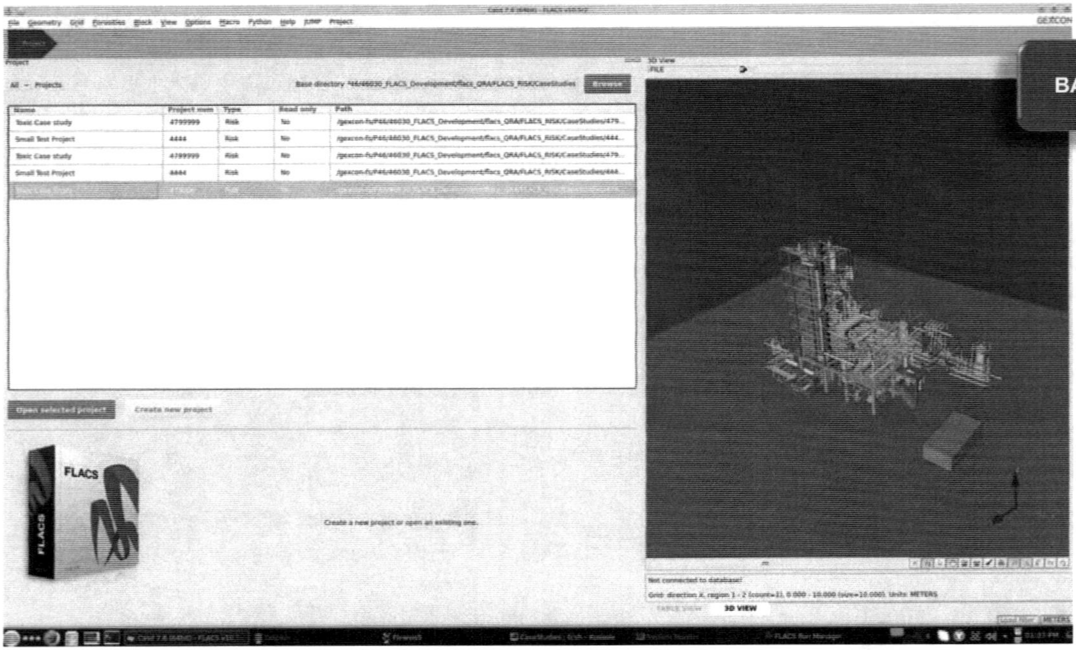


BACK

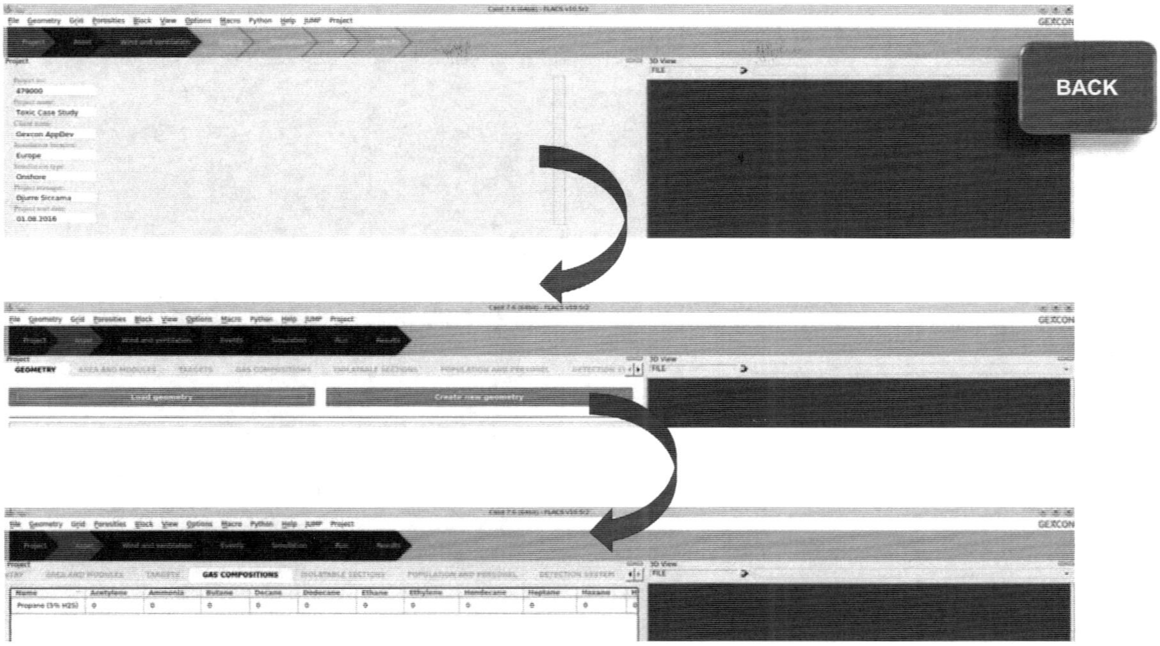
BACK

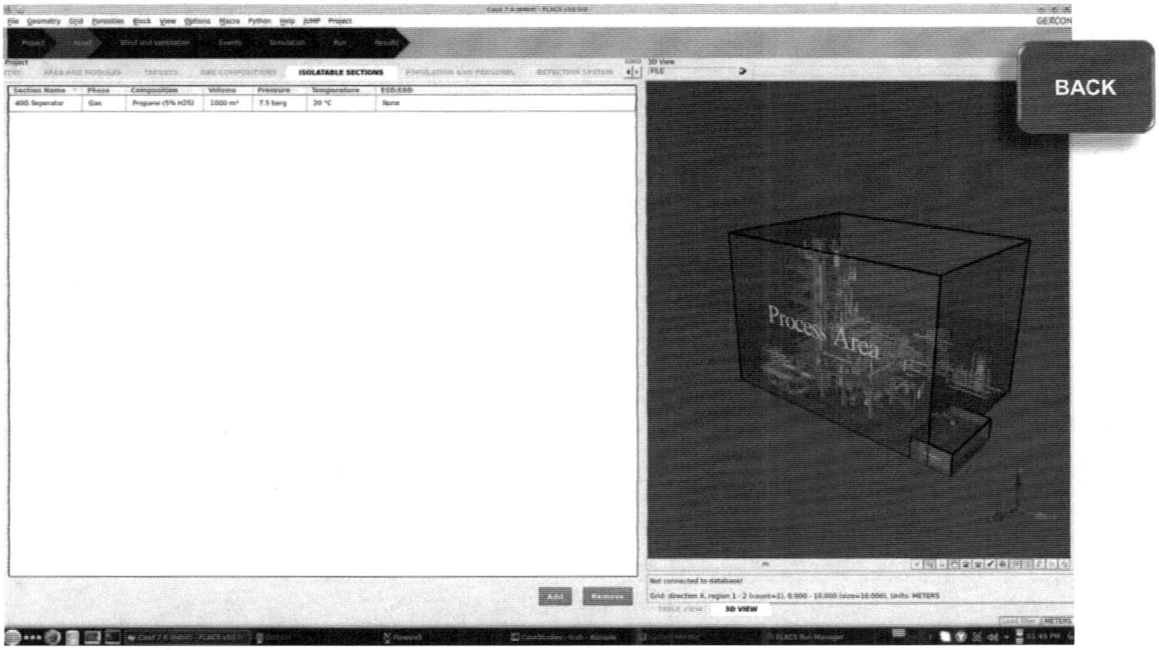
BACK
Process Area

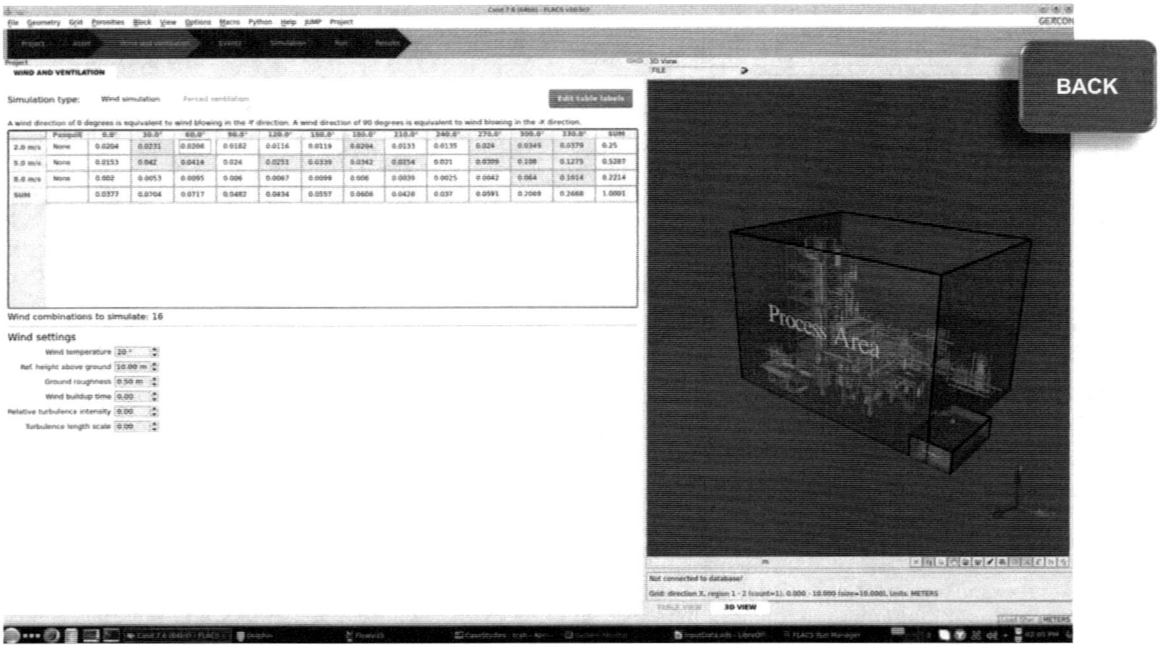


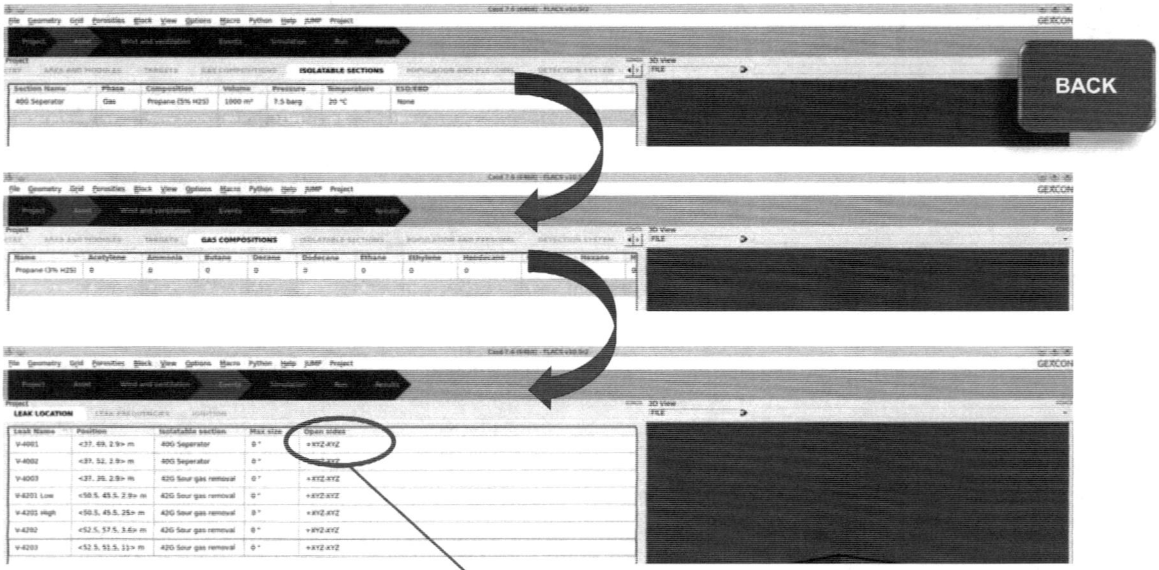


6 release directions for each along x, y and z in both positive and negative directions

GEXCON
SOFTWARE

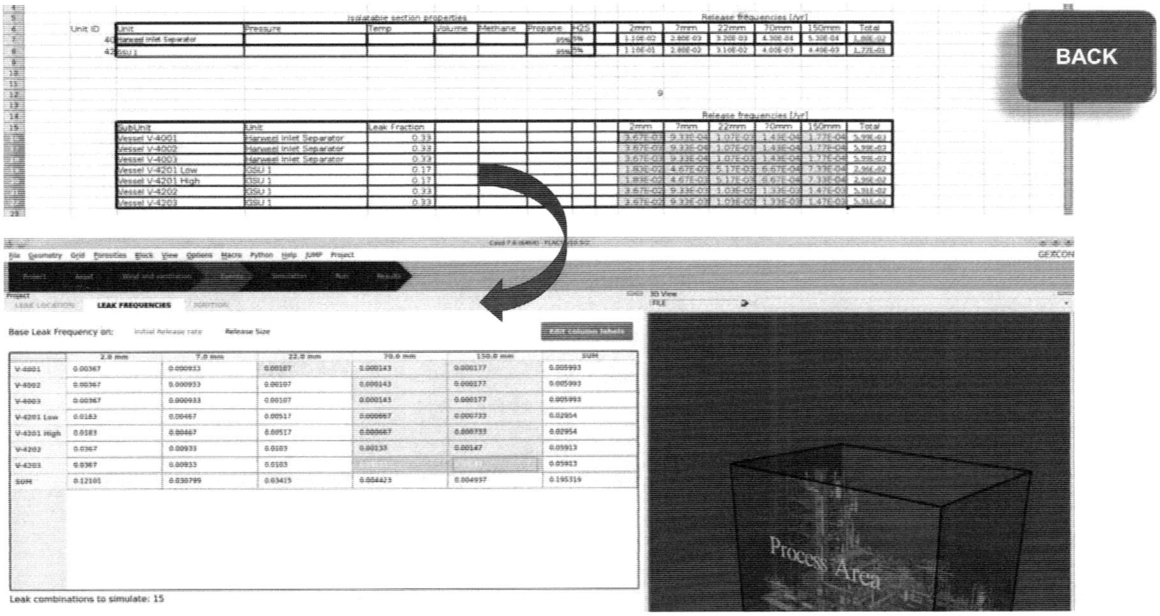
BACK
Base Leak Frequency on:
Leak combinations to simulate: 15
Process Area

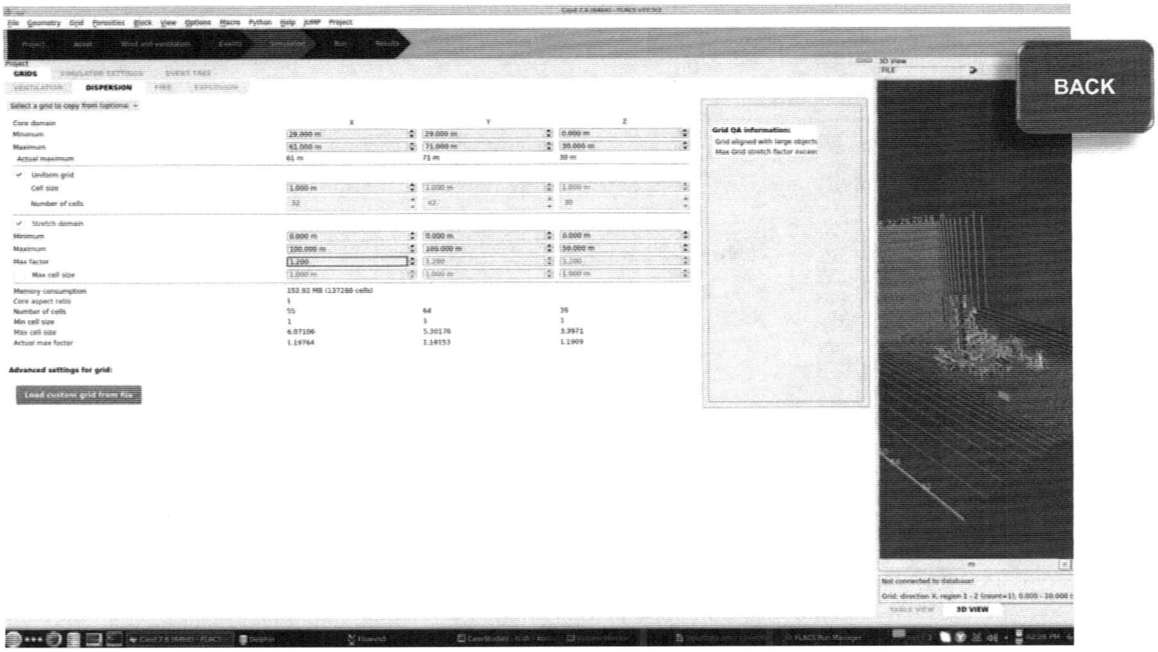
BACK

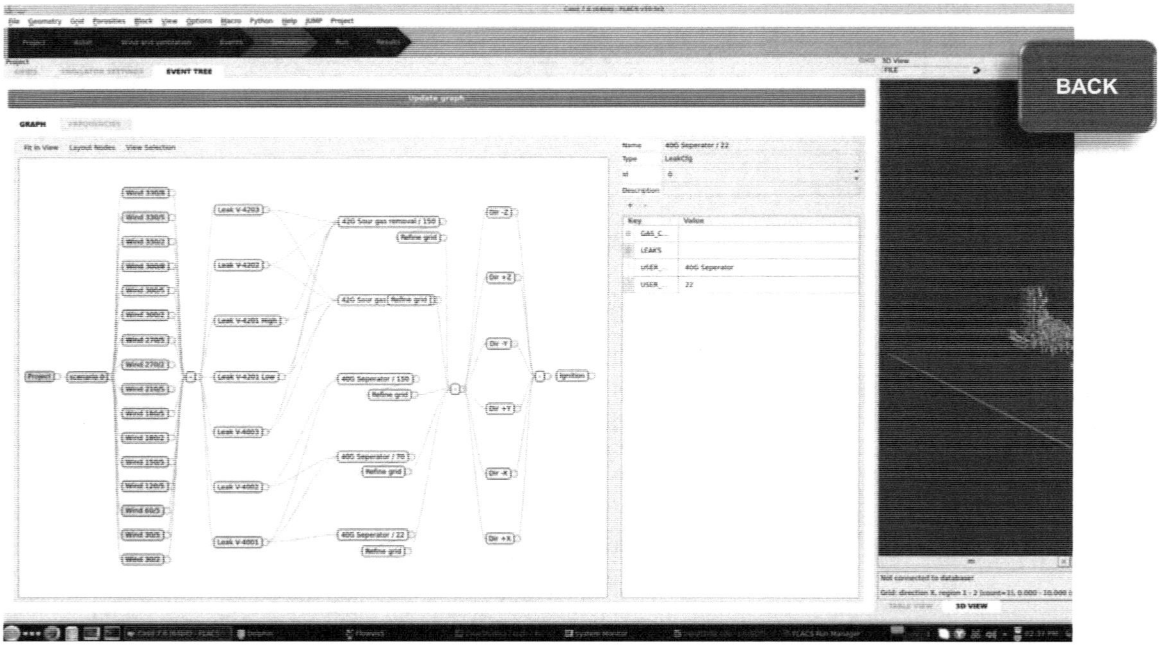


## Likelihood of death (/year) in open space

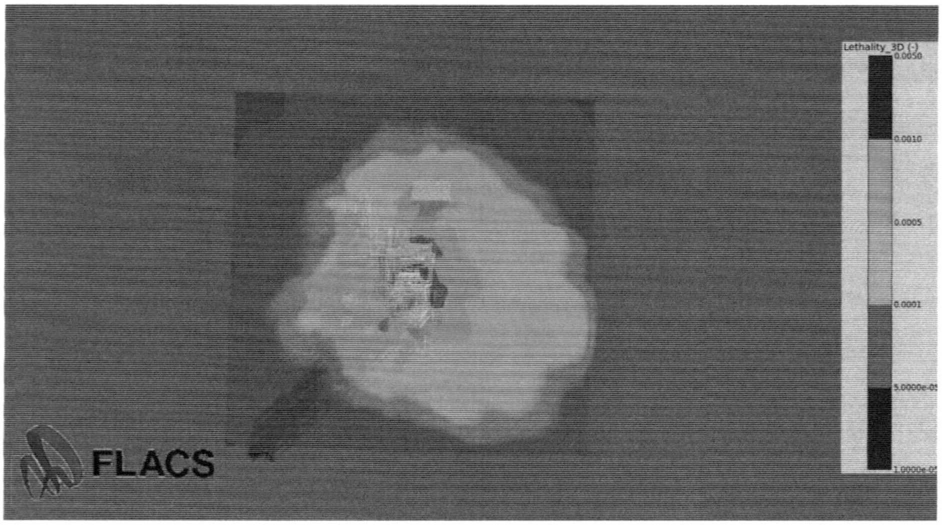

## 2D cut plane representing likelihood of fatality due to H2S

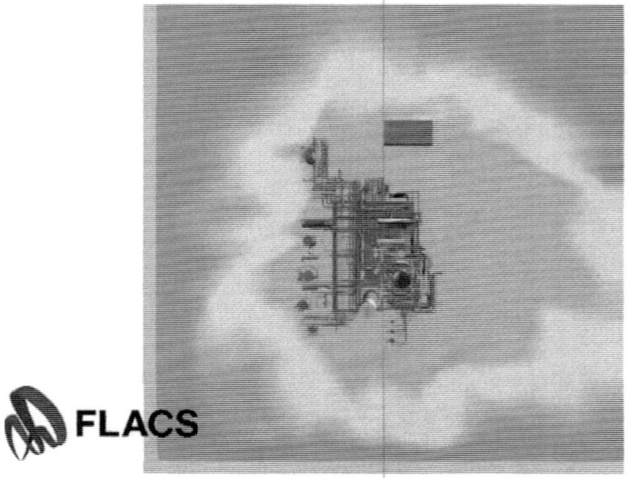


Semi-transparent cut planes showing exceedance contours results using FLOWVIS existing colour palette

Red for >= 1e-3/year
Amber for >= 1e-4/year
Yellow for >= 1e-5/year
Green & blue for >=1e-6/year

## 2D cut plane representing likelihood of fatality due to H2S

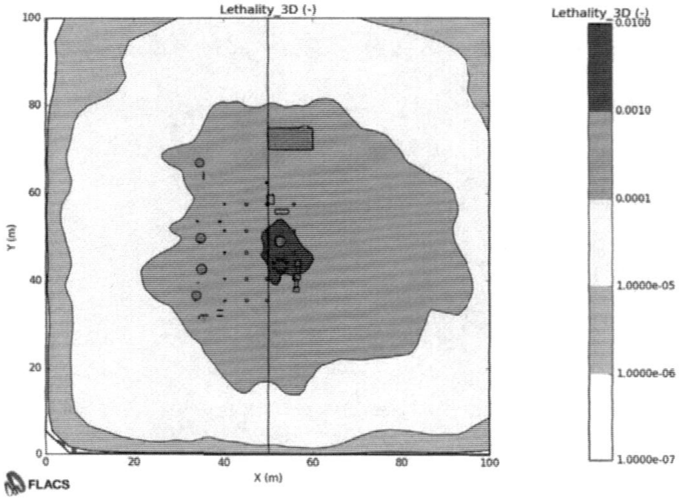


Horizontal cut planes showing exceedance contours

Red for >= 1e-3/year
Amber for >= 1e-4/year
Yellow for >= 1e-5/year
Green for >=1e-6/year

## Red zone classification

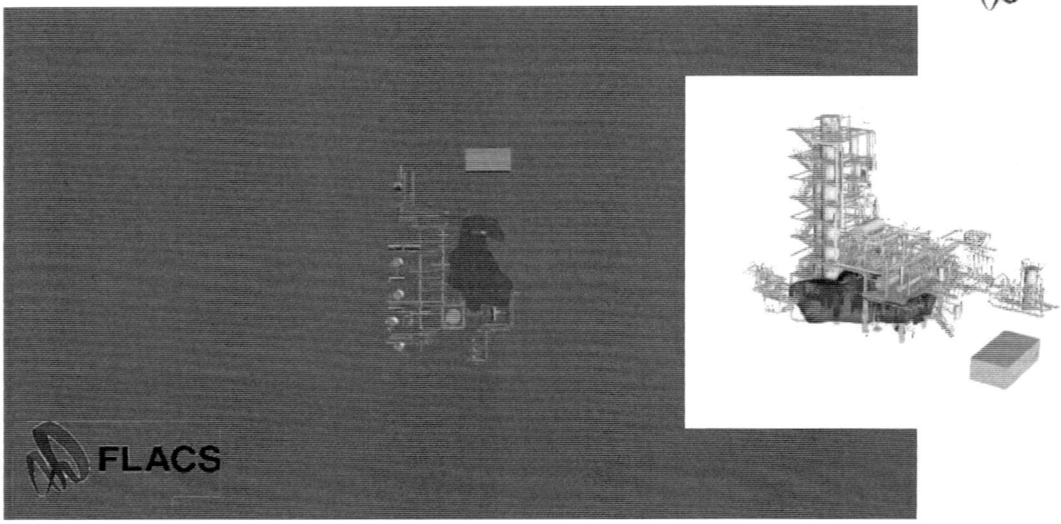


## Amber zone classification

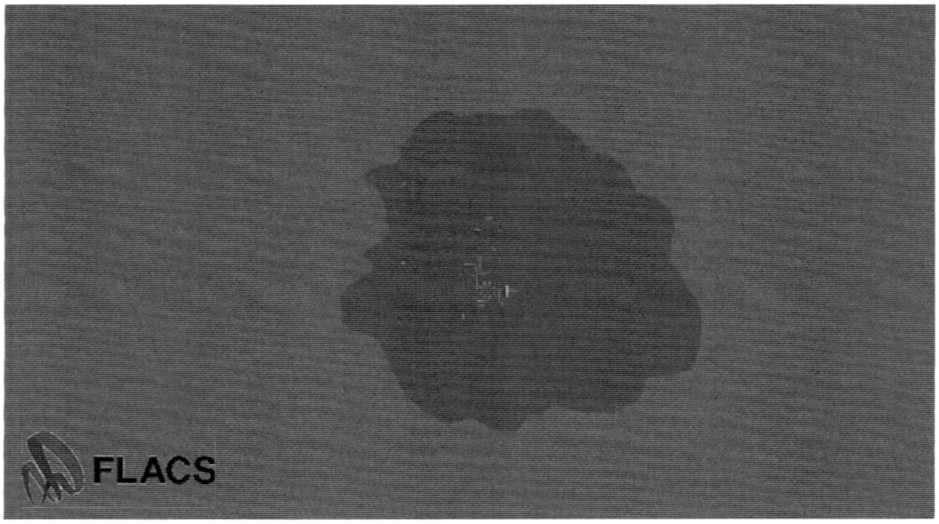

## Yellow zone classification

## Green classification zone

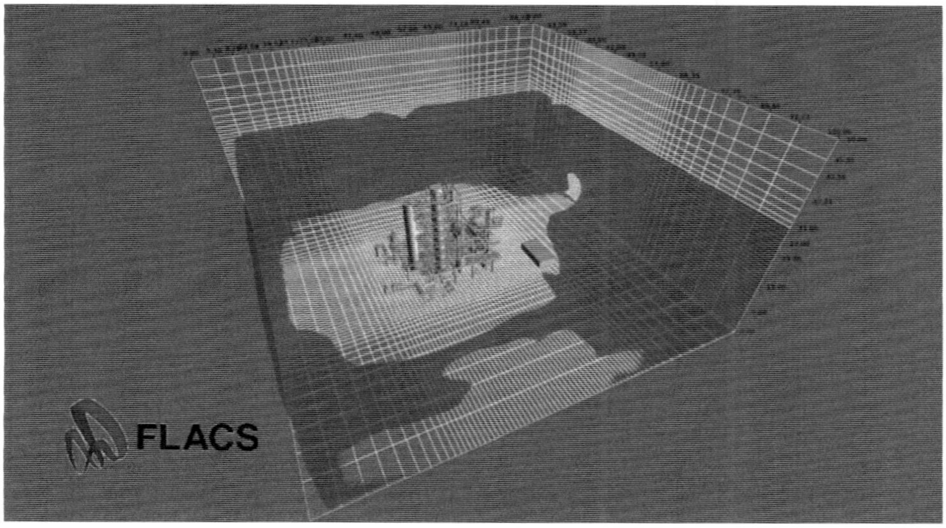

## H2S concentration measured at HVAC inlet

## H2S concentration measured at HVAC inlet

GEXCON
SOFTWARE

## H2S concentration exceedance curves at HVAC inlet

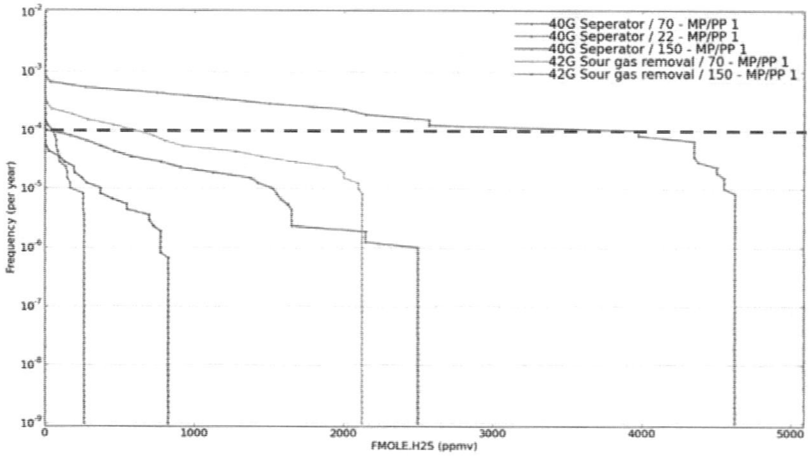


Split by isolatable sections (5 in total)

## H2S concentration exceedance curves at HVAC inlet

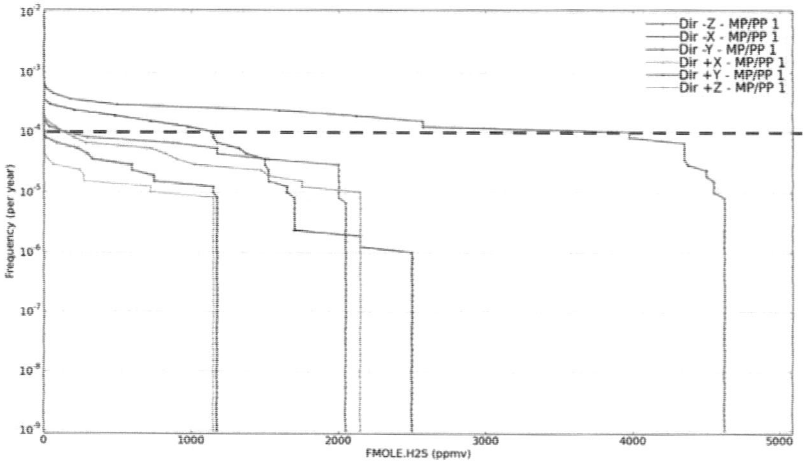


Split by direction of releases
+Y direction releases are the worse offenders
Which makes sense as they are the ones directed towards the building

FLACS

# FLACS-VR

GEXCON
SOFTWARE

## Gexcon Virtual Reality (VR) technology and experience

- 2009 development by CMR for Statoil
- 3DRM framework
- 2015 Stormfjord acquisition
- VR development using Unity3D game engine

## Example of application of VR using Unity3D

FLACS

## Example of application of VR using Unity3D

FLACS

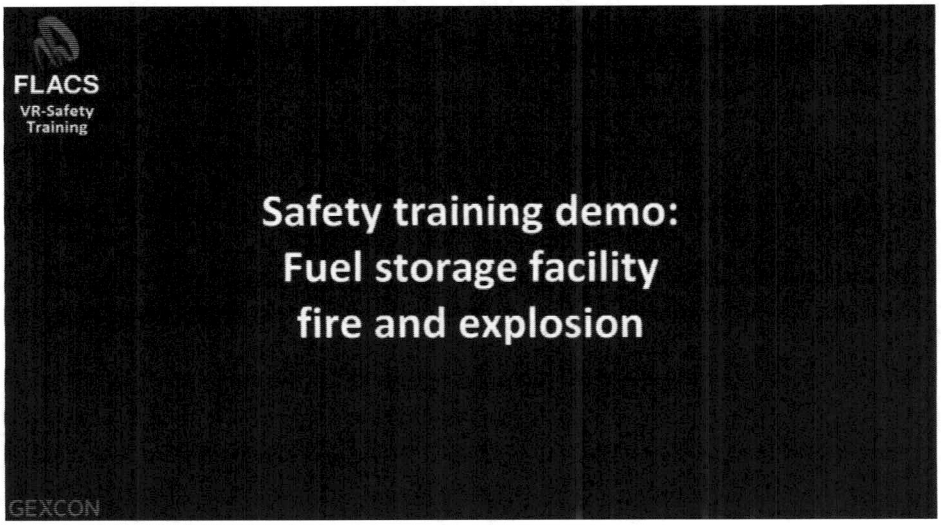

## Conclusions

- FLACS is the most suitable tool to predict accurately the ventilation field and the consequences of a gas (toxic and/or flammable) or liquid release, leading to a fire or an explosion
- FLACS has been extensively validated over the past 20 years and has become the standard worldwide for explosion analysis and 3D consequence simulation software for process safety
- FLACS-Risk builds on FLACS to combine frequency and probability data to predict the risk to assets and people and will help quantify the impact of control and mitigation measures
- FLACS-Risk can be used to perform probabilistic analysis to design effectively inherently safe assets and protect workers
- FLACS and FLACS-Risk can be combined with dynamic structure response analysis techniques (e.g. Non- Linear Finite Element Analysis) to assess the feasibility of design options based on correct prediction of blast loads for instance
- A FLACS analysis can help optimization PFP thanks through a Fire Risk Analysis (FRA) using FLACS-Risk, which will determine accurately the fire threats to the facility
- GEXCON has to technology to include the effects of the assessment software FLACS into a virtual environment to promote the communication of risk to personnel

GEXCON
SOFTWARE

## Thank you
## Any Questions?

**Dr Pascal Le Gal**
**+971 50 666 7935**
Pascal.legal@gexcon.com

GEXCON GEXCON GEXCON GEXCON FLACS

Gexcon Middle East
Dubai
United Arab Emirates
Pascal.legal@gexcon.com

Gexcon.com

## Contents

- Binh Son Refinery (BSR) Introduction
- KPI Implementation – BSR Approach
- Definitions of Key Terms
- Overview of Process Safety Indicator Pyramid
- Progress/ Outcome So Far
- Main Challenges
- Next Steps
- Q&A

2

# Binh Son Refinery (BSR)

3

# BSR Introduction

- Binh Son Refining and Petrochemical Co Ltd. (BSR) has a Dung Quat Refinery and Polypropylene Plant.
- The plants have been Commissioned 2009 and 2010 respectively.
- Refinery Capacity is 6.5 million tons/yr and Polypropylene Plant capacity is 150k tons PP/yr.
- We are 30 % domestic supplier.
- Expansion project is in progress (FEED stage).

4

## Why PS KPI in BSR

- We Measure our performance and Improve by our own:
  - ✓ Self-assessment and self-Improvement. A proactive approach.
- Preventing major incidents.
  - ✓ Recording of major incidents and careful analysis of their root causes can provide lessons to prevent recurrence.
- Improving reliability:
  - ✓ The quality and productivity of a company's operations

5

## Why PS KPI in BSR

- Avoiding complacency:
  - ✓ KPIs provide a constant reminder of asset integrity, the attention needed on process safety management systems and the warnings from near misses and less severe incidents.
- Communicating process safety performance:
  - ✓ Inform management to focus on right path and show commitment ,transparency and progress to work force.

A Constant Reminder

6

## Process safety KPI – BSR Approach

- The four tier approach:
  - ✓ **Use of API 754 for Tier 1 and 2**
  - ✓ **Selected elements by our own experience for tier 3 and 4**
  - ✓ **Transparent to the Stakeholder (Backer's report guideline)**
  - ✓ **Communicate to the management and workforce about the safety system in the plant.**
  - ✓ **Systematic tactic for process safety violations**
  - ✓ **Appropriate corrective actions for Tier 3 and 4 Incidents**

We get what we inspect, not what we expect

7

## Process safety KPI – BSR Approach

- ✓ Objective, few and simple
- ✓ Well defined, capable of being applied consistently across the plant
- ✓ Useful to drive process safety performance improvement and learning.
- ✓ Useful to all stakeholders and allow internal and external benchmarking
- ✓ Statistically valid

Our Goal is Continuous Improvement and Zero Tier 1 and 2 Events in BSR

8

# Indicators

**Lagging indicators:**

Events that that have already occurred and may indicate potential recurring problems and may include fires, releases, and explosions.

**Leading indicators:**

They are forward-looking and indicate the performance of the key work processes; they are intended to give an indication of potential problems or deterioration in key safety systems early enough that corrective action may be taken.

Our Concept Is Keep It In The Pipes

9

# Containment

**Primary Containment:**

A tank, vessel, pipe or equipment designed to serve as primary container or used for transfer of material.

**Secondary Containment:**

Secondary containment systems include tanks, dykes, curbing around process equipment, drainage collection systems into segregated oily drain systems, the outer wall of double walled tanks, etc.

10

## Process Safety Indicator Pyramid

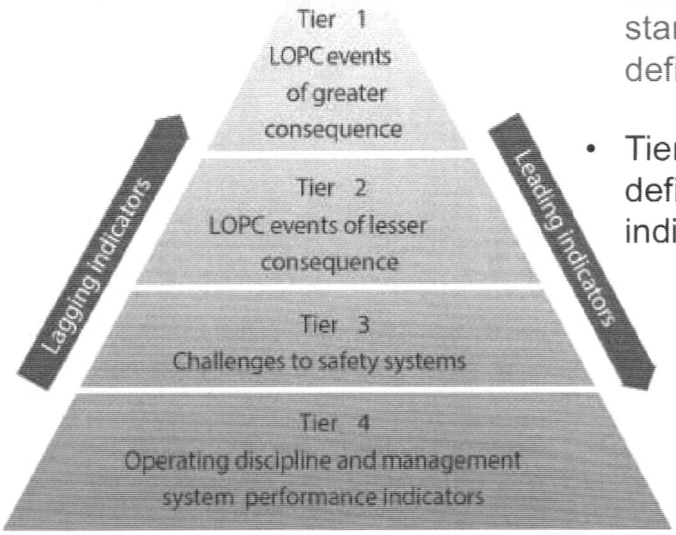


- Tiers 1 & 2 are RP- 754 standardized definitions
- Tiers 3 & 4 are BSR defined performance indicators.

11

## API 754 – Process Safety KPIs

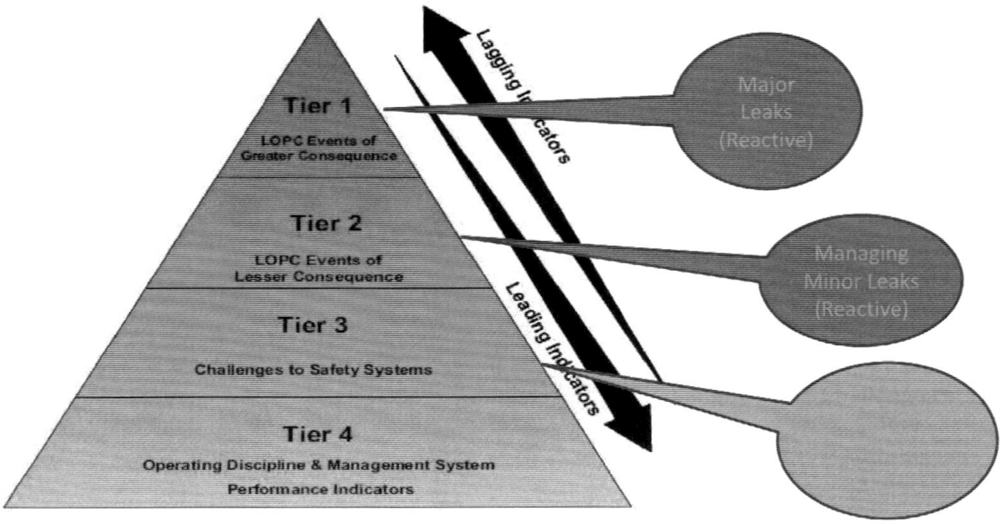


12

# BSR Process Safety KPI – 2016

**Tier 1**

- Nil

**Tier 2**

- Nil

**Tier 3**

- Operating Limit Excursions –
- Minor LOPC –
- Relief valve fail to Danger -
- No of Repeated Incidents -
- Trip Over rides / Off ( MOS) -
- GDs / Deluge / FM 200 failure -
- No of Temporary MOC Overdue -

**Tier 4**

- Relief valve Calibration overdue -
- No of overdue Hi Risk / Hi priority actions -
- No. of management Safety Walks -
- % of Preventive Maintenance Compliance –
- % of ERP conducted vs plan –
- % of Contractor management audit conducted vs plan
- % of Alarm management compliance - %
- % of Operation Procedure reviewed vs plan-
- % of Hot work permit audit vs plan-
- % of PSM training vs plan –

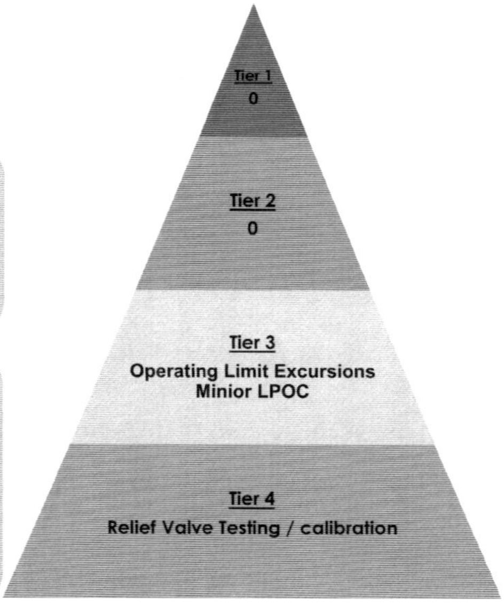


# PS KPI – The Case for Change

- Significant progress shown on leading KPIs ( Zero tier 1 and 2 Incidents) Great achievement so for
- Growing set of 'leading' KPIs in 2017
- Positive feedback from recent PSM audit
- Cheering remarks from Insurance risk surveyors (premium reduced about 10%)
- Leading Indicators reporting improved
- Emergency preparedness well experienced

Process Safety Effectiveness

# PS KPI – The Case for Change

- PSM visibility dramatically improved
- Predictive monitoring and identification of performance trends enhanced.
- Noble tool to report management and workforce
- MOC effectiveness / control in good shape
- Contractor compliance Improved
- Repeated incidents reduced
- Learning enriched

Process Safety Effectiveness

15

# PS KPI Implementation – Key Challenges

- Collecting Data from various departments
- Tracking recommendations
- Follow-up Implementations status
- Incident reporting

16

# Next Steps in KPI Implementation

- **KPI awareness sessions/posters**
- **Dedicated person for KPI tracking from PSM team**
- **KPI tracking for concept of acute release**
- **Create a 'chronic unease' with the workforce**
- **Train PSM champions**
- **Promote appreciation awards**

"Create a chronic unease"

17

# Questions

18

ENGINEERING SAFETY CONSULTANTS

*The Global Provider of Functional Safety Expertise and Technical Consultancy*

# 3rd CCPS Global Summit on Process Safety

## Key Performance Indicators in the Management of Functional Safety


**Ron Bell**

**Engineering Safety Consultants Ltd**

ron.bell@esc.uk.net

www.esc.uk.net




## KPI's for Functional Safety

- Functional Safety: The basics
- Key Performance Indicators (KPIs)
- Functional Safety KPIs
- Concluding comments

## KPI's for Functional Safety

- Functional Safety: The basics
- Key Performance Indicators (KPI's)
- Functional Safety KPIs
- Concluding comments




## What is functional safety?

Safety: freedom from unacceptable risk

Functional safety: part of the overall safety that depends on a system or equipment operating correctly in response to its inputs.

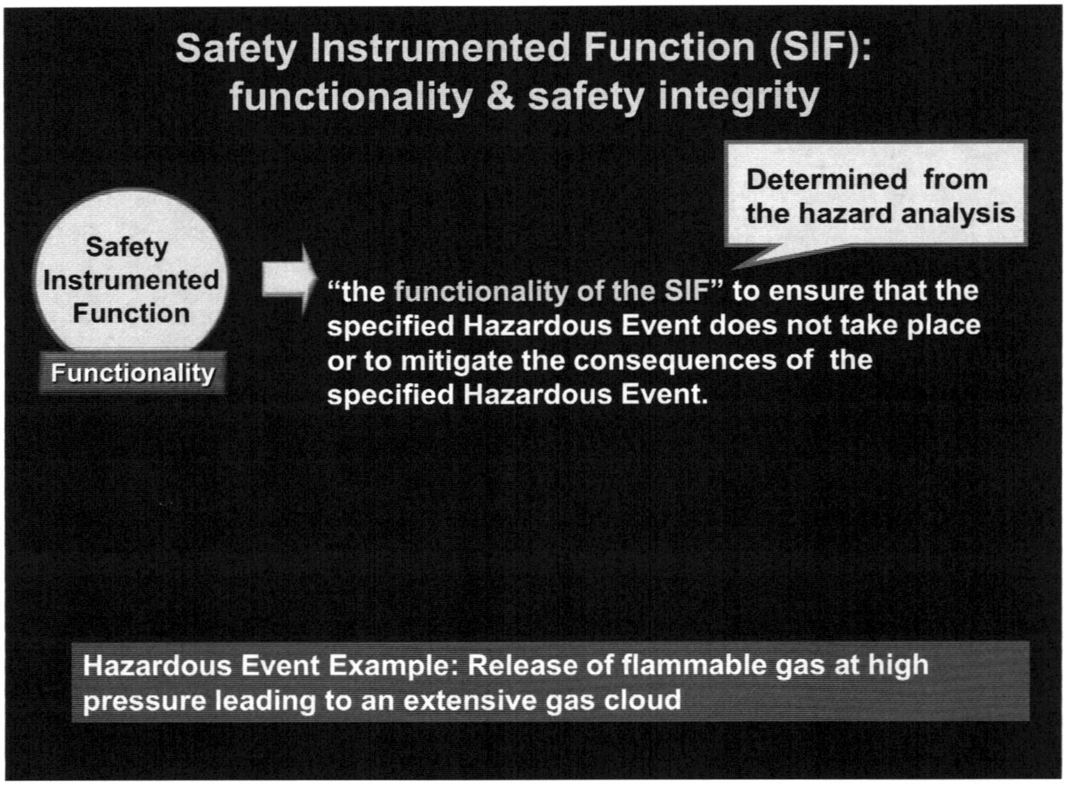
Safety Instrumented Function (SIF):
functionality & safety integrity
Determined from
the hazard analysis
Safety
Instrumented
Function
Functionality
"the functionality of the SIF" to ensure that the specified Hazardous Event does not take place or to mitigate the consequences of the specified Hazardous Event.
Hazardous Event Example: Release of flammable gas at high pressure leading to an extensive gas cloud

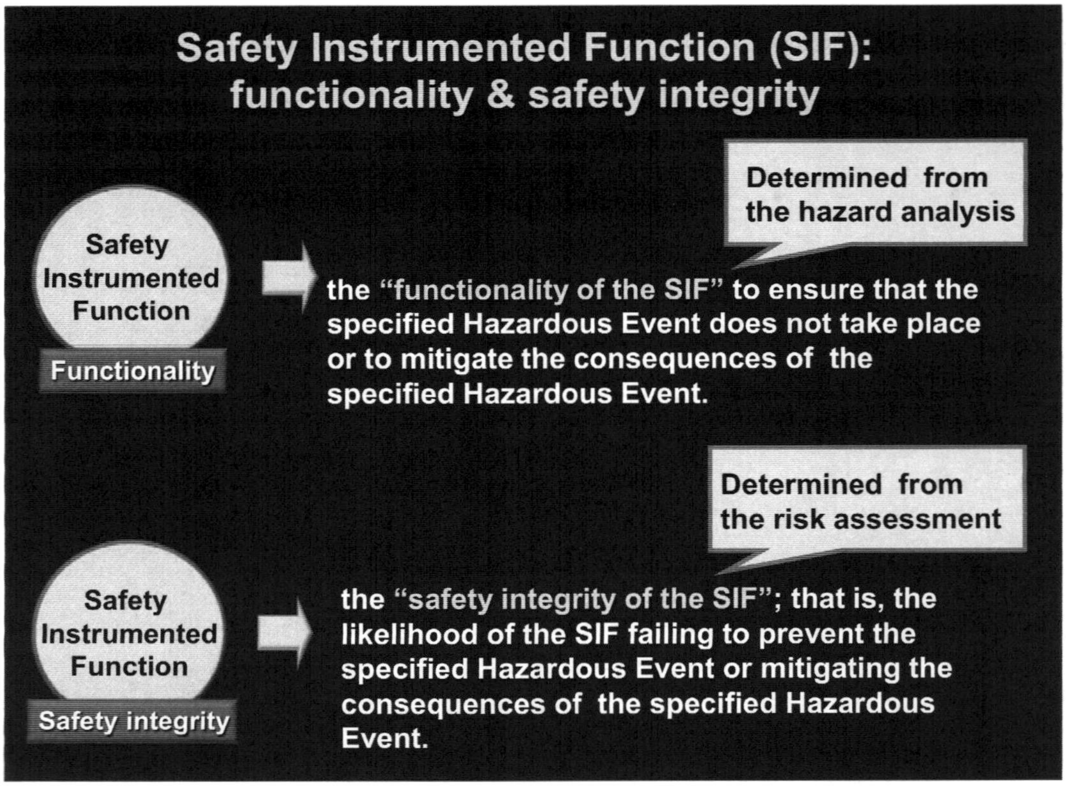
Safety Instrumented Function (SIF):
functionality & safety integrity
Determined from
the hazard analysis
Safety
Instrumented
Function
Functionality
the "functionality of the SIF" to ensure that the specified Hazardous Event does not take place or to mitigate the consequences of the specified Hazardous Event.
Determined from
the risk assessment
Safety
Instrumented
Function
Safety integrity
the "safety integrity of the SIF"; that is, the likelihood of the SIF failing to prevent the specified Hazardous Event or mitigating the consequences of the specified Hazardous Event.

## Safety Integrity Level

| Safety Integrity Levels (SILs) |
| --- |
| 4 |
| 3 |
| 2 |
| 1 |

**One of four possible discrete levels for specifying the safety requirements of the safety functions to be allocated to the safety-related systems**

**Safety Integrity Level 2 = SIL 2**

- ***4 Highest***
- ***1 Lowest***

## Safety Instrumented Function (SIF): functionality & safety integrity

**Safety Instrumented Function — Functionality**

**the "functionality of the SIF" to ensure that the specified Hazardous Event does not take place or to mitigate the consequences of the specified Hazardous Event.**

**Safety Instrumented Function — Safety integrity**

**the "safety performance of the SIF"; that is, the likelihood of the SIF failing to prevent the specified Hazardous Event or mitigating the consequences of the specified Hazardous Event.**

**Example**

**SIF (Functionality):** In order to prevent the rupture of pressure vessel "X", valve "Y" should open in 2 seconds when the pressure in the vessel rises to 2.6 bar.
**SIF (Safety integrity):** The safety integrity of the safety function shall be Safety Integrity Level 2 (SIL 2)

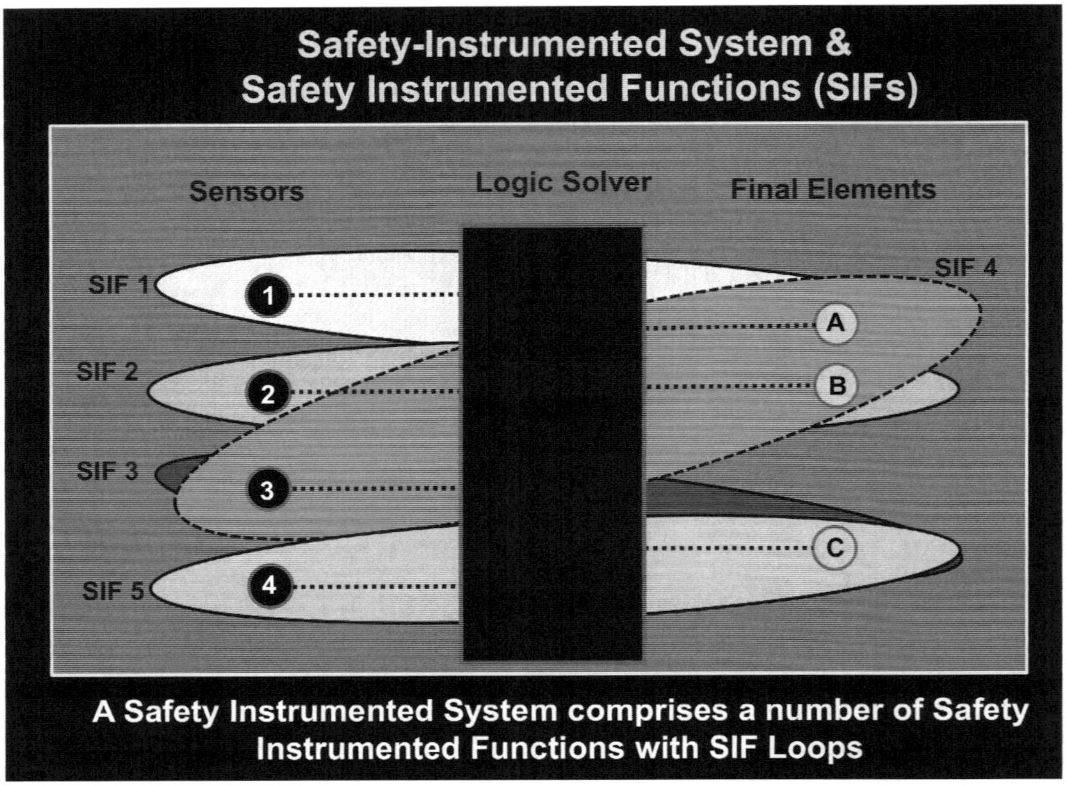
Safety-Instrumented System &
Safety Instrumented Functions (SIFs)
Sensors
Logic Solver
Final Elements
SIF 1
SIF 2
SIF 3
SIF 4
SIF 5
1
2
3
4
A
B
C
A Safety Instrumented System comprises a number of Safety Instrumented Functions with SIF Loops

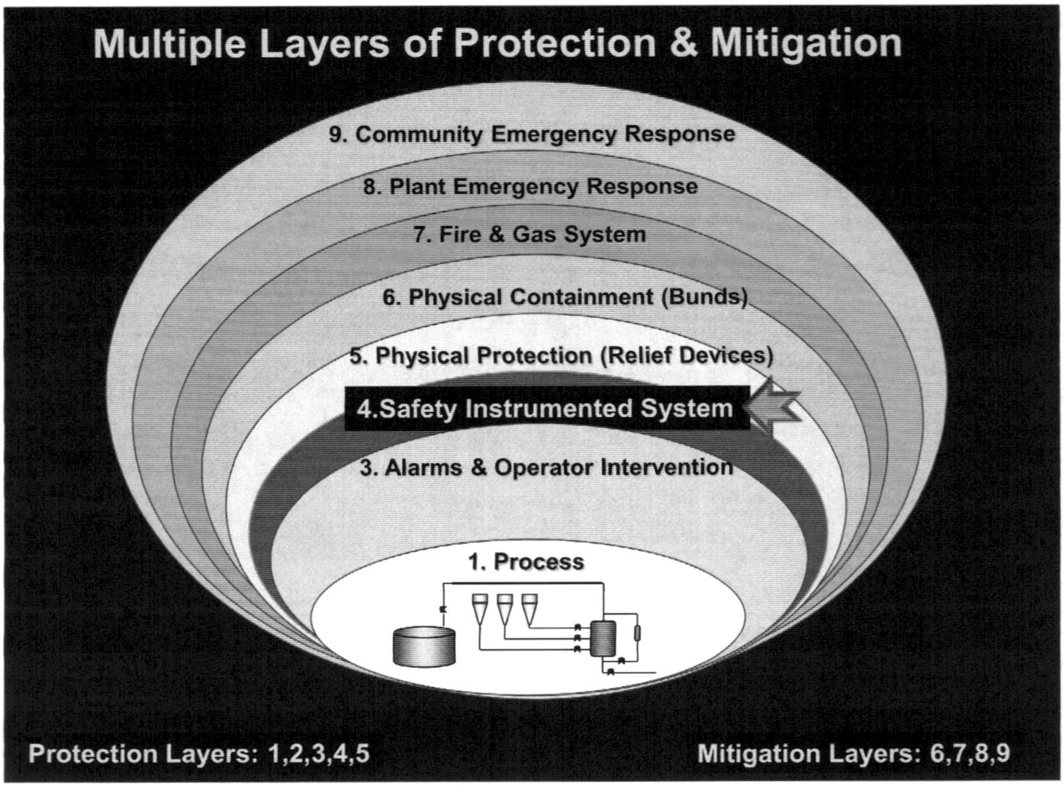
Multiple Layers of Protection & Mitigation
9. Community Emergency Response
8. Plant Emergency Response
7. Fire & Gas System
6. Physical Containment (Bunds)
5. Physical Protection (Relief Devices)
4.Safety Instrumented System
3. Alarms & Operator Intervention
1. Process
Protection Layers: 1,2,3,4,5
Mitigation Layers: 6,7,8,9

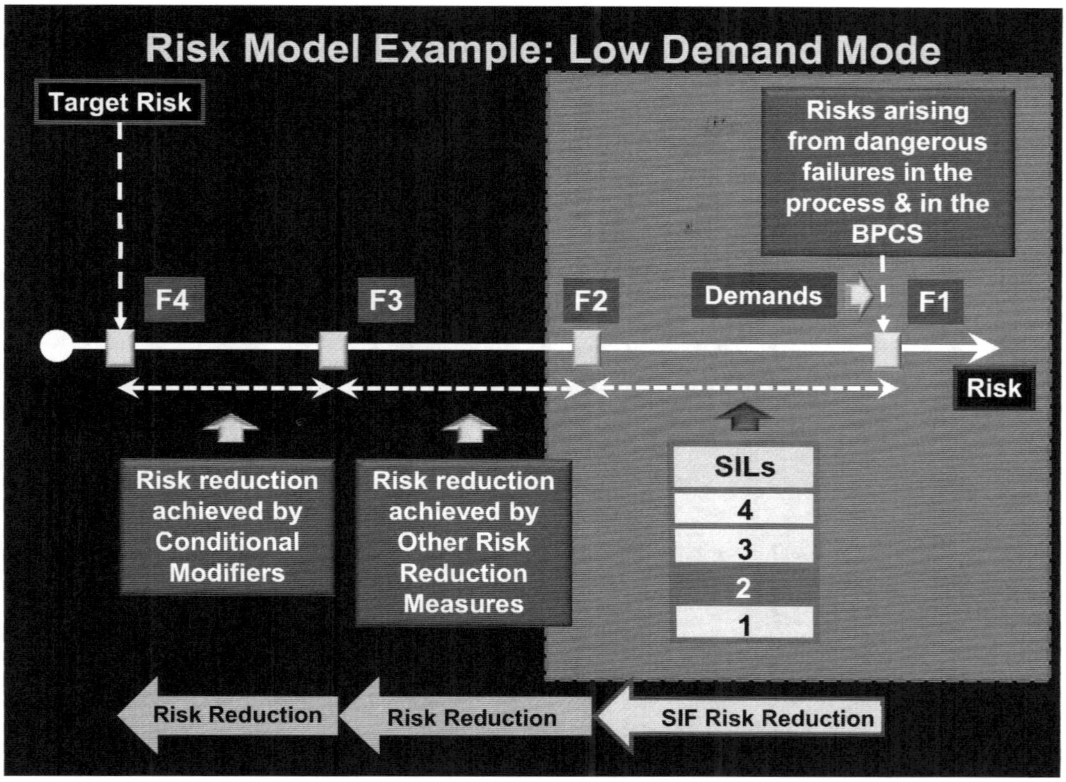
Risk Model Example: Low Demand Mode
Target Risk
Risks arising from dangerous failures in the process & in the BPCS
Demands
F4
F3
F2
F1
Risk
Risk reduction achieved by Conditional Modifiers
Risk reduction achieved by Other Risk Reduction Measures
SILs
4
3
2
1
Risk Reduction
Risk Reduction
SIF Risk Reduction

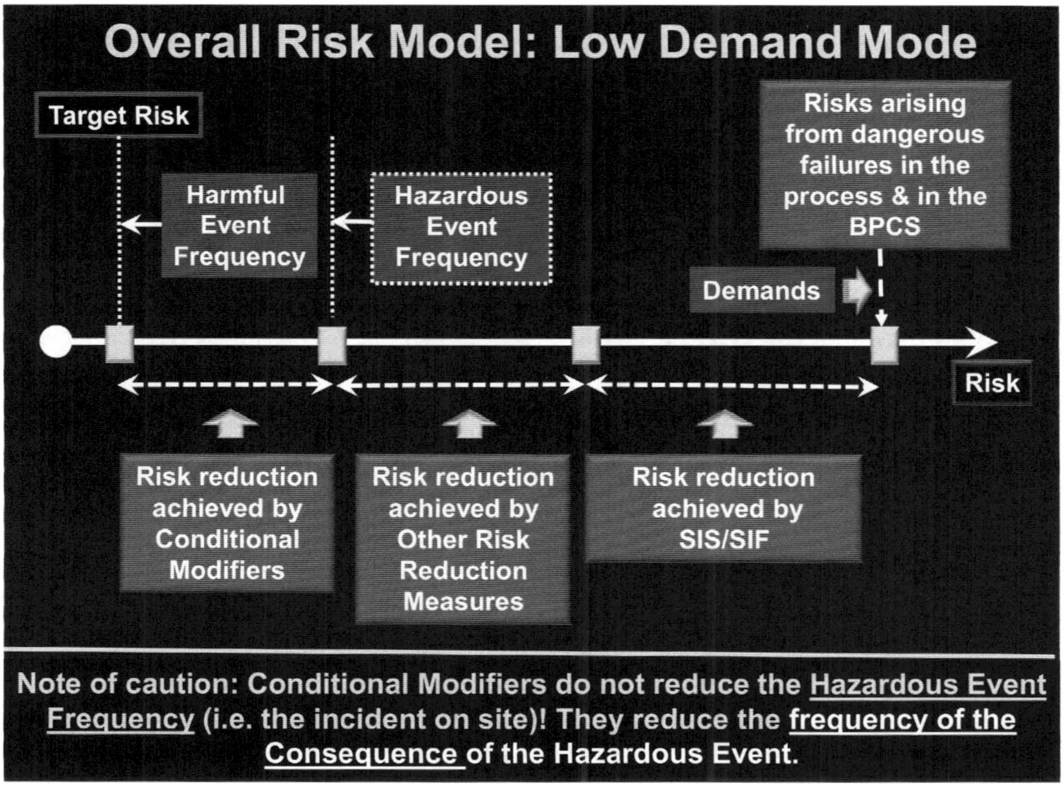
Overall Risk Model: Low Demand Mode
Target Risk
Harmful Event Frequency
Hazardous Event Frequency
Risks arising from dangerous failures in the process & in the BPCS
Demands
Risk
Risk reduction achieved by Conditional Modifiers
Risk reduction achieved by Other Risk Reduction Measures
Risk reduction achieved by SIS/SIF
Note of caution: Conditional Modifiers do not reduce the Hazardous Event Frequency (i.e. the incident on site)! They reduce the frequency of the Consequence of the Hazardous Event.

## KPI's for Functional Safety

- Functional Safety: The basics
- Key Performance Indicators (KPI's)
- Functional Safety KPIs
- Concluding comments




ESC

## Leading Indicators

Leading indicators – monitoring deviations to expected practices and/or procedures (inputs) potentially causing loss of one or multiple protection layers

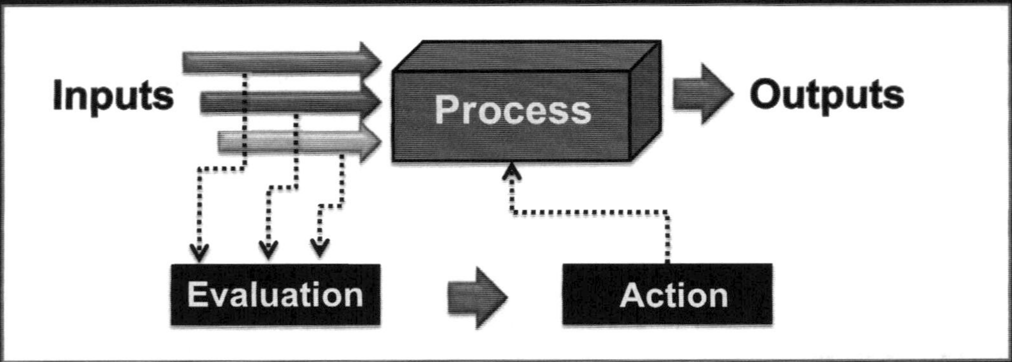

## Lagging Indicators

Lagging indicators – reviewing failures and non-compliances which resulted in loss of a layer of protection and/or the hazardous event (outputs) being protected against

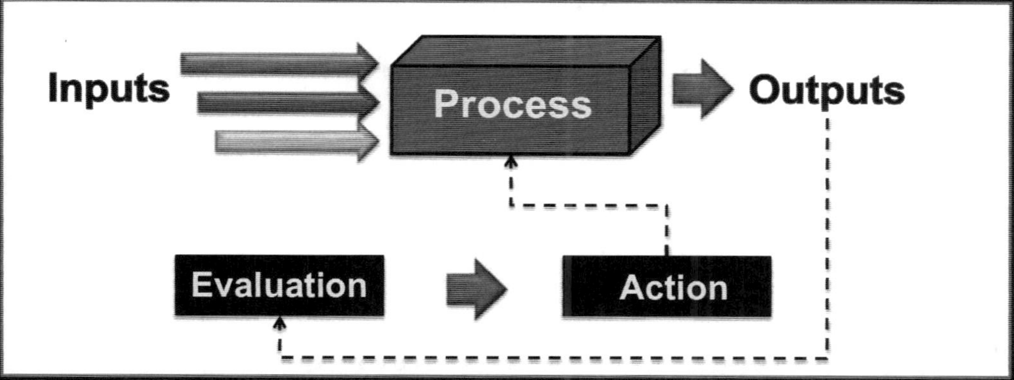





## Full Picture

Leading & Lagging Indicators

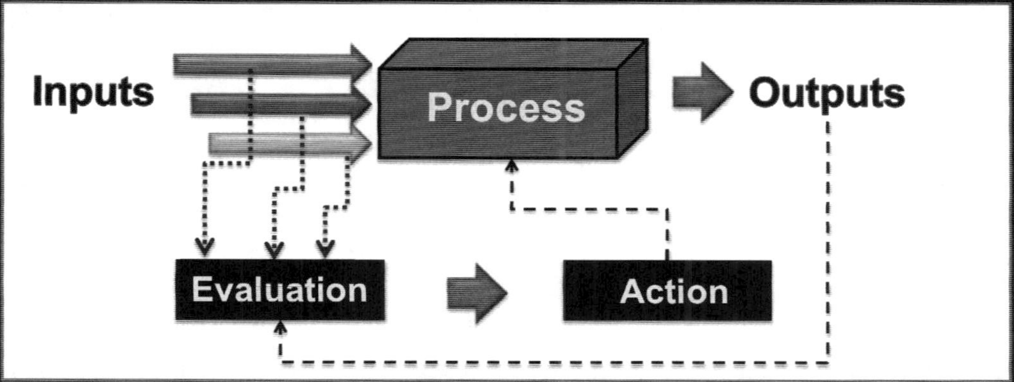

## Leading & Lagging Indicators

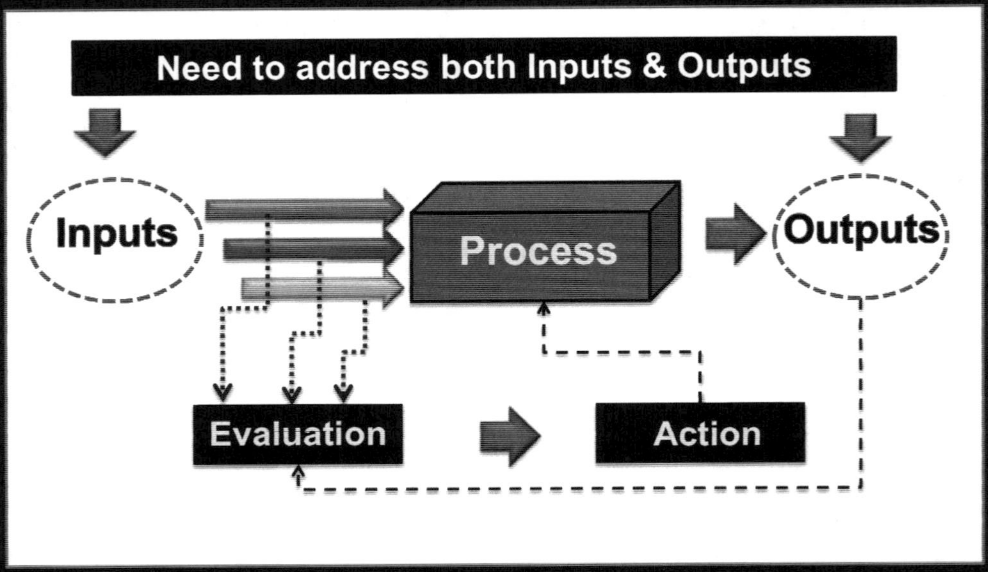





## KPI's for Functional Safety

- Functional Safety: The basics
- Key Performance Indicators (KPI's)
- <u>Functional Safety KPIs</u>
- Concluding comments

ESC

## Functional Safety: Inputs & Outputs

**Inputs**

- Competence
- Hazard and Risk Assessment
- Assurance Measures:
  - Functional Safety
  - Assessment;
  - Functional Safety Audit;
  - Verification;
  - Validation
- Proof Testing Procedures
- Bypass/Overrides
- Conditional Modifiers





ESC

## Functional Safety: Inputs & Outputs

**Inputs**

- Competence
- Hazard and Risk Assessment
- Assurance Measures:
  - Functional Safety
  - Assessment;
  - Functional Safety Audit;
  - Verification;
  - Validation
- Proof Testing Procedures
- Bypass/Overrides
- Conditional Modifiers

**Outputs**

- Dangerous Failures on demand
- Testing & Inspection backlog
- Protection Layer failures
- Dangerous Failures during testing
- Dangerous Failures during normal operation
- Dangerous Failure Rates achieved

## SIS safety lifecycle: IEC 61511

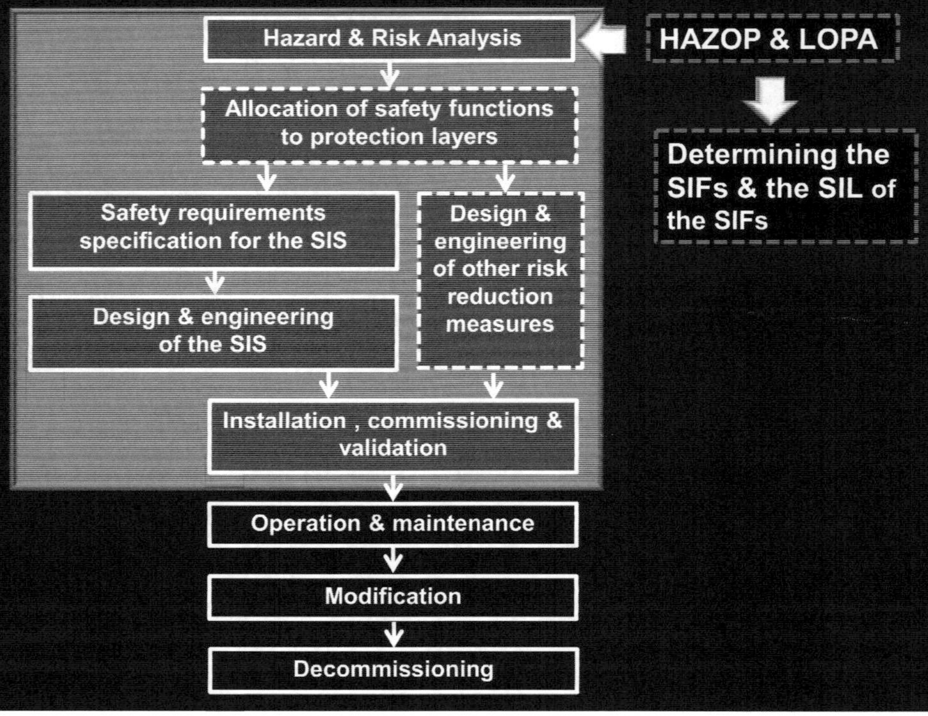


## Hazard & Risk Assessment

- Risk analysis is the building block for any Risk Control System
- Up to date studies (e.g. HAZOP, LOPA) are key to understand the ongoing role of protective systems [SIFs] and layers of protection [relief systems, critical tasks, BPCS...]
  - Revalidation cycle.... on what it is based?
  - Changes in plant design...are they captured as part of Management of Change?
    - Are the risk analyses (identification and assessment) updated as a result?

**POTENTIAL IMPACT**

Could result in poor identification of new requirements for SIF's in modified or new equipment/operational practices and reduced protection against Loss of containment

 

## Hazard & Risk Assessment

**Extract from IEC 61511-1 2016: Functional Safety - Safety Instrumented Systems for the Process Industry Sector:**

> The Hazard & Risk Assessment shall be recorded in such a way that the relationship between the [specified requirements listed above] are clear and traceable.

**Traceability**




## Functional Safety KPI's: Inputs

- Competence
- Hazard and Risk Assessment
- Assurance Measures:
  - Functional Safety Assessment;
  - Functional Safety Audit;
  - Verification;
  - Validation
- **Proof Testing**
- Bypass/Overrides
- Conditional Modifiers
- ……………………
- ……………………

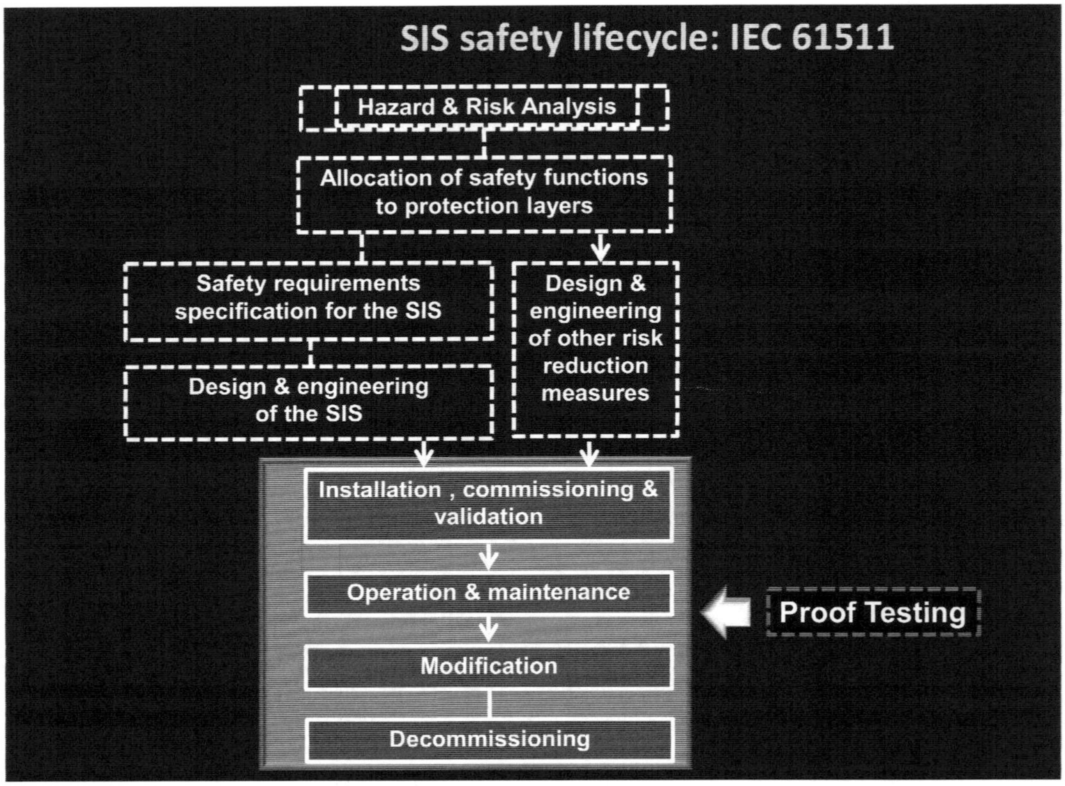


# Proof Test & Functional Test

## Proof Test:

Periodic test performed to detect dangerous hidden failures in a safety-related system so that, if necessary, a repair can restore the system to an "as new" condition or as close as practical to this condition IEC 61508-4 / 3.8.5 /Edition 2].

## Functional Test:

Usually referred to the testing of a safety-related system to ensure that the specified function is working correctly.

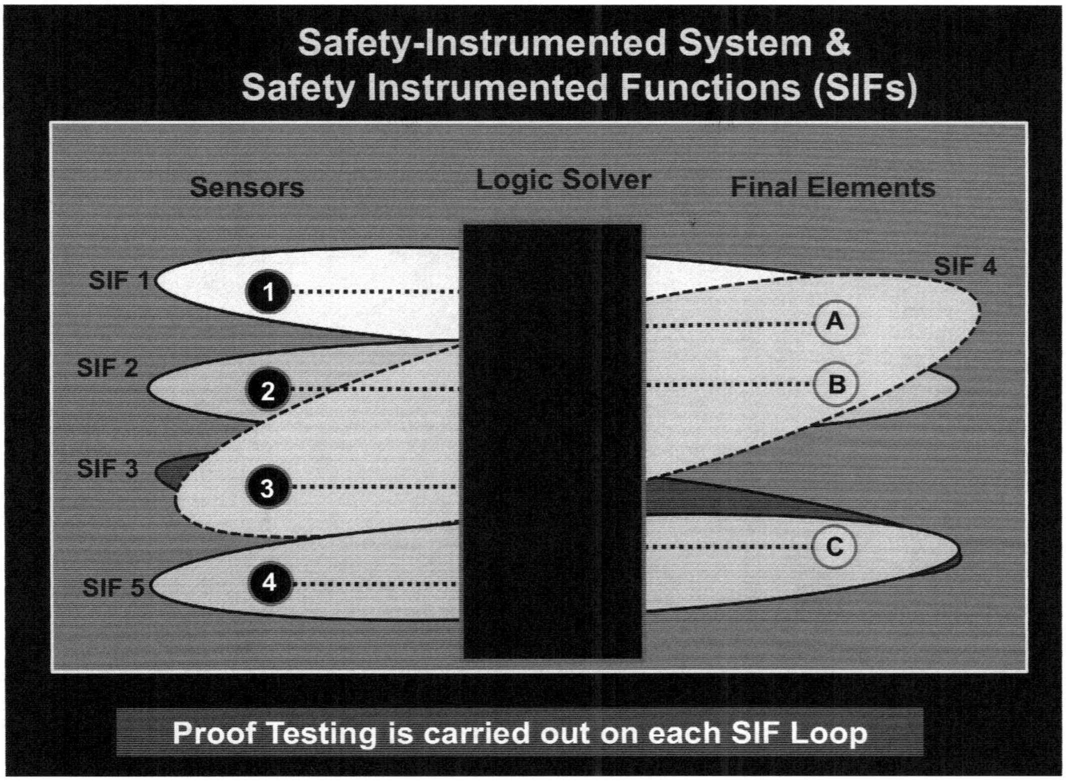
Safety-Instrumented System &
Safety Instrumented Functions (SIFs)
Sensors
Logic Solver
Final Elements
SIF 1
SIF 2
SIF 3
SIF 4
SIF 5
1
2
3
4
A
B
C
Proof Testing is carried out on each SIF Loop

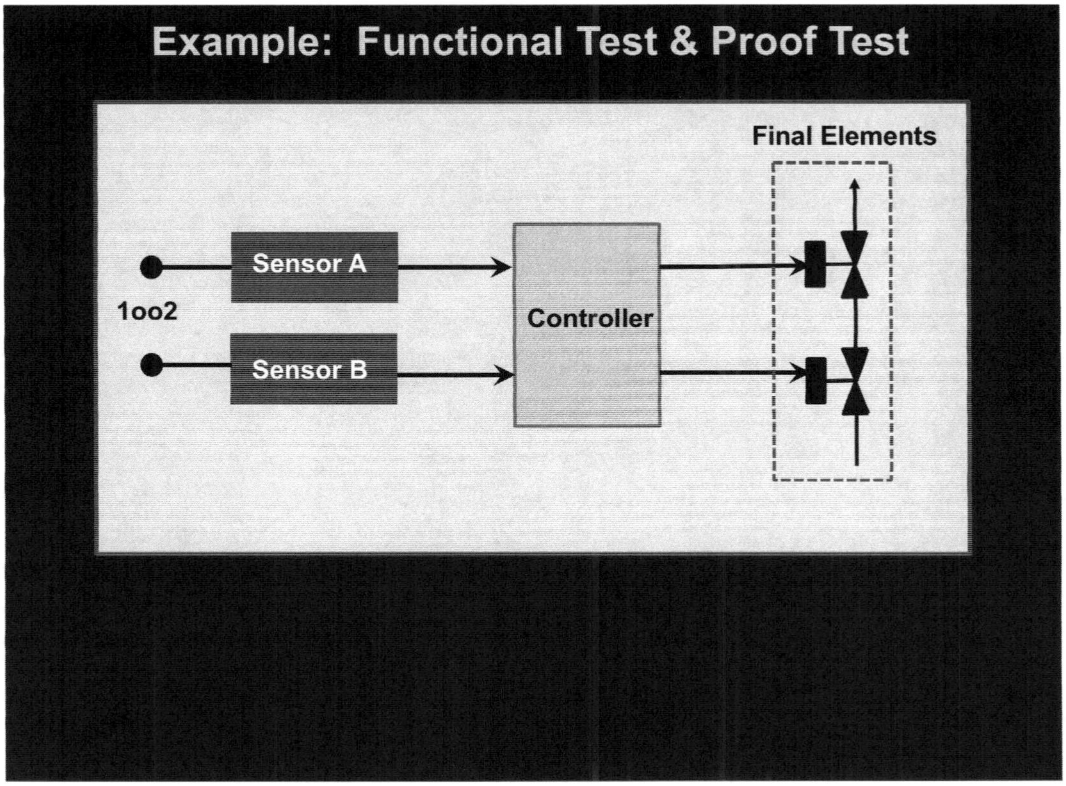
Example: Functional Test & Proof Test
Final Elements
Sensor A
Sensor B
1oo2
Controller

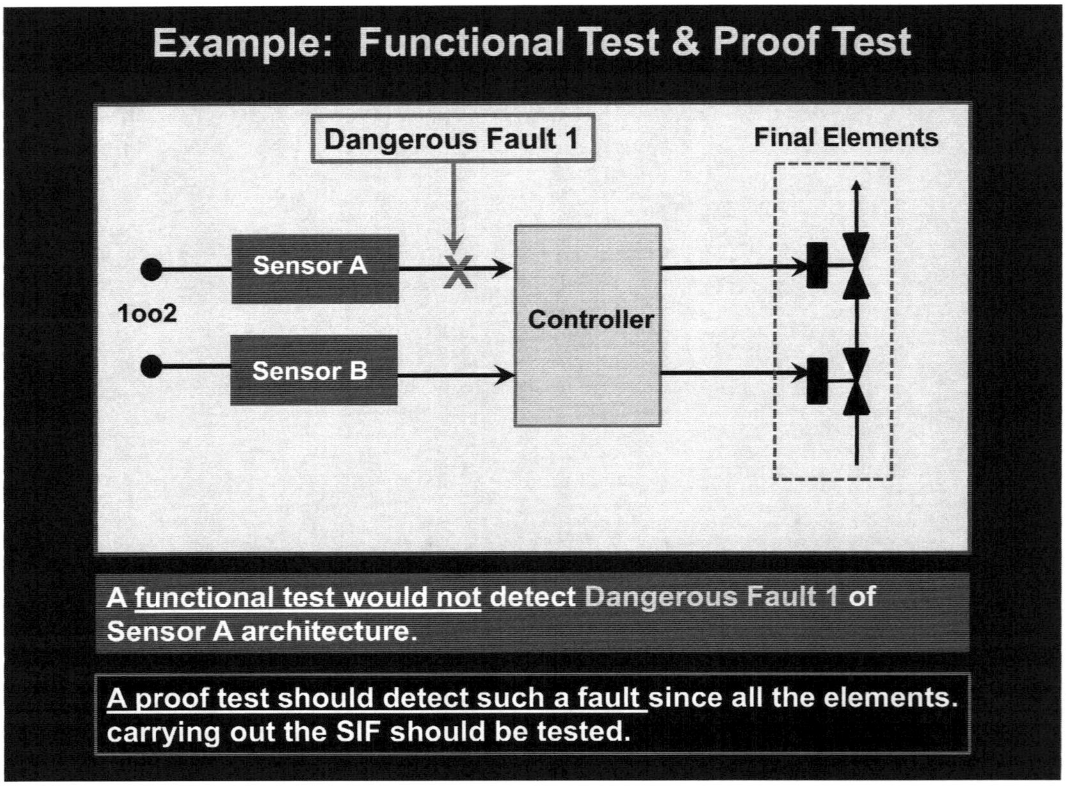
Example: Functional Test & Proof Test
Dangerous Fault 1
Final Elements
Sensor A
1oo2
Controller
Sensor B
A functional test would not detect Dangerous Fault 1 of Sensor A architecture.
A proof test should detect such a fault since all the elements. carrying out the SIF should be tested.

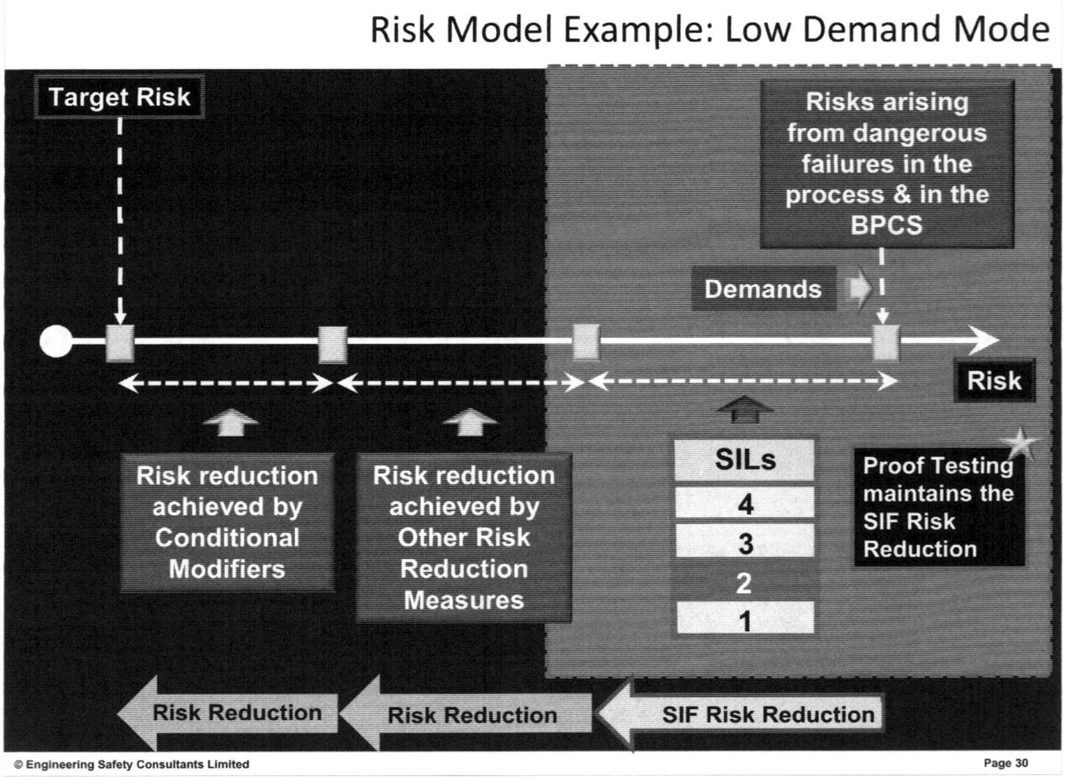
Risk Model Example: Low Demand Mode
Target Risk
Risks arising from dangerous failures in the process & in the BPCS
Demands
Risk
Risk reduction achieved by Conditional Modifiers
Risk reduction achieved by Other Risk Reduction Measures
SILs
4
3
2
1
Proof Testing maintains the SIF Risk Reduction
Risk Reduction
Risk Reduction
SIF Risk Reduction
© Engineering Safety Consultants Limited
Page 30

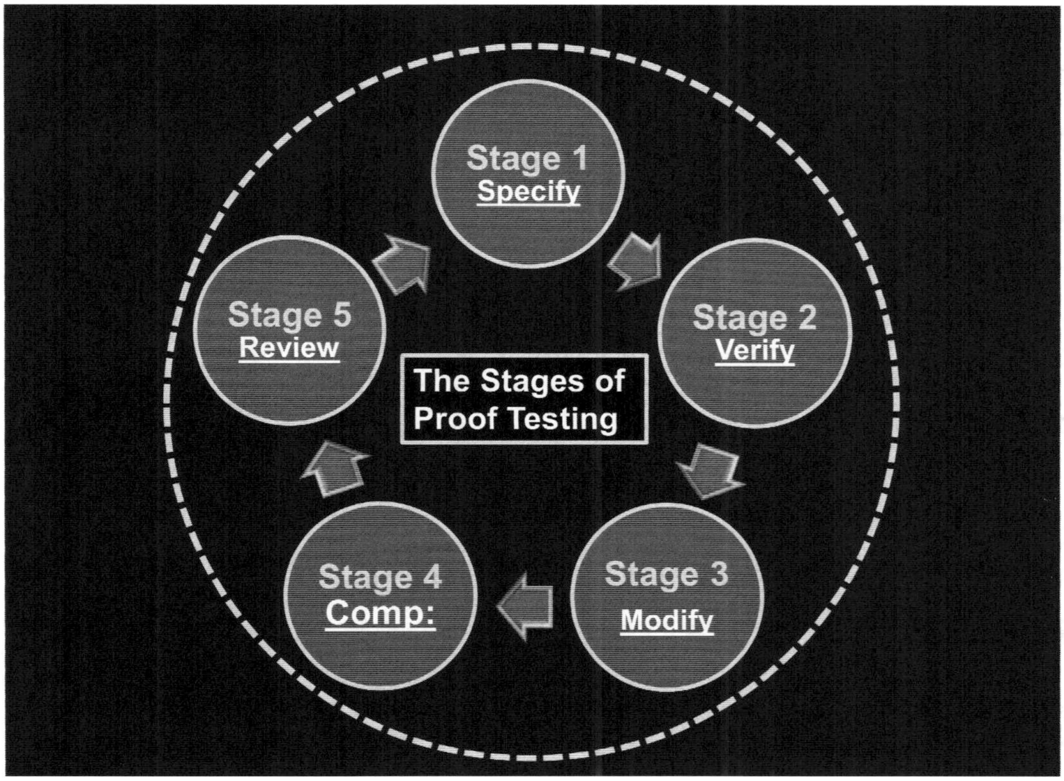


## Importance of Recording Proof Test Results

Recording results from proof test procedures is of utmost importance:

- Allows evaluation of the failure modes being experienced on your plant;
- Allows failure rate analysis in order to determine if the assumptions made in design are correct;
- Allow demonstration that the proof tests are being completed for internal and external audits.

## Proof Testing

- Where the Proof Test procedures based on a systematic analysis?
- Are proof tests controlled in the maintenance system?
- Are test failures understood and documented in test results?
- Are the failures documented "as found" and any repairs documented as "as left"?
- Is there an effective Failure classification for "as found dangerous failures" that is based on a risk assessment and analysis of the specific SIF architecture.
- Are Proof Test results analysed to enable evaluation of:
  - the failure dangerous modes being experienced on your plant;
  - the failure rate of the dangerous modes in order to determine if the assumptions made in design are correct……this has operational and functional safety implications.




## Proof Testing

POTENTIAL IMPACT

- If the Proof Test is not undertaken at the prescribed frequency the residual risk will increase.
- Inappropriate Proof Testing could result in undetected failures which would potentially stay latent in the system.

## KPI's for Functional Safety

- Functional Safety: The basics
- Key Performance Indicators (KPI's)
- Functional Safety KPIs
- <u>Concluding comments</u>

 

## Concluding Comments

**In the context of Functional Safety:**

- KPIs need to be addressed for all safety lifecycle phases;
- Recording, Traceability, Justification and Documentation are of paramount importance;
- Without Traceability, effective Management of Functional Safety is not possible.

**Recording... Traceability... Justification... Documentation!**

 

ENGINEERING SAFETY CONSULTANTS

*The Global Provider of Functional Safety Expertise and Technical Consultancy*

Thank you

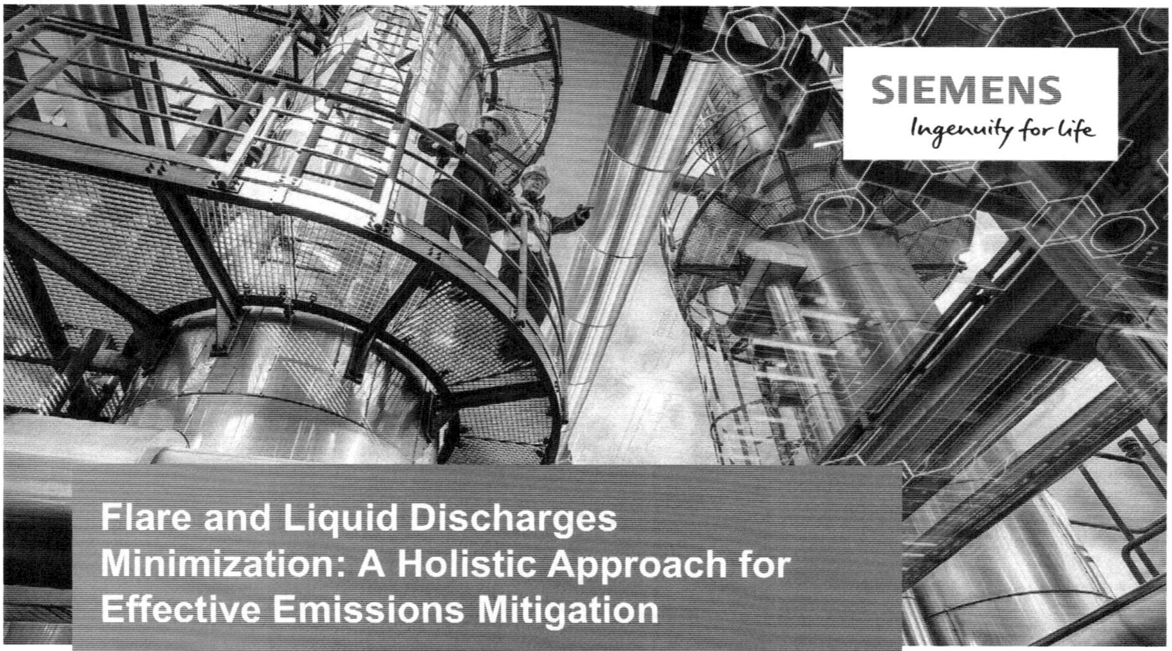

SIEMENS

*Ingenuity for life*

# Flare and Liquid Discharges Minimization: A Holistic Approach for Effective Emissions Mitigation

3rd Global Summit on Process Safety
Dammam, Saudi Arabia – 4 December 2016



Dr. Abdulrehman Aldeeb

## Disclaimer

SIEMENS

The information contained in this presentation represents the current view of the authors at the time of publication.

Process safety management is a complex subject and this document cannot embody all possible scenarios or solutions related to compliance.

This document is for informational purposes only.

Siemens makes no warranties, express or implied, in this paper or presentation.

## Today's Agenda

SIEMENS

- Relief and flare systems basics
- Strategy for flaring and liquid discharges minimization
- Risk-based based approach
- Feasibility of minimization opportunities
- Integration into plant lifecycle
- Summary



## Barriers-based Process Safety

SIEMENS

Application of James T Reason's Barrier Model to Process Safety

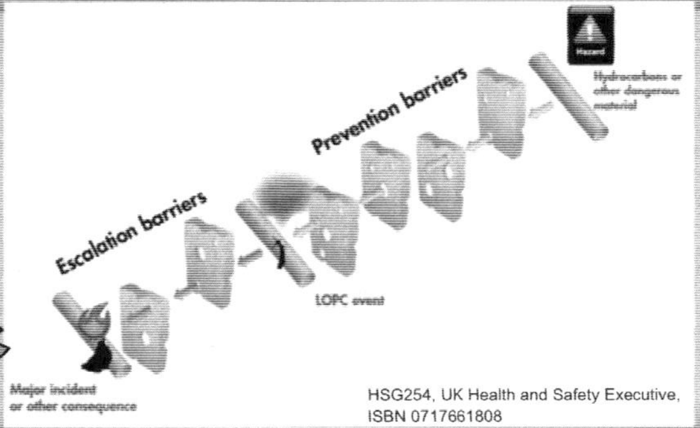


HSG254, UK Health and Safety Executive, ISBN 0717661808

- Determine the **critical process safety barriers** in your facility
- Barriers are "**hard**" and "**soft**": Plant Safety Systems, HSE Management Processes and People
- Barrier deterioration is **dynamic** (holes should never align)
- We want a program to systematically manage hard and soft barrier **performance and drive continuous improvement** (i.e., learning from experience)

*HSG254, UK Health and Safety Executive, ISBN 0717661806

## Process Safety Hard and Soft Barriers

SIEMENS

Safety System (Hard) Barriers

Management System (Soft) Barrier

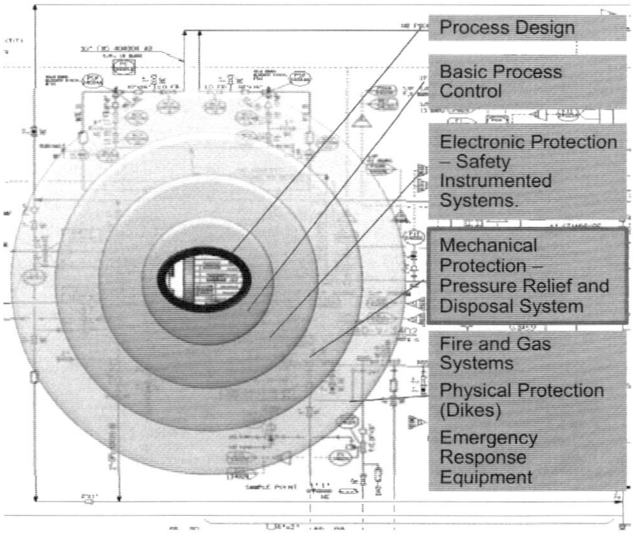


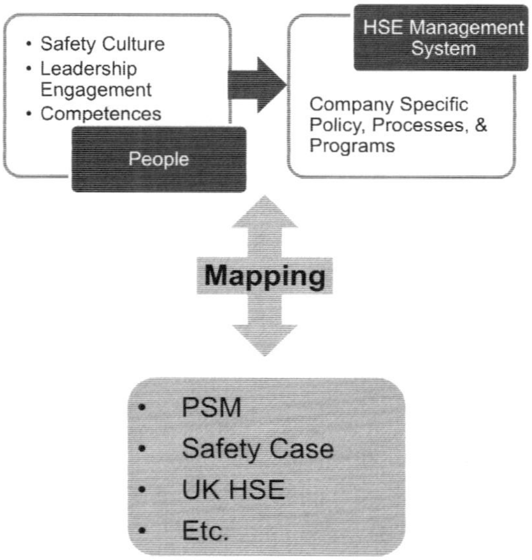




## Relief Disposal System - Simplified

SIEMENS

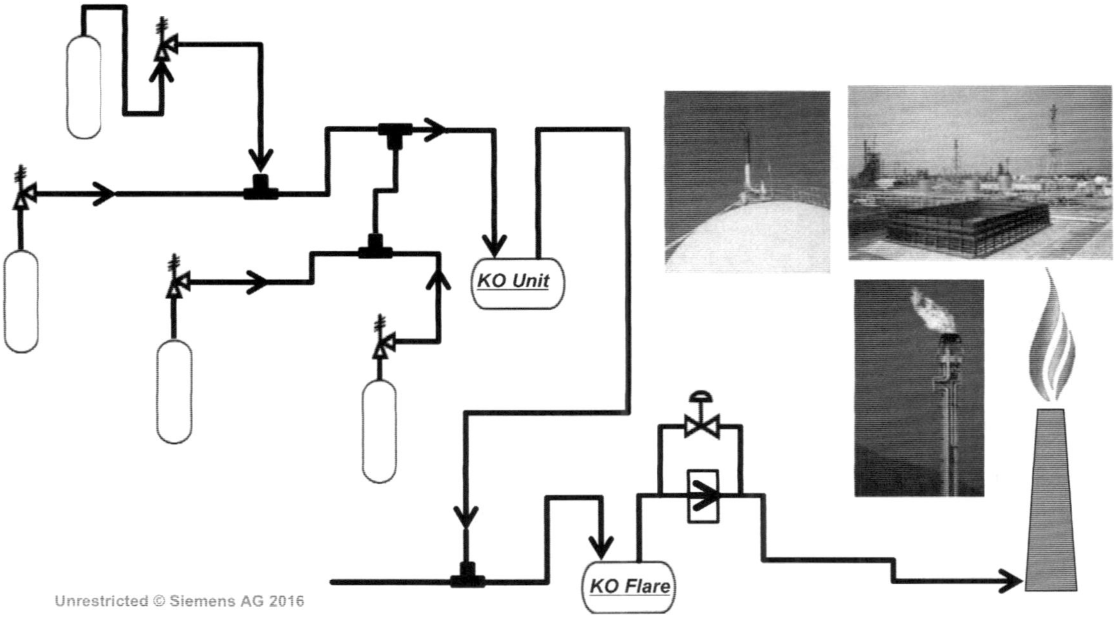

## Relief Disposal System Systems – Hard and Soft Barriers

SIEMENS

Safety System (Hard) Barriers

Relief Devices & Disposal System

Management System (Soft) Barrier

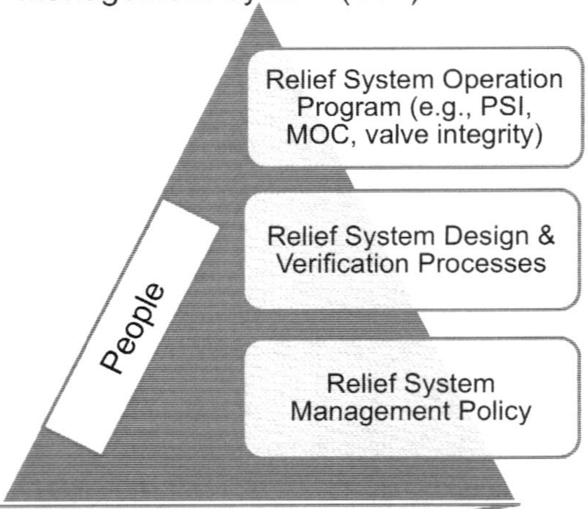


Performance Assurance & Management

**Protect equipment from over or under pressure and ensure that the barriers are strong and reliable so that they do not fail on demand.**





## Industry Trends and Impact on Process Safety

SIEMENS

- Tremendous oil, gas, and petrochemical industry expansion over the past 20 years
- Increasing HSE regulations
- Decreasing returns on investment (ROI)
- Aging facilities
- Introduction of more complex new technology

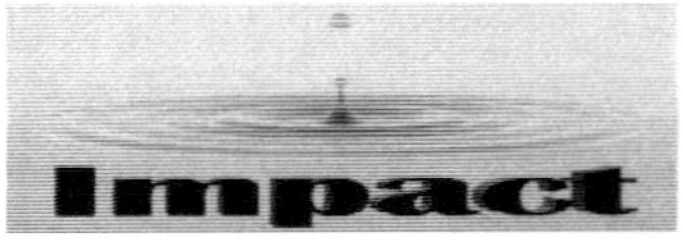


- Pressure to cut cost impacting safety and environment protection
- Lack of comprehensive visibility on process safety performance indicator
- Increasing incidents frequency
- Decreased public tolerance for industry risk
- Introduction of new hazards with new technology

***All driven by the desire to improve performance, sustainability, and reduce carbon dioxide footprint***

## Relief and Disposal Systems Challenges

- Continuous process expansions and changes results complex system arrangements, multiple disposal systems, operation modes, etc.
- Managing relief system design data through process lifecycle.
- Flaring systems and burn pits are major sources of emissions.

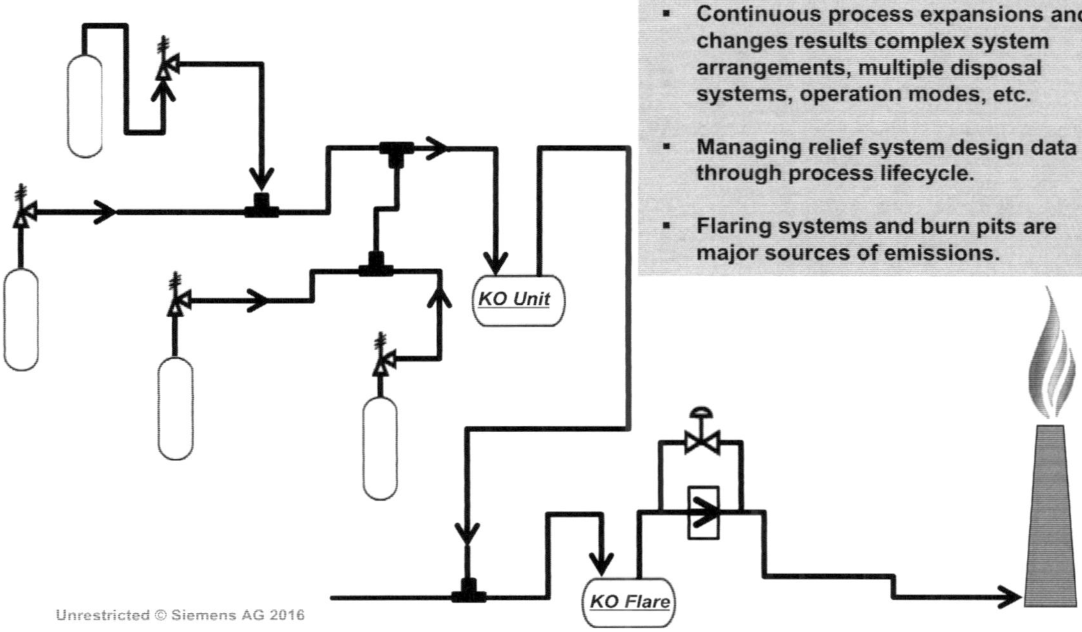








## Strategy for Flare & Liquid Discharges Minimization

**Quantify**
- Systematic analysis and quantification of vapor and liquid discharges
- Apply risk-based analysis for emergency discharging contingencies

**Identify**
- Identify flaring and liquid discharging minimization opportunities
- Desktop safety review: Engineering, safety, and operations
- Evaluate economical feasibility

**Improve**
- Effective integration of relief and disposal system analysis with MOC and PHA
- Maintain system design up-to-date
- Continuously identify further improvement opportunities

## Strategy for Flare & Liquid Discharges Minimization

SIEMENS

**Quantify**
- Systematic analysis and quantification of vapor and liquid discharges
- Apply risk-based analysis for emergency discharging contingencies

**Identify**
- Identify flaring and liquid discharging minimization opportunities
- Desktop safety review: Engineering, safety, and operations
- Evaluate economical feasibility

**Improve**
- Effective integration of relief and disposal system analysis with MOC and PHA
- Maintain system design up-to-date
- Continuously identify further improvement opportunities




## Systematic Pressure Relief Systems Analysis

SIEMENS

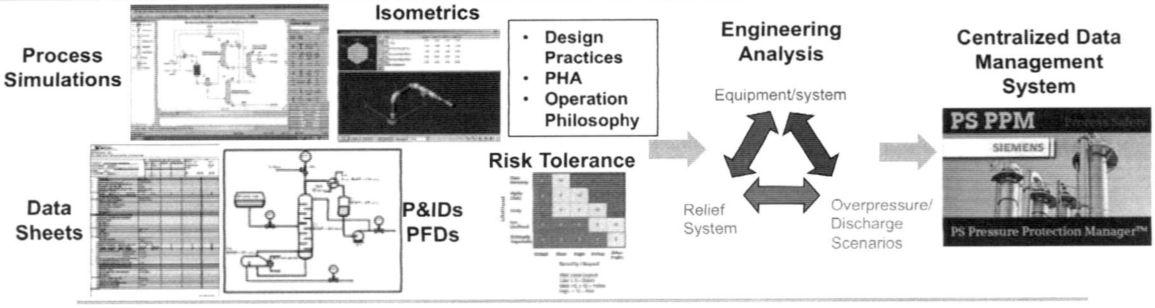


| Identify overpressure and discharging scenarios | Applicability & calculation discharge relief rates | Size relief & disposal systems |
|---|---|---|
| Emergency and upsets<br>Start-up and shut-down<br>Purging<br>Non-routine line-ups<br>Maintenance | Steady vs dynamic discharge loads<br>Simultaneous discharge contingencies<br>No credit for other than relief systems | Piping network hydraulic analysis<br>Flare radiation & dispersion modeling<br>Minimum purging requirements<br>Adequacy for protection |

SIEMENS

## Risk-based Emergency Relief Requirements

What are the conditions of using instrumentation in emergency overpressure protection – to eliminate and/or reduce discharges?

**Independent Protection Layers (IPL)**

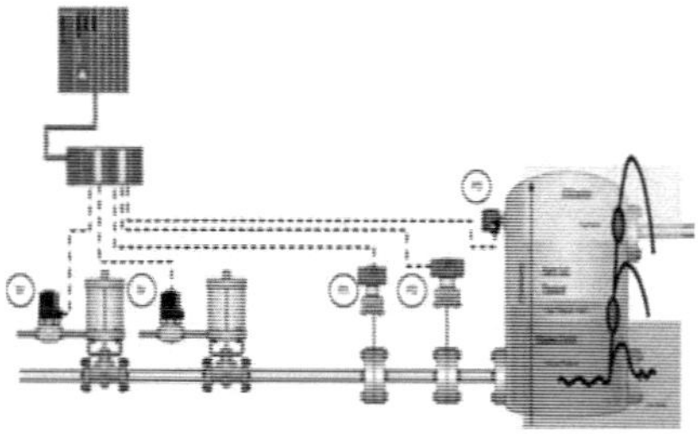

» Credit for IPL to reduce emergency pressure relief requirements.

**Risk Tolerance Criteria = Consequence X Frequency**

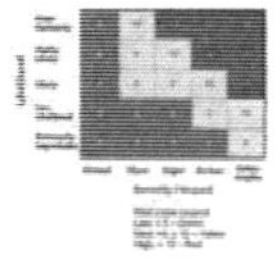

» Such applications must fit within plant risk tolerance criteria.





SIEMENS

## Examples of IPL's

| Examples of IPL | Typical PFD | Passive/ Active IPLs |
|---|---|---|
| Relief Valve | $1\times10^{-1}$ to $1\times10^{-3}$ | Active |
| Rupture Disk | $1\times10^{-1}$ to $1\times10^{-3}$ | Active |
| Basic Process Control System | $1\times10^{-1}$ | Active |
| Safety Instrumented System | SIL-1: $\geq 10^{-2}$ and $< 10^{-1}$<br>SIL-2: $\geq 10^{-3}$ and $< 10^{-2}$<br>SIL-3: $\geq 10^{-4}$ and $< 10^{-3}$ | Active |
| Dike | $1\times10^{-2}$ to $1\times10^{-3}$ | Passive |
| Open Vent | $1\times10^{-2}$ to $1\times10^{-3}$ | Passive |
| Fireproofing | $1\times10^{-2}$ to $1\times10^{-3}$ | Passive |
| Blast Wall/Bunker | $1\times10^{-2}$ to $1\times10^{-3}$ | Passive |
| Flame/ Detonation Arrestors | $1\times10^{-1}$ to $1\times10^{-3}$ | Passive |

IPL: Independent protection layers
PFD: Probability of failure on demand

# Determine a Risk-based Emergency Relief Requirements: Flare QRA

SIEMENS

- Flare QRA Objective is to develop a relationship between scenario likelihood and consequence while taking into account all available ILP.
- Outcome is a system (vessels and piping) hydraulic pressure profile vs. time intervals, i.e., risk criteria.

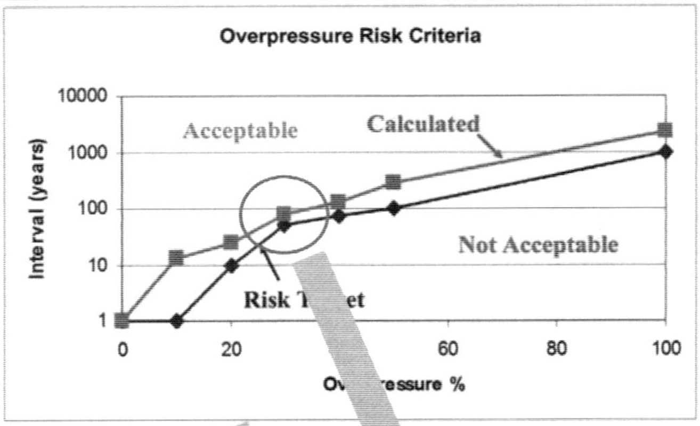


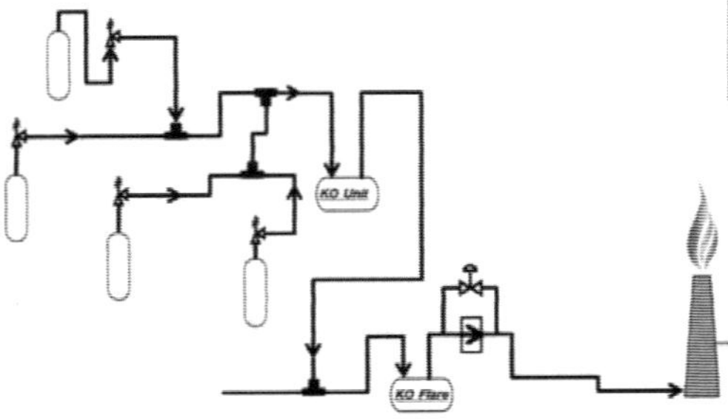


- Discharge load can be calculated based on risk tolerance criteria.




# Strategy for Flare & Liquid Discharges Minimization

SIEMENS

**Quantify**
- Systematic analysis and quantification of vapor and liquid discharges
- Apply risk-based analysis for emergency discharging contingencies

**Identify**
- Identify flaring and liquid discharging minimization opportunities
- Desktop safety review: Engineering, safety, and operations
- Evaluate economical feasibility

**Improve**
- Effective integration of relief and disposal system analysis with MOC and PHA
- Maintain system design up-to-date
- Continuously identify further improvement opportunities

## Flaring and Liquid Discharging Minimization Opportunities

SIEMENS

- Process modifications
- Process line-up improvement
- Flare gas recovery systems (FGRS)
- Operating procedures modifications
- Safety instrumentation systems (SIS)
- Flare consolidation

+

**Desktop Safety Review**
Engineering, safety, and operations

**Economical Feasibility**




## Strategy for Flare & Liquid Discharges Minimization

SIEMENS

**Quantify**
- Systematic analysis and quantification of vapor and liquid discharges
- Apply risk-based analysis for emergency discharging contingencies

**Identify**
- Identify flaring and liquid discharging minimization opportunities
- Desktop safety review: Engineering, safety, and operations
- Evaluate economical feasibility

**Improve**
- Effective integration of relief and disposal system analysis with MOC and PHA
- Maintain system design up-to-date
- Continuously identify further improvement opportunities

## Continuous Improvement – Process Lifecycle Integration

SIEMENS

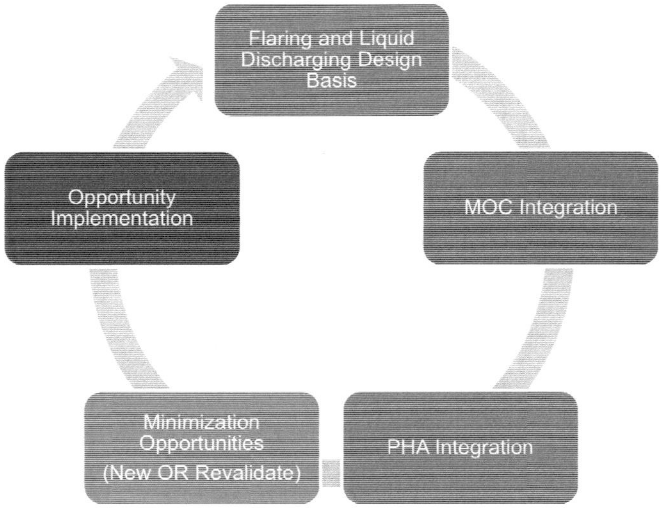




## Conclusion: From Compliance to Excellence

SIEMENS

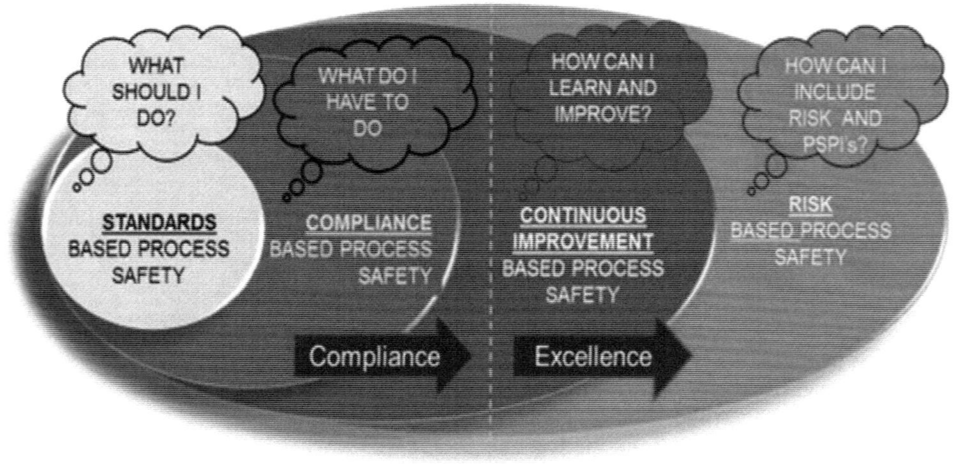

## Thank You. Questions?

SIEMENS

**USA:**
Siemens Energy, Inc.
4615 Southwest Freeway
Houston, TX 77027, USA
Phone: +1 (713) 570-2900
E-mail: support.energy@siemens.com

**KSA:**
Al-Raja Tower
Khobar-Dammam Expressway
Al-Khobar 31952, Saudi Arabia
Phone: +966 (13) 865-7748

**UAE:**
Siemens Building
Unit No B-05, Masdar City
P.O. Box 47015, Abu Dhabi, UAE
Phone: +971 (55) 200-3380
abdul.aldeeb@siemens.com

# PHA/HAZOP Implementation - the Good, the Bad, the Ugly

(optimizing PHA/HAZOPs to address all scenarios, efficiently)

3rd Global Process Safety Summit
Dammam, December 4-5, 2016

William (Bill) Bridges & Adel Dakheel
Process Improvement Institute

PROCESS IMPROVEMENT INSTITUTE
Risk and Reliability Specialists

3RD CCPS GLOBAL SUMMIT ON PROCESS SAFETY

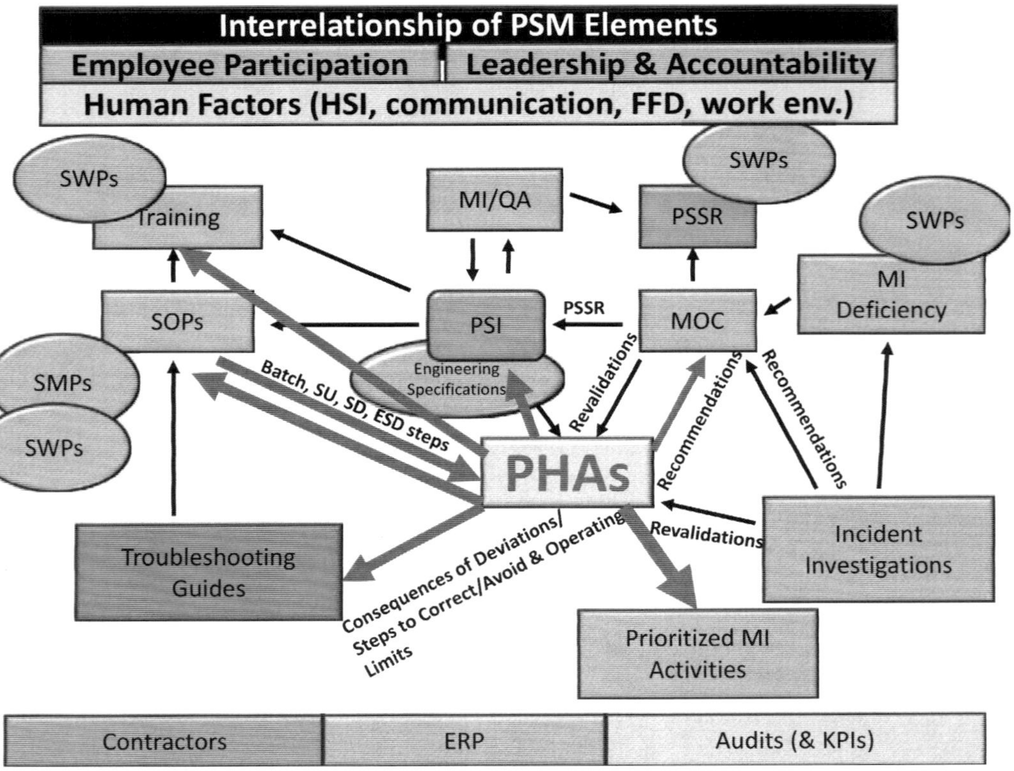

# Elements for Optimizing PHAs

- **Best Practices for Scope of Analysis**
- **Best Practices for PHA Team Membership and Leadership**
- **Best Practices for use of Meetings Rules & Methodology and for PHA Documentation**
- Best Practices for Follow-through on PHA recommendations
- Best Practices for Implementation of PSM Management Practices that support or derive from PHAs

PROCESS IMPROVEMENT INSTITUTE
Risk and Reliability Specialists



# Experiments in Optimizing PHAs

- During >5000 unit-size PHAs, **With** and **Without**:
  - Projection of meeting notes (100s of examples with)
  - Scribe (more than 1000 experiments without)
  - Using a Risk Matrix in meetings (more than 1000 with)
  - LOPA use in meetings (within meetings for 50+ PHAs and outside for 500+ PHAs)
  - Various meeting hours per day, per week, etc.
  - Step-by-step (HAZOP) or What-If of Startup & Shutdown & Online maintenance procedures
  - Variations on rules for documentation of analysis
  - Various Software (LEADER, PHA Pro, PHA Works, Excel)
  - Multiple PHAs of the same system by different teams

PROCESS IMPROVEMENT INSTITUTE
Risk and Reliability Specialists

4

# GOAL - GOOD PHA/HAZOPs

- Identify > 95% of major accident scenarios *(perfection not attainable)*
- Produce good decisions on control of risk of each major scenarios
- Are documented to clearly convey the findings of the PHA/HAZOP team
- Are efficient at completing the goals above
- Are followed through on by management

5

# Bad/Ugly PHA – Deficient SCOPE

- PHA (e.g., using HAZOP/What-if methods) fails to cover All Operating Modes
- PHA fails to cover All Types of Damage Mechanisms
- PHAs for major projects not performed during multiple phases

PROCESS IMPROVEMENT INSTITUTE
Risk and Reliability Specialists

6

# Bad/Ugly PHA – Deficient SCOPE

**POP QUIZ: During which mode of operation do MOST Major Process Safety Accidents Occur?**

- ❑ Normal (continuous) Mode of Operation?
- ❑ Startup Mode?
- ❑ Online Maintenance?
- ❑ Shutdown?

PROCESS IMPROVEMENT INSTITUTE
Risk and Reliability Specialists

7

# When do most Major Process Accidents Occur?

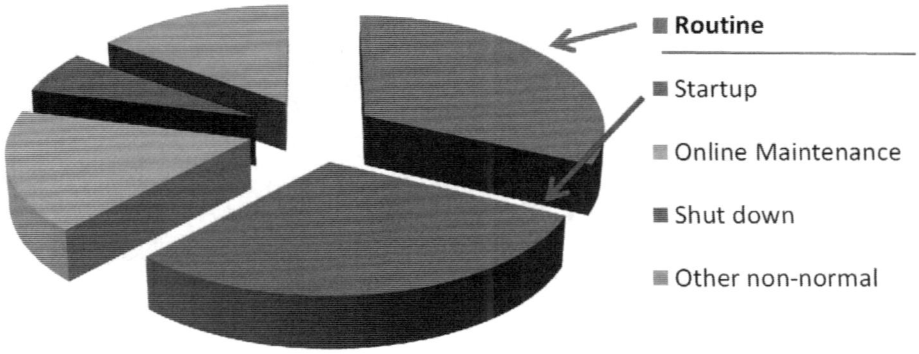

8

## Current Status of PHAs - related to PHA of ALL modes of Operation

- **In a majority (>80%) of both older operations and new plants/projects, the non-routine modes of operations are not analyzed at all**, or else not as recommended in **Chapter 9.1**, 3rd Ed, *Guidelines for Haz Eval Procedures*, CCPS/AIChE (2008)
- Instead of focusing on the most hazardous modes of operation, most PHAs focus on normal operations (e.g., HAZOP of equipment nodes)
- *NOTE: In some PHAs, the team discuss startup, shutdown, and online maintenance as a "deviation" during HAZOP of equipment nodes; but this method only uncovers 10% of the scenarios for these modes of operation*

PROCESS IMPROVEMENT INSTITUTE
Risk and Reliability Specialists

9

## PROPER Approach for PHA/HAZOP of ALL Modes of Operation for a CONTINUOUS Process

(Per CCPS Haz. Eval. textbook)

1. HAZOP or WHAT-IF of normal mode of operation (HAZOP of nodes, etc.)
2. 2 Guideword HAZOP of Critical Procedures (rarely 7 guidewords)
3. WHAT-IF of less critical procedures
4. Checklist of Global Issues (facility siting, human factors, utility failures)

PROCESS IMPROVEMENT INSTITUTE
Risk and Reliability Specialists

10

## *RETURN ON INVESTMENT IN PHA OF NON-ROUTINE MODE*

***Detailed PHA of Procedures does not take much investment, yet can save millions to billions USD***

Chemical-related companies estimate the return on investment at 1000 to 1, by avoiding Newly discovered accident scenarios that would likely occur over the next 40 years

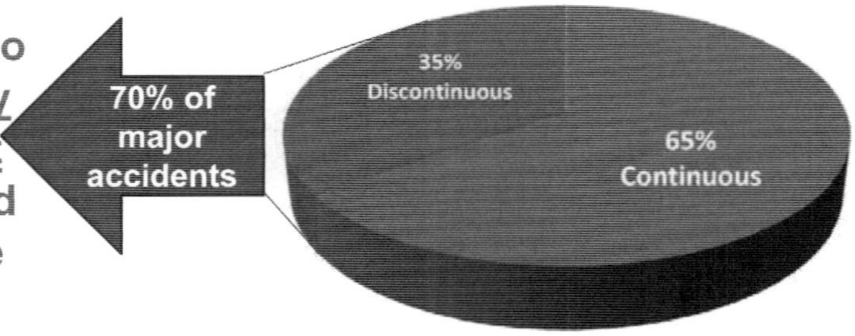


PROCESS IMPROVEMENT INSTITUTE
Risk and Reliability Specialists

14

11

## Current Status of PHAs - related to PHA of ALL Damage Mechanisms

- Most PHAs (roughly 70%) reviewed by PII do NOT cover damage mechanisms at each node (each equipment section) as recommended in CCPS's *Guidelines for Haz. Eval. Procedures*, 2nd and 3rd editions, 1991 & 2008. Should cover:
  - External Impact
  - External Fire
  - Corrosion/Erosion
  - Improper Maintenance
  - Wrong alloy/material
  - Detection of releases
  - Response to releases
- Some companies discuss once per unit; should be per node

PROCESS IMPROVEMENT INSTITUTE
Risk and Reliability Specialists

12

## Current Status of PHAs – related to PHA (e.g., HAZOP) during Multiple Project Phases

- Most Companies do ONE major HAZOP for each Major Project; near design "freeze" ....
  - But the Operating procedures are not completed by then and so how will they do a PHA of deviations of the steps for startup, shutdown, and online maintenance?
  - And a project typically has hundreds of changes from design freeze until startup... how are these changes captured in the PHA finally delivered to the new operating staff?
- Multiple PHAs/HAZOPs must be done during major projects to (1) find all hazards during all modes, (2) find hazards in designs that are not truly final at design freeze, and (3) to provide a good basis for future revalidations

PROCESS IMPROVEMENT INSTITUTE
Risk and Reliability Specialists

13

## Number of PHAs/HAZOPs necessary versus project size

| Project Size | Example Project Scope | Example Project Length/Duration (concept→ commissioning) | Number of Risk Reviews |
|---|---|---|---|
| Major | Major projects handled external to an affiliate/plant, such as expansions and new facilities | 12-36 months | 4-7 |
| Medium | Works engineered by an affiliate/ plant (installing a new design of knockout pot for a feed to a unit) | 6-9 months | 2-3 |
| Small | Minor affiliate/plant works (installing piping to bypass a control valve) | 1-2 months | 1 |

PROCESS IMPROVEMENT INSTITUTE
Risk and Reliability Specialists

## Current Status of PHAs - related to Team Membership and PHA Leadership

- Most of the PHA leaders (more than 80%) we have met at sites or projects have good inherent abilities but are NOT competent because they have not been properly trained & coached in the right methods and rules (this leads to the same deficiencies in Scope described earlier)

- Many teams ( > 30% ) of the PHAs we have audited have deficient team members; the staff who are "available" are assigned to the team instead of the assigning the most expert staff. As well, PII fights this issue on many PHAs we lead (many times resulting in PII stopping the PHA)

PROCESS IMPROVEMENT INSTITUTE
Risk and Reliability Specialists



## Current Status of PHAs - Methods and Rules/Approach for Good Efficiency

- **Why Optimize?**
  - Inefficient PHA/HAZOP burns out the team – reduces brainstorming / more scenarios are missed
  - You can lose management support if PHAs are taking too long for the benefit achieved
  - Wasting of time during PHA of normal mode of operation will drain budget and leave less or no time for analysis of non-routine modes
- **Status: Due to inefficient rules, most PHAs spend 2 to 3 times of the effort actually needed**

PROCESS IMPROVEMENT INSTITUTE
Risk and Reliability Specialists

16

## Meeting Time – For PHA of a Continuous Flow Unit

With LOW/NO coverage of Startup, Online Maintenance, and Shutdown modes of operation

**Not Optimized**

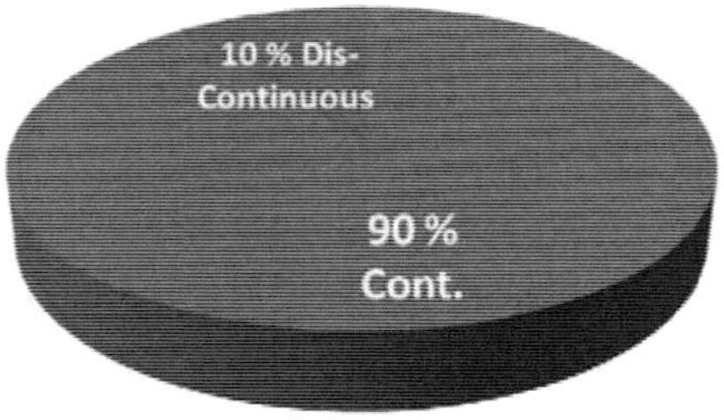


**Optimized**

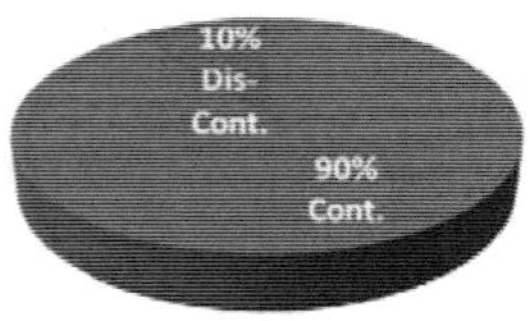


**1/3 original meeting time**

With FULL Process Hazard Analysis of ALL Modes of Operation

**Optimized**

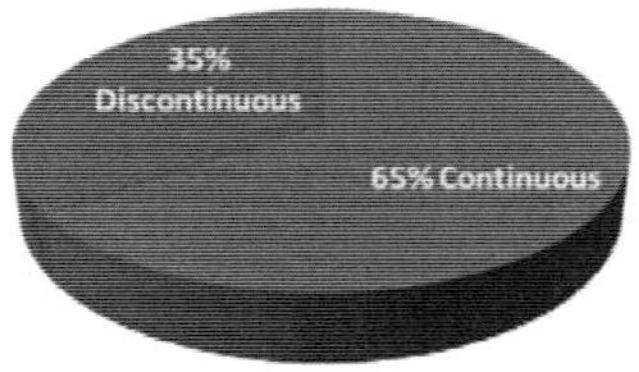


**1/2 original meeting time**

17

## Best Practices (for efficiency): Preparation, Meetings, and Documentation

- **Surprising Rule:** Do NOT project the analysis worksheets during team meetings

It detracts from BRAINSTORMING – this rule saves 30% of meeting time as well

"There are pluses and minuses to projecting notes live in the meeting, but actual results show that the negatives far outweigh positives"

PROCESS IMPROVEMENT INSTITUTE
Risk and Reliability Specialists

18

## Best Practices: Preparation, Meetings, and Documentation (cont'd)

- Create appropriate Nodes and Deviations (in software) BEFORE the meetings – saves 10% of meeting time
- Use dedicated PHA software if doing more than 2 weeks of PHA per year – saves 20% meeting time; saves 80% preparation time. *(Also ensure that Node and Deviation selection matches software defaults)*
- Use a dedicated Scribe for meetings longer than 4 hours – allows Leader to focus on FACILITATION; saves 30-50% of PHA meeting time and increases brainstorming (so allows more scenarios to be found)

PROCESS IMPROVEMENT INSTITUTE
Risk and Reliability Specialists

19

## Best Practices: Preparation, Meetings, and Documentation (cont'd)

- Make definite risk decisions as soon as possible after the scenario is understood – 10-20% less meeting time
- Identify which Safeguards will *likely* meet Independent Protection Layer (IPL) rules:

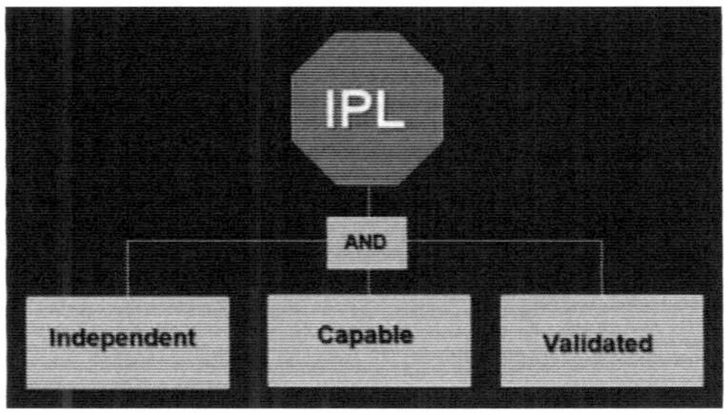


- Identify if SIF is needed and determine What SIL is needed for the SIF (see POSTER in exhibition hall)

PROCESS IMPROVEMENT INSTITUTE
Risk and Reliability Specialists

| **No.:** 2 XXXX storage spheres xxx-T-XX A/B/C/D/E/F/G/H/I/J/K/L (1 of 12) | | | | | |
|---|---|---|---|---|---|
| **#** | **Dev.** | **Causes** | **Consequences** | **Safeguards** | |
| 2.1 | High level | Too much flow to one sphere from XX Plant (through their pump; about 40 bar MDH) | High pressure (see 2.5) | High level SIF with level sensors voted 2oo2, to close inlet valve - **SIL 1**<br>Overflow thru pressure equalization line to other spheres (through normally open [NO] valve) - **IPL** | |
| 2.2 | Low level | Failing to switch from the sphere with low level in time (based on level indication) | Low/no flow - Liquid from spheres through high pressure product pumps to the vaporizer (see 4.2) | Level indication and low level alarm, inspected each year, per government regulation **(not IPL; part of the cause)**<br>9 other spheres with possibly enough level to switch to<br>Feeding from two spheres at all times, so unlikely for BOTH spheres to have low level at the same time - **IPL** | R H tc sp d sl S o o m v |
| | | | Low/no flow - Unqualified liquid from spheres back to Plant (see 6.2) | Two level indication from SIS level transmitter, with low level alarm, with more than 60 min available to switch tanks (SIF driven alarm and response) - **possible IPL, if action of the operator is quick enough** | |

23

# CLOSING

- **A majority of the currently unpredicted Major Process Safety Accidents can be prevented if companies close the Gap in the SCOPING of PHAs:**
  - **Complete a PHA/HAZOP of non-routine modes of operation**
  - **Discuss all damage mechanisms at each equipment section**
  - **Perform multiple (3-4) PHAs/HAZOPs in major projects**
- **Demand competency of Team leaders; enforce use of expert Scribes; ensure the most experienced staff are available for teams**
- **Follow the Best Practice Rules for leading and documenting PHAs/HAZOPs (to eliminate inefficiencies)**

**RESULTS: Close the gaps and find the scenarios you are currently missing (that lead to +80% of major accidents), and do PHA/HAZOP in 50% of the time currently being invested**

# Questions or Comments?

**Bill Bridges**
**PII**
wbridges@piii.com

**Adel Dakheel**
**PII**
adakheel@piii.com

PROCESS IMPROVEMENT INSTITUTE
Risk and Reliability Specialists

3RD CCPS GLOBAL SUMMIT ON PROCESS SAFETY

أرامكو السعودية saudi aramco

# Process Focused Reviews Our Journey to Safe Operations

By Bouziane KHLOUF

12/04/2016 - Session 2B - Process Safety Excellence / Leadership

where energy is opportunity™

## Agenda

1. What is a Focused Review?
2. Methodology
3. Process Safety Improvement Approach
4. Sample of Conducted Focused Reviews
5. Tracking System Overview

# What is a Focused Review?

1





## What is a Focused Review?

### Definition

Systematic field survey, narrow in scope, to identify the degree to which a Department is meeting the applicable Saudi Aramco Engineering Standards, Procedures, Reports, Standard Drawings, and related OIMs.

### Objectives

- Assess plant compliance against applicable standards
- Identify areas of improvements
- Technical Skills Development tool
- Support customers to identify and manage hazards and related risks
- Strengthen Process Safety focus among customers

## What is a Focused Review? Cont.

### Topic Selection

Topics are selected based on findings from:

- Incident reports
- Trend analysis reports
- Insurance reports
- Required assessments for new/revised standard requirements

### Report

The Focused Report includes:

- Executive summary
- Objective of the survey
- Methodology
- Findings with recommendations for corrective actions



# Methodology

2

## Methodology

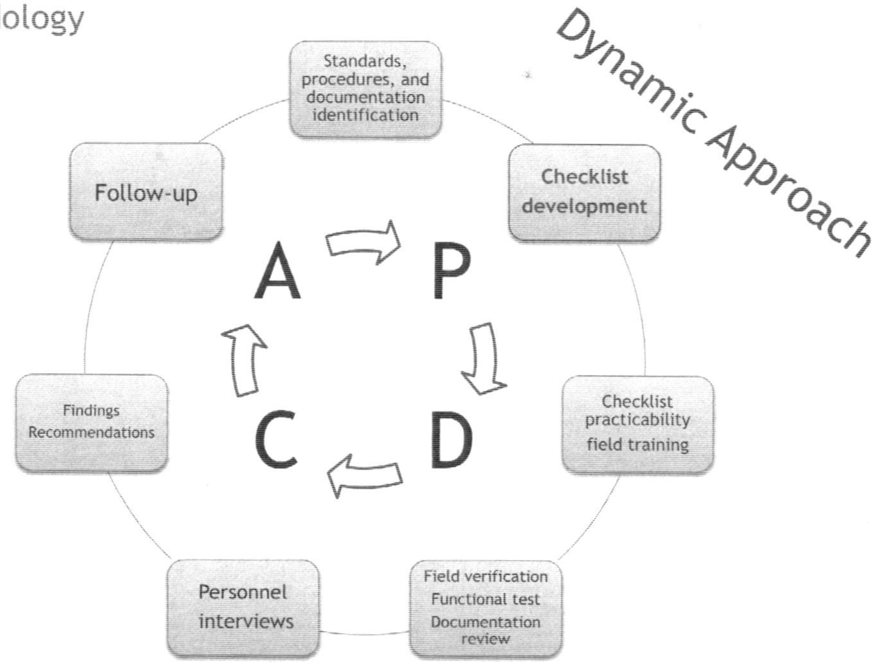




## The way we work together

# Four Critical Behaviors

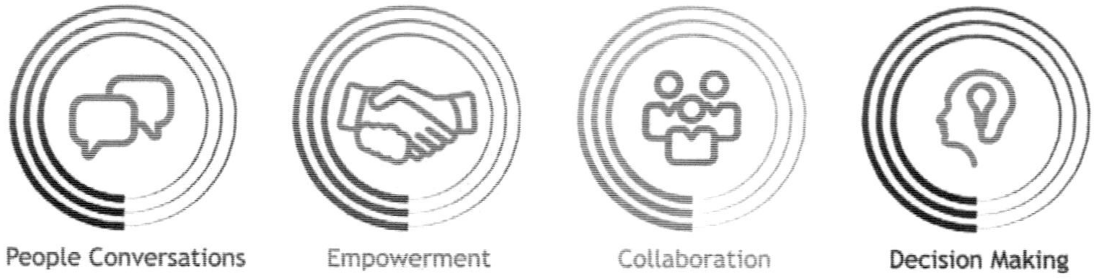

# Process safety improvement approach

# 3



## Process safety improvement approach

Focused review and layers of protection relationship

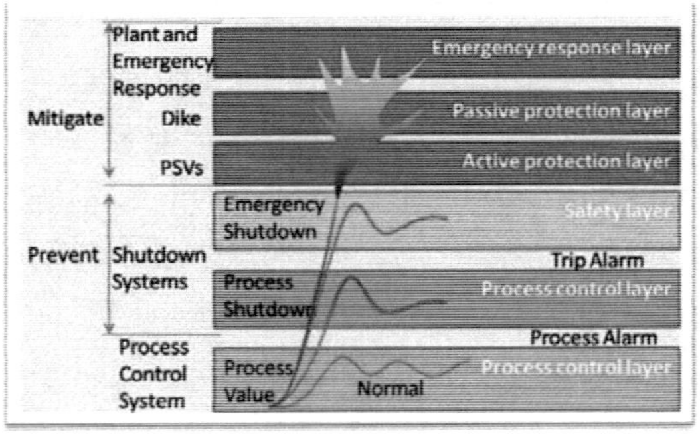

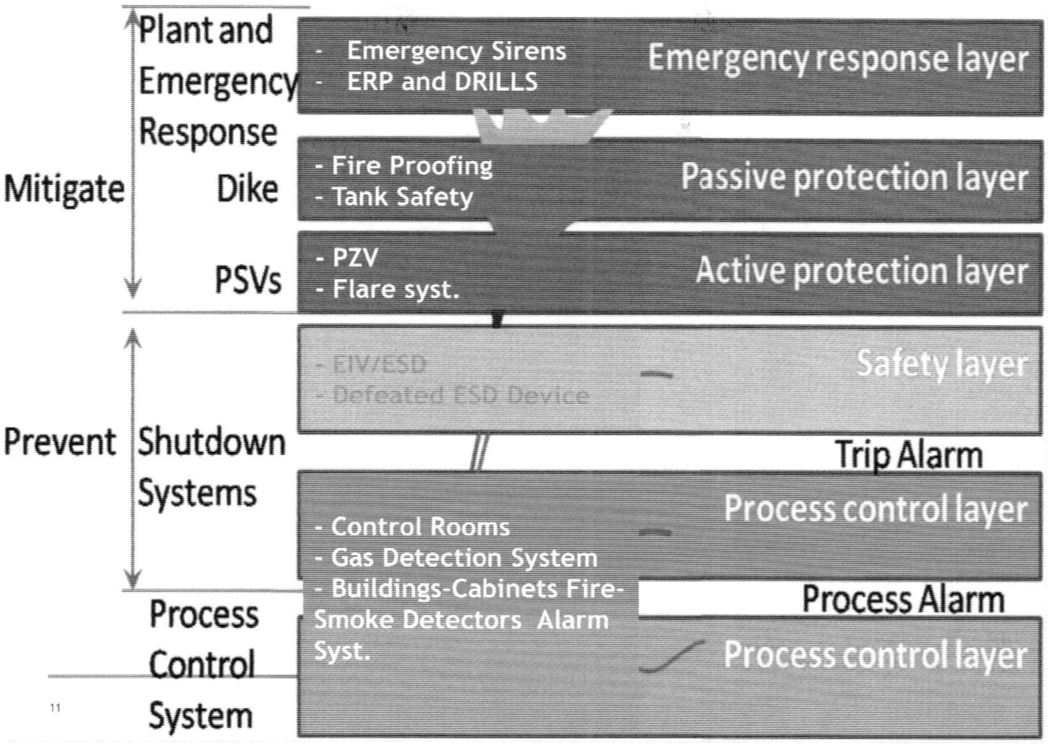


# Sample of Conducted Focused Reviews

4

12

## Sample of Conducted Focused Reviews

- PZV and flare system
- ESD system
- Battery rooms
- Isolation, lockout, and use of hold tags
- Fire protection system preventive maintenance program
- Waste water treatment plant
- Oily water lift station
- Transformer yards
- Etc.



# Tracking System Overview

5

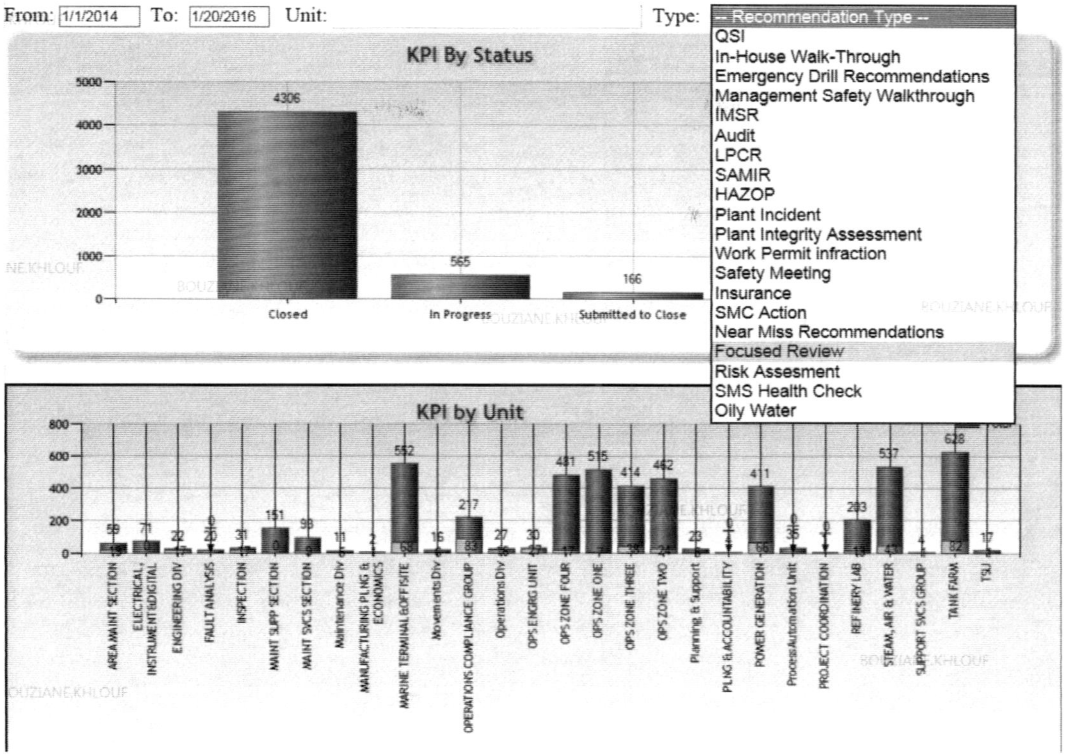


# JRD Safety Recommendation Tracking System

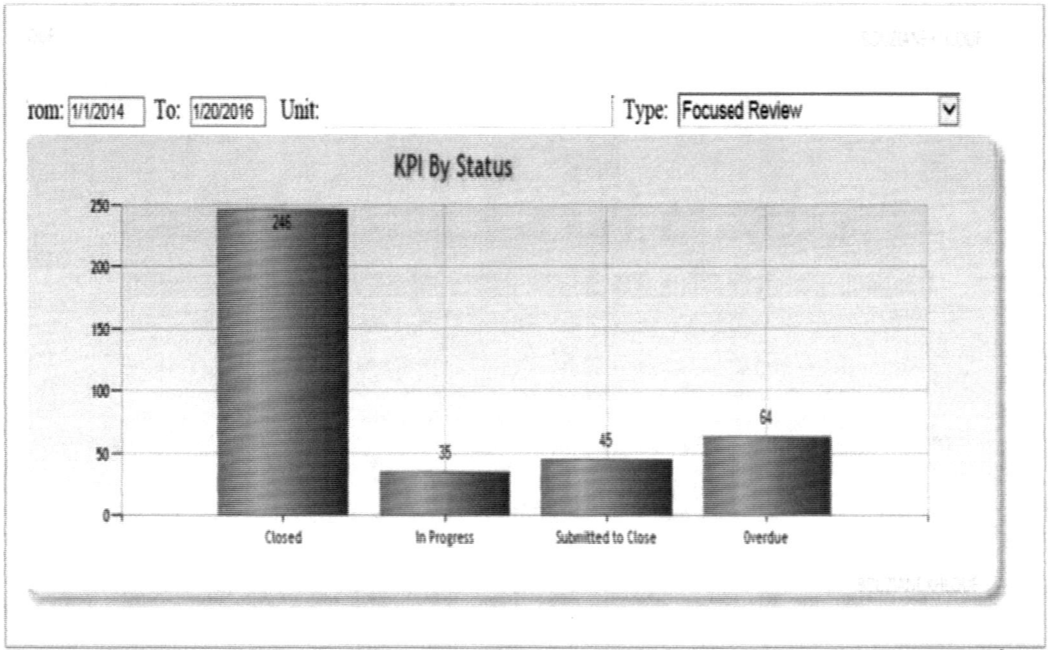


16

Saudi Aramco: Public

أرامكو السعودية
saudi aramco

## Summary

- Collaborative tool with great impact on leadership development & demonstration
- Recommendations addressing both Behavior & System safety to fix identified RC
- Dynamic and systematic Inspection preventing complaisance







3RD CCPS GLOBAL SUMMIT ON PROCESS SAFETY

أرامكو السعودية
saudi aramco

# Thank You

where energy is opportunity

saudi aramco

entrepreneurship

# An Approach to Assess, Manage and Treat Major Industrial Incidents

December 2016

CCPS

where energy is opportunity™

## Major Accidents

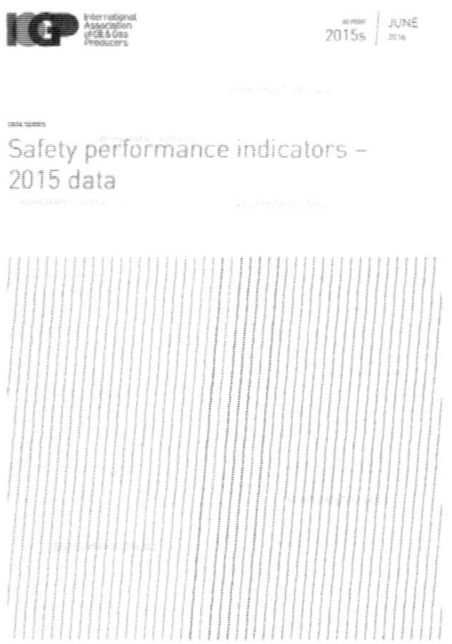


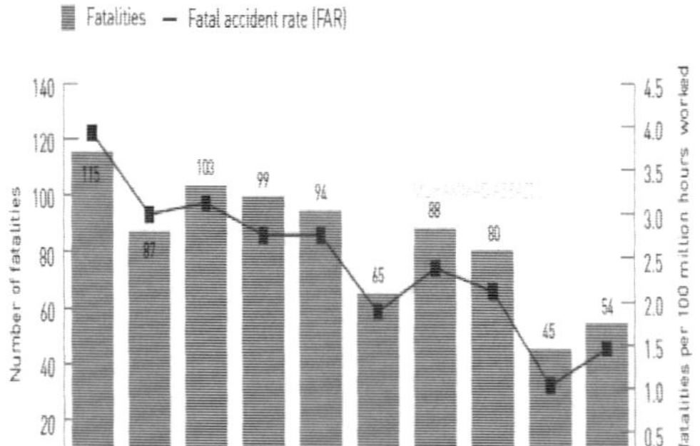


- PROCESS (CONDITIONS): Organizational: Inadequate training/competence
- PEOPLE (ACTS): Inattention/lack of awareness: Improper decision making or lack of judgement
- PROCESS (CONDITIONS): Organizational: Inadequate work standards/procedures
- PROCESS (CONDITIONS): Organizational: Inadequate supervision

- **PROCESS (CONDITIONS): Organizational: Inadequate hazard identification or risk assessment.**

# Major Accidents

The publication summarized the 100 largest property-damage losses occurred in the hydrocarbon extraction, transport, and processing industry from 1974 to 2013

**The total accumulated value of the 100 largest losses is over US$34 billion!!**



# Major Accidents

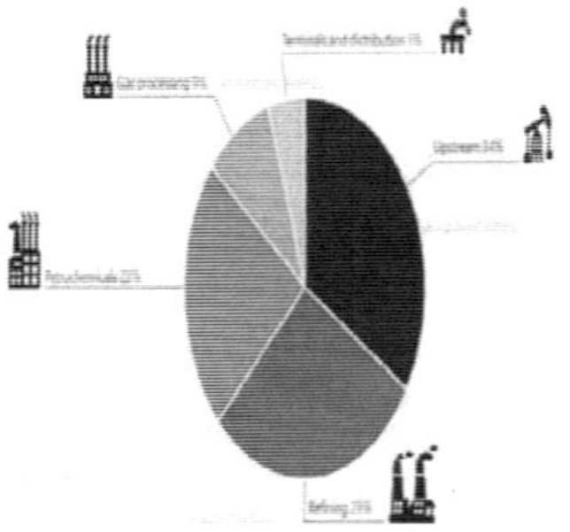

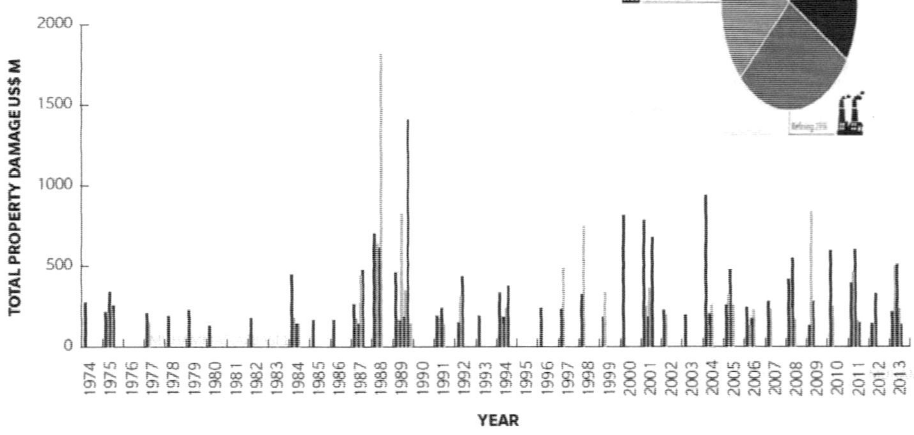

# ERM Process

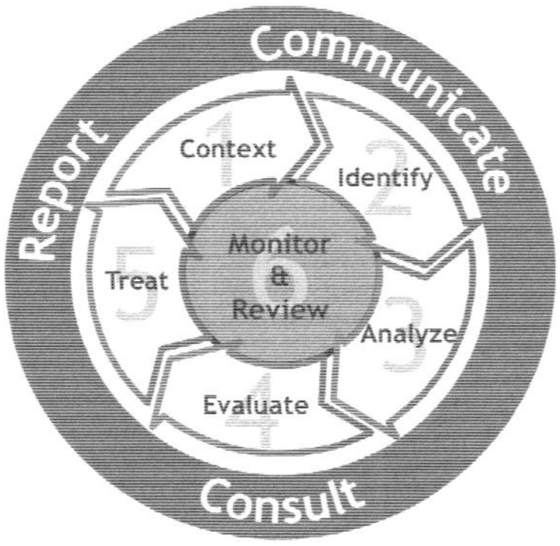


A structured approach to Enterprise Risk Management (ERM) and the requirements of ISO 31000



# Activities - Step 1

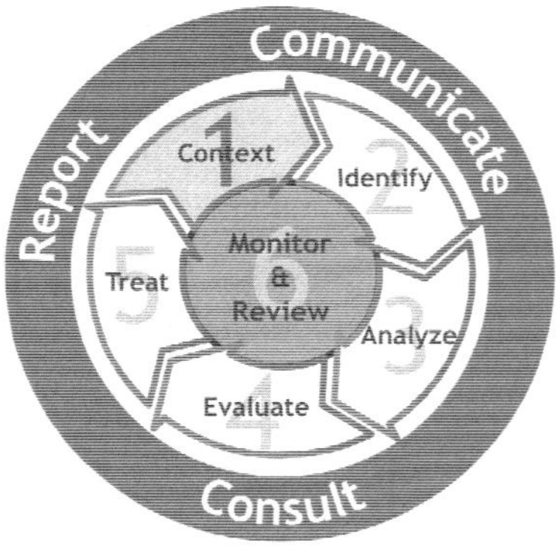


A coordinator, senior consultant level assigned from Consulting Services Department

1. A team established with diverse representation from Loss Prevention, Process and Control System, Environmental Protection, Facilities Planning, Inspection and Consulting Services Departments.
2. Team has >100 yrs combined industry experience and strong backgrounds (4 PhDs)

# Scope

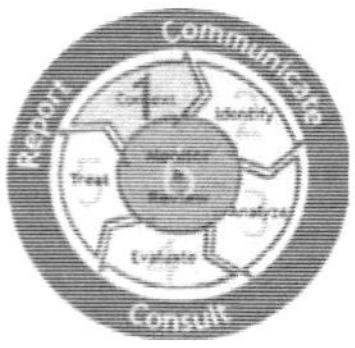


## Risk Description

Identify the risks that lead to an incident resulting in:

- Damage to a major industrial facility or
- Loss of life



# Governance and reporting

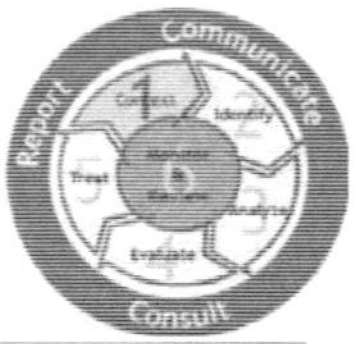


**Head of Risk (Corporate Planning VP) Review**

- Head of Risk
- ERM Group
- Risk Champions/ Cos

**Management Committee – Annual**

- Management Committee
- Risk Champions/ Cos

**Risk Champion & Team**

- Risk Champion/ Co-champion
- Coordinator
- Taskforce
- Risk Professional

## Activities - Step 2

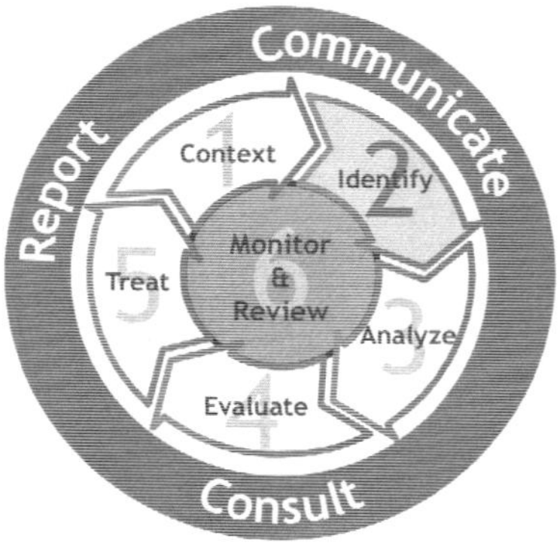


1. Complete background research to review relevant existing or previous initiatives/studies
2. Conduct facilitated sessions with cross functional teams
3. Determine an adequate Risk representation methodology



## Risk Identification Approach

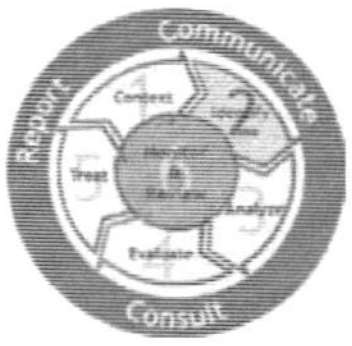

**5**

Workshops

**8**

Admin Areas

**241**

Admin Area Risks

**8**

Consolidated Sub Risks

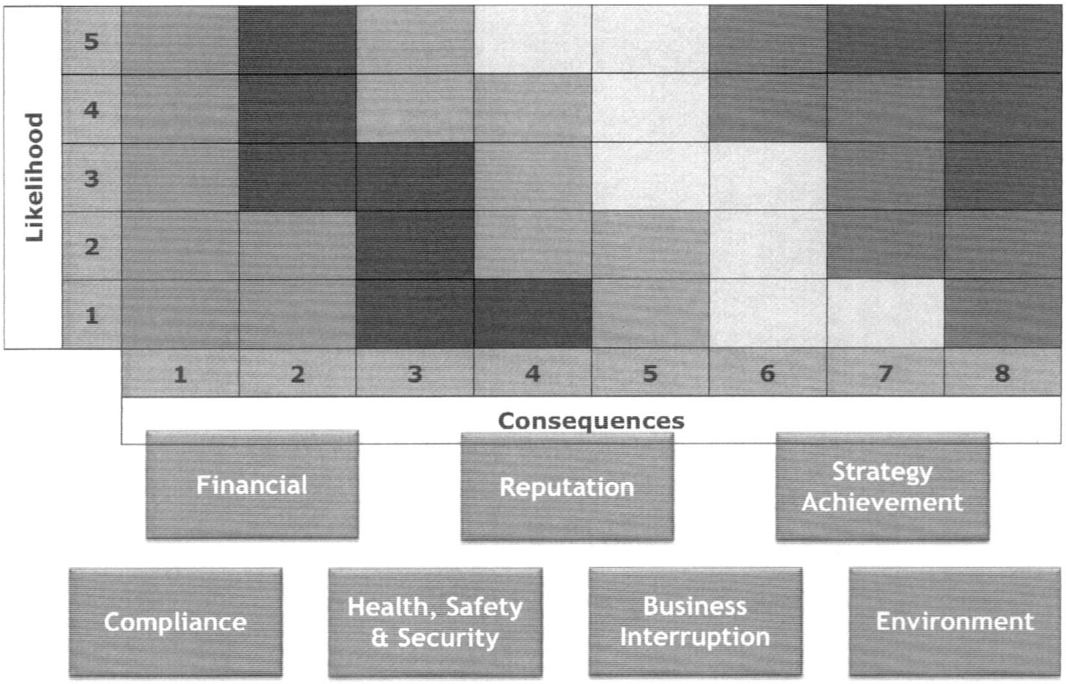




# Activities - Step 3

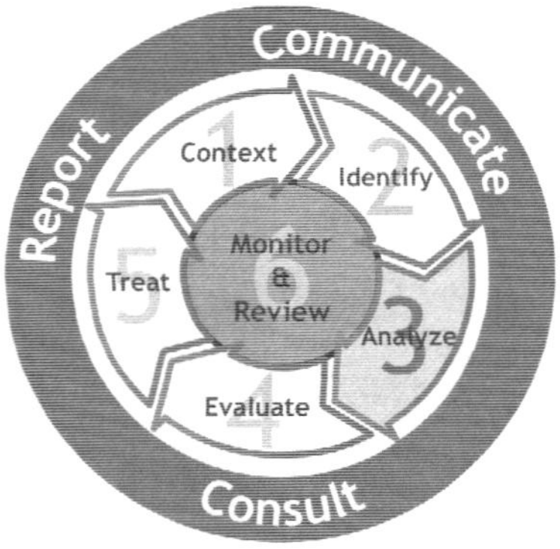


1. Assess the consequence and likelihood of the Risk Events identified in the previous step based on credible scenarios
2. Prioritize causes

# Bow-Tie Approach

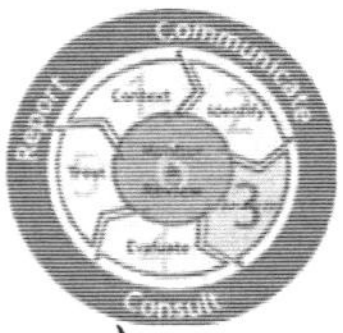


Multiple Causes and Consequences for each Risk (Toxic Release)

# Detailed Activities - Step 4

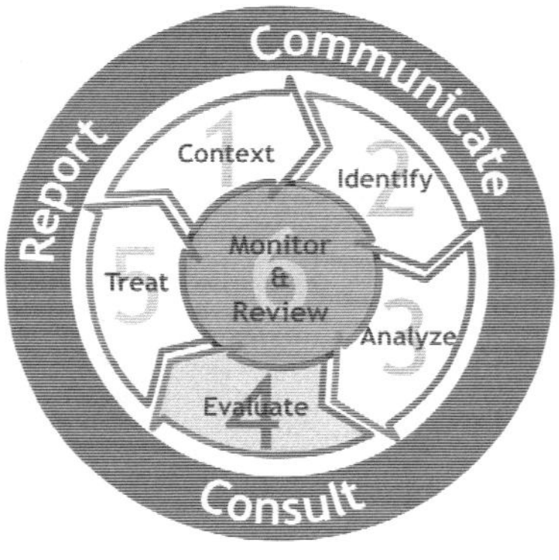


1. Identify existing controls, at appropriate level
2. Assess effectiveness of existing controls
3. Document weaknesses in controls

# Control Identification

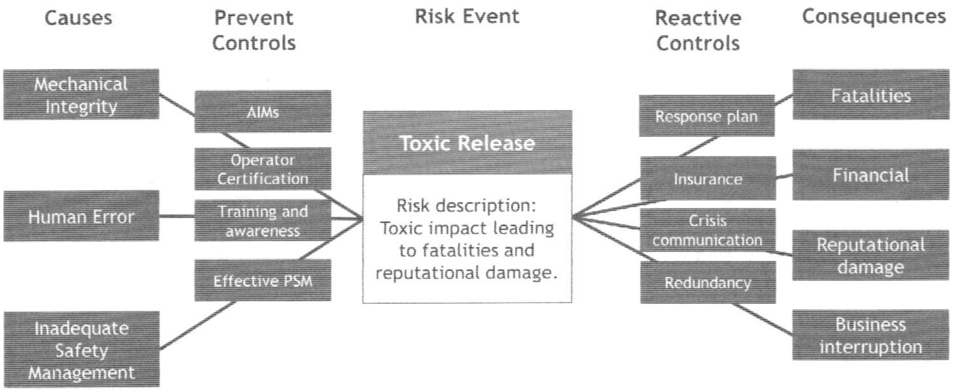




# Activities - Step 5

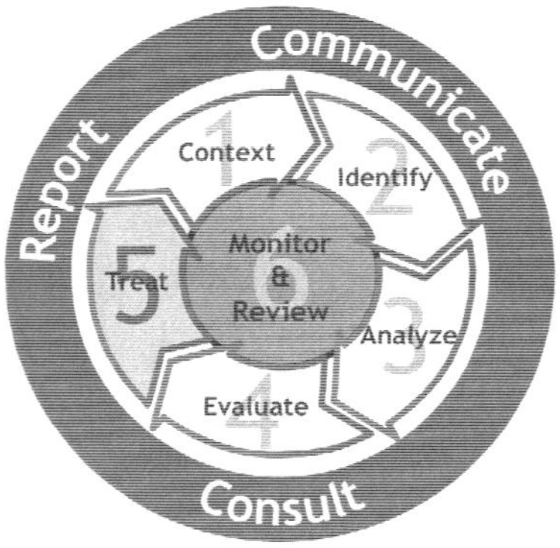


1. Define treatment type and plan, if existing controls are not adequate
2. Develop a method of Treatment governance and implementation by assigning actions and review date
3. Identify targets for treatment implementation

# Proposed Treatments

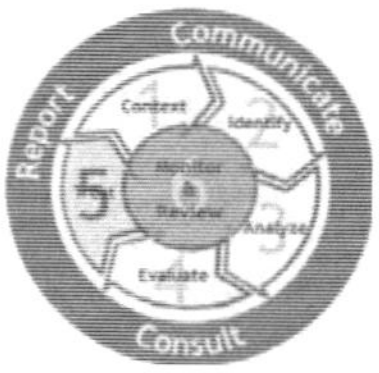


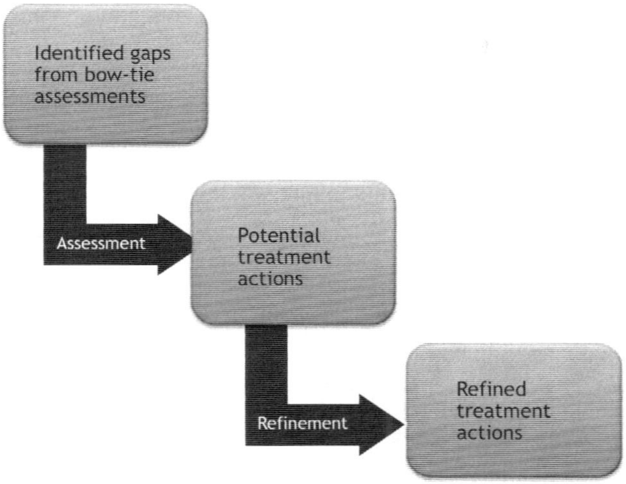




# Risk-Cause-Control-Treatment Relationships/Linkage

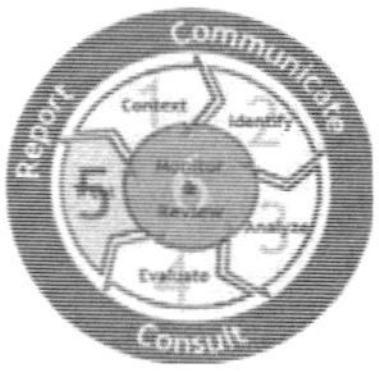

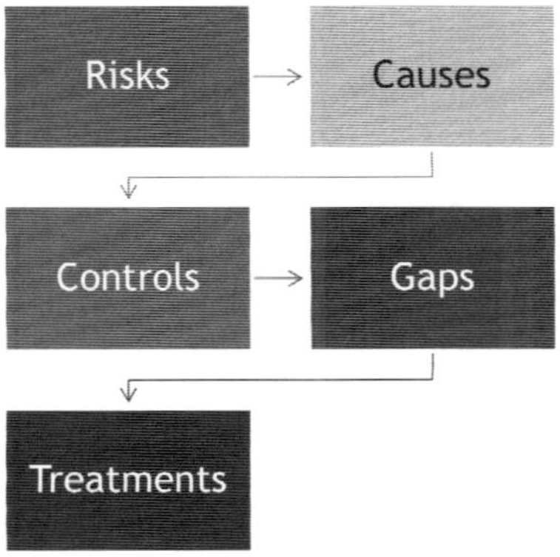

## Treatment Awareness Workshops

Conducted a number of Workshops;

- To Engage Stakeholders
- Objective: Raise Awareness

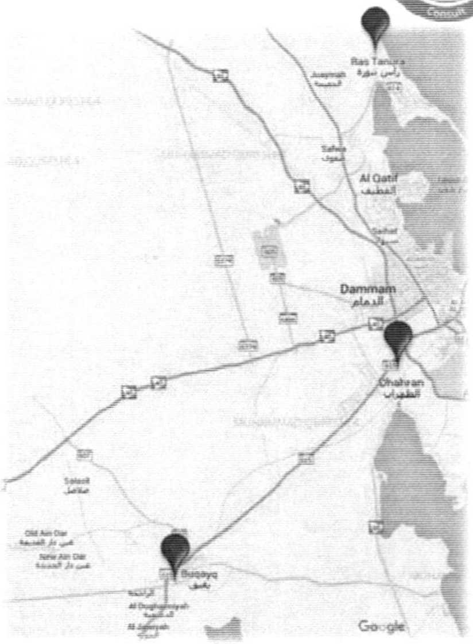








## Treatment Evaluation

Evaluation Matrix

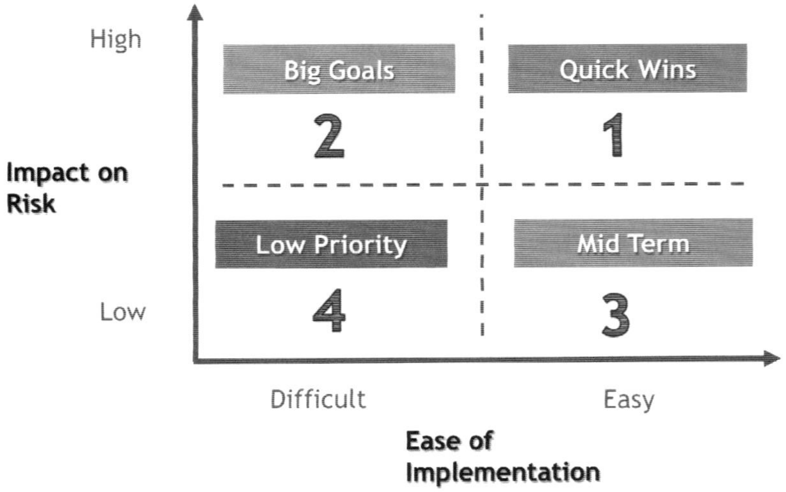

## External Review Post Treatment Step

- Concurred with the team approach and methodology.
  - ✓ No major risk has been overlooked.
  - ✓ No mitigation measure has been missed.
- Provided input on improving the execution plan.



## Detailed Activities - Step 6

1. Monitor implementation of treatment
2. Deploy metrics or indicators for monitoring
3. Report to appropriate level
4. Initiate steady state review

# Key Performance Indicators

A number of leading and lagging indicators were identified

- **Lagging Indicators**

Monitor failures or weaknesses of a barrier

- **Leading Indicators**

Monitor input activity to maintain a strong and healthy barrier



# Summary

- Major industrial incidents remain a challenge for the oil and gas industry
- Saudi Aramco has adopted the ERM 6-step process and is proactively mitigating potential major industrial incidents in the company

## Questions?

كيميا
kemya

# PROCESS SAFETY ACADEMY(PSA)- FOR FRONT LINE SUPERVISORS

## A Step Towards Building A Process Safety Culture

By: Hesham K. Al-Subait
Sr. Manager, EHSS (Former PSE-Lead)

Seju Davis
Lead Engineer, Safety

Al-Jubail Petrochemical Company (Kemya) - A SABIC Affiliate

3rd CCPS Global Summit, Dammam, 04/05 Dec 2016

A SABIC and ExxonMobil Joint Venture

كيميا
kemya

## WHY PROCESS SAFETY !

"There's an old saying that if you think safety is expensive$$, try an accident. Accidents cost a lot of money. And, not only in damage to plant and in claims for injury, but also in the loss of the company's reputation

No. 1

kemya

## CONTENTS

- KEMYA OVERVIEW
- PROCESS SAFETY ACADEMY (PSA) OBJECTIVES
- PSA STRUCTURE & CONTENTS
- CONCLUSION

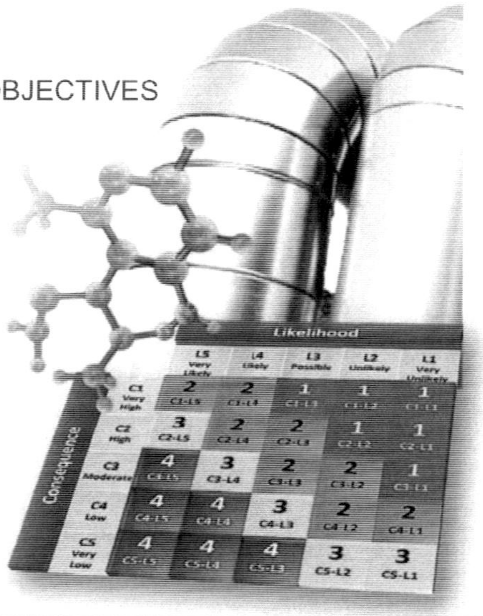


No. 2

kemya

## CASE STUDY PLANT - KEMYA OVERVIEW

Al-Jubail Petrochemical (KEMYA) was formed on 26th April 1980 as an equally owned partnership through a joint venture agreement between Saudi Basic Industries Corporation (SABIC) and Exxon Chemical Arabia, Inc., an affiliate of Exxon Chemical Company.

**POLYMER**

LLDPE
LLDPE : 900 KMTPY, Prod. Started: 1984

LDPE
LDPE : 303 KMTPY, Prod. Started: Aug 2000

**HYDROCARBON**

KEMYA OLEFINS PLANT
ETHYLENE : 760 KMTPY, Prod. Started: Nov 2000
PROPYLENE : 200 KMTPY, Prod. Started: Nov 2000

**ELASTOMER**

CARBON BLACK
CB : 50 KMTPY, 2015

PBR/SBR
PBR/SBR : 100 KMTPY, 2016

EPDM
EPDM : 110 KMTPY, 2016

HALOBUTYL
HB : 110 KMTPY, 2016

Our values

Integrity

Inspire Engage Create Deliver

- 25+ Years / 29+ MMhrs w/o employee LTI (since Oct 90)
- 29+ Years w/o environment violation or exceedance
- No major process safety accidents
- Healthy management system rating
- Many safety awards

Direct Hire Employees ~ 1200

No. 3

## PROCESS SAFETY ACADEMY (PSA) FOR FLS : WHY ?

- Safety leadership training feedback: Need improvement in process safety knowledge and leadership attributes.
- **New resources:** Plants, process and new employees.
- Learning from major process safety accidents : More effectiveness of tools/management system = people focus.
- FLS is Key: Front line supervisors can make or break an organization's safety culture.
- Beyond Personal Safety: Enabling FLS will help them to develop and lead a culture of process safety within their shift teams.
- Improve organizational safety management capability: Develop FLS with process safety mindset and behaviors; Align FLS with professionals/ leadership, creates synergy.

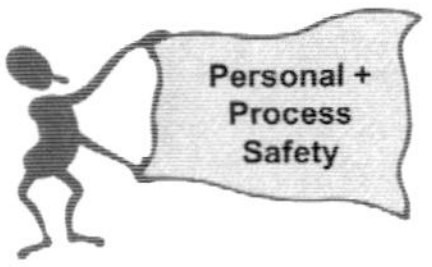

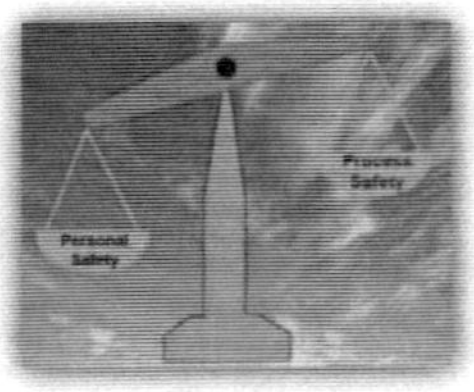



## PROCESS SAFETY ACADEMY OBJECTIVES

- Develop process safety knowledge of tools, management system & behaviors for FLS. Integrate with SHEMS* application.
- Enhance process safety leadership and culture of FLS, complementing key EHSS leadership behaviors, leading to safety excellence.
- Practice safety leadership in the field and proactively resolve process safety issues
- Act as body of knowledge and role models, for future FLS

**FLS**

- Ensure compliance
- Encourage the team
- Promote risk awareness
- Involve the team

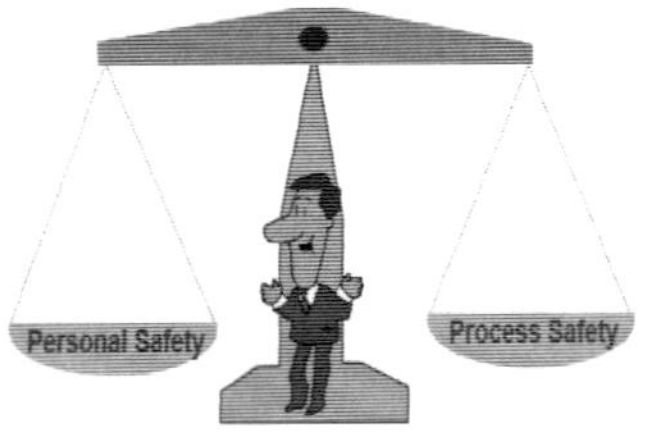


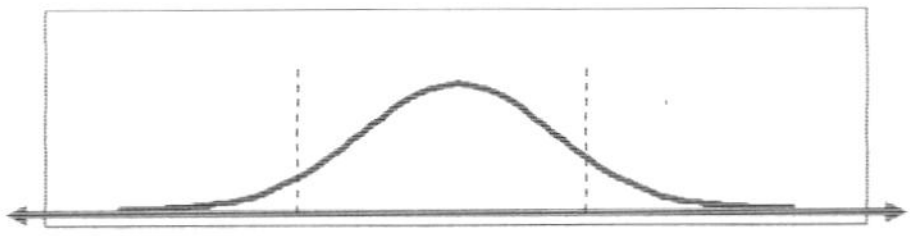

## PROCESS SAFETY ACADEMY – HOW?

- An in-house team developed modules based on existing process safety tools and management system, with inputs also taken from SABIC/ExxonMobil.
- PSA workshop of six- including one day filed assessment - days were tailored to suit adult learning with plenty of exercises, role plays, videos and field visits.
- Senior Management involvement and support assured in each step and throughout workshop.
- Periodic process safety walkthrough and metrics ensures continuous benefit.

No. 6

## PROCESS SAFETY ACADEMY – HOW?

- 28 FLS from operation & maintenance functions selected under the guidance of senior management from different plants, based on experience and passion towards safety. [16 Kemya + 12 from SABIC].
- Participants divided into groups of 5 members each, for effective learning.
- Each group has a leader selected for the day and leadership rotated daily.
- Participants list their expectations from PSA and compared at the end of PSA.
- More focus on process safety films (e.g. CSB case studies) and practical examples, from site and high learning value incidents from outside.
- Participants list three major process safety risk in their area and ended up finding mitigations – Problem solvers!; Also practiced process safety walkthrough (PSW)...

No. 7

كيميا
kemya

## CONTENTS OF PSA : WHAT?



No. 8

كيميا
kemya

## PROCESS SAFETY ACADEMY – WHAT?

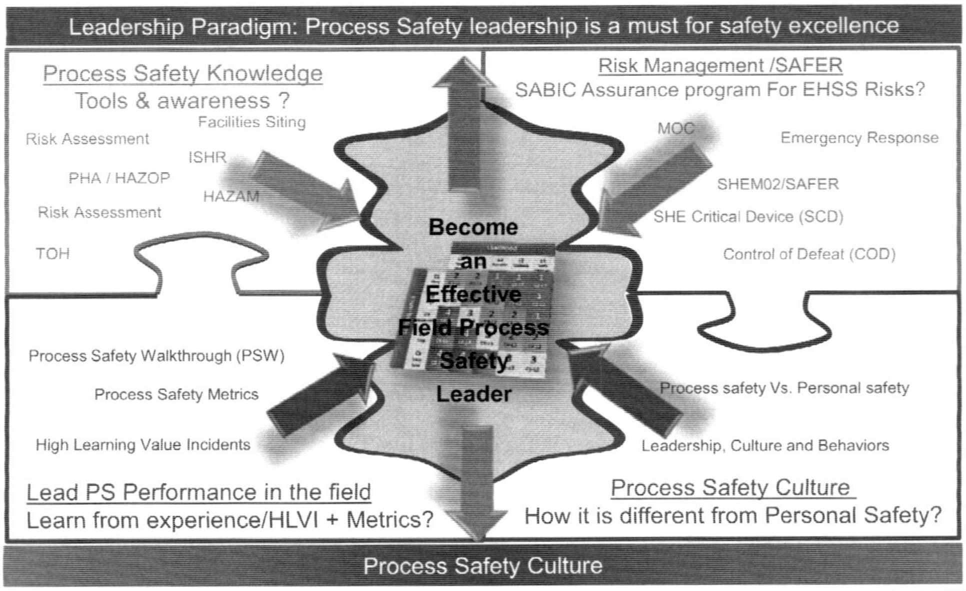


No. 9

كيميا
kemya

## PROCESS SAFETY ACADEMY – PS CULTURE

PROCESS SAFETY CULTURE

1. Process safety Vs. Personal safety
2. Leadership, Culture and Behaviors

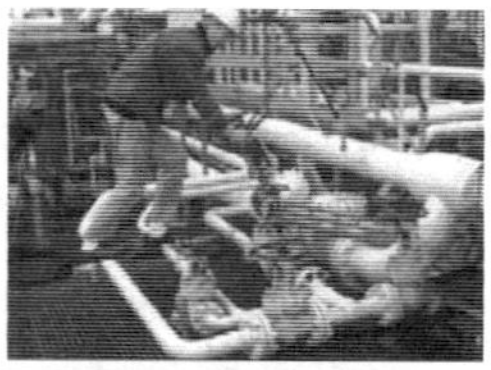

Observable Behaviors ?
- Why?
- Barriers?

**Personal Safety** **Process Safety**

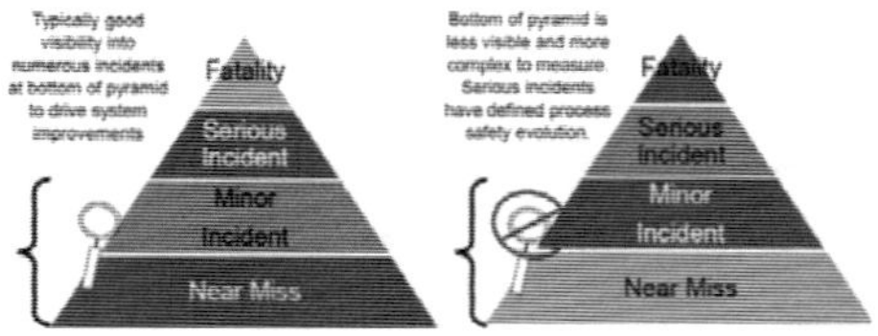


**Personal Safety**
- Prevention of a series of incidents

Incident A
Incident B
Incident C
Incident D

**Process Safety**
- Prevention of the catastrophic incident

Incident A

No. 10

كيميا
kemya

## PROCESS SAFETY ACADEMY – PS KNOWLEDGE

PROCESS SAFETY KNOWLEDGE

1. PHA/ HAZOP/ HAZAMS*
2. Risk Assessments
3. ISHR/ TOH*
4. Safety Engineering Technology for FLS
5. Facilities Siting

Confusing Manifold –wrong connection?

The Car Seal Open valves are not painted yellow.

No. 11

# PROCESS SAFETY ACADEMY – RISK MANAGEMENT

RISK MANAGEMENT

1. SAFER
2. Management of Change
3. Stewardship

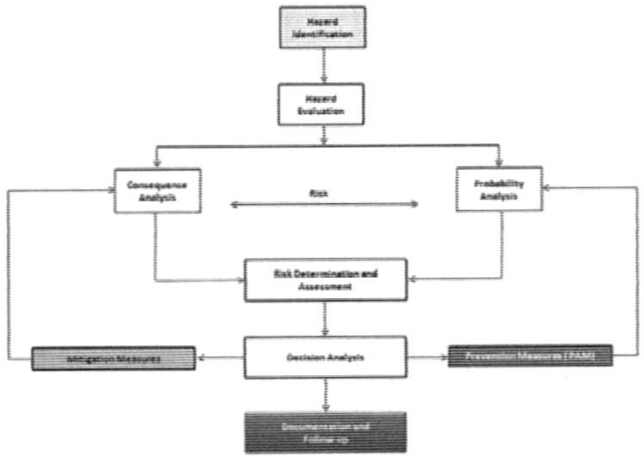


| Risk Class Description |
|---|
| High Risk |
| Moderate Risk |
| Low Risk |
| Acceptable Risk |

| Consequence | | Likelihood: L5 Very Unlikely | L4 Unlikely | L3 Possible | L2 Likely | L1 Very Likely |
|---|---|---|---|---|---|---|
| C1 | Very High | 2 | 2 | 1 | 1 | 1 |
| C2 | High | 3 | 2 | 2 | 1 | 1 |
| C3 | Moderate | 4 | 3 | 2 | 2 | 1 |
| C4 | Low | 4 | 4 | 3 | 2 | 2 |
| C5 | Very Low | 4 | 4 | 4 | 3 | 3 |

Based on risk level, stewardship mechanism and approval level for continued operation defined in SAFER!

No. 12

# PROCESS SAFETY ACADEMY – LEAD PS PERFORMANCE

LEAD PROCESS SAFETY PERFORMANCE

1. Process Safety Metrics (lagging & Leading)
2. High Learning Value Incidents
3. Process safety walk through (PSW)

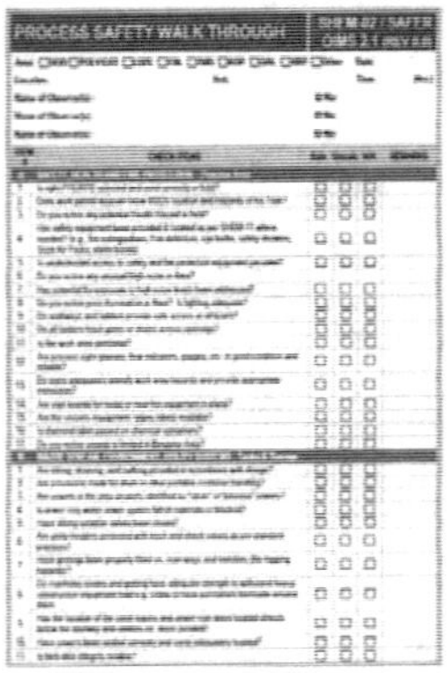


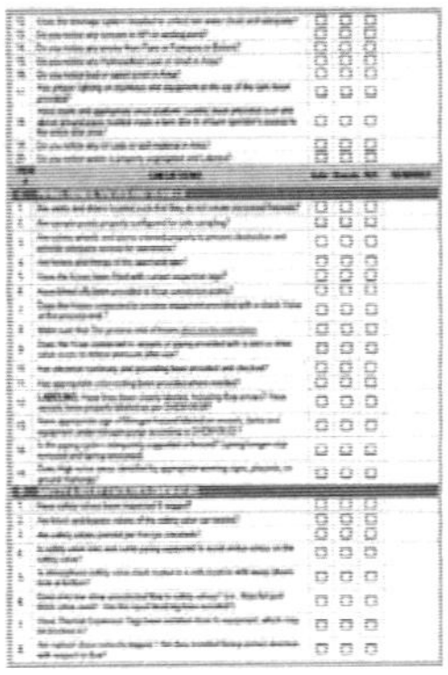

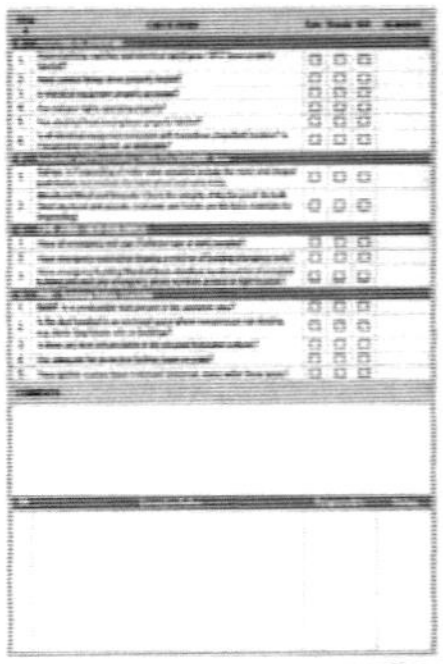

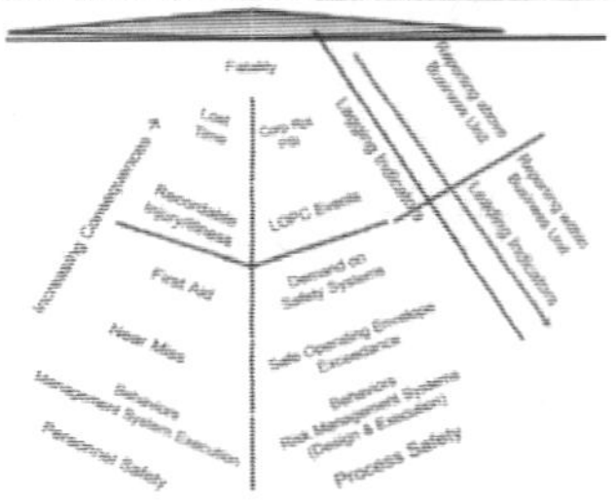


No. 13

كيميا
kemya

## PROCESS SAFETY ACADEMY – METRICS (PART)

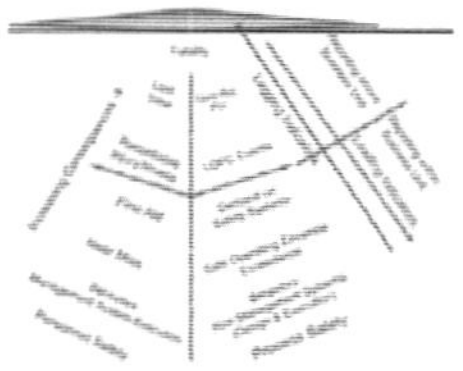

| KEY PERFORMANCE INDICATORS (KPI) | THIS MONTH Plan/Actual | YTD Actual | YTD Target | TL |
|---|---|---|---|---|
| 2) No. of Cat.1 & 2 Risk Items Closure Overdue | 0 / 0 | 13 | 0 | ● |
| 3) No. of Open Non-T/A Cat 1 & 2 Risk Items Age > 2 yrs. | 0 / 0 | 24 | 0 | ● |
| 4) No. of PS Linked OCVM FV and OKV Audits Conducted | 47 / 60 | 360 | 277 | ● |
| 5) % of PS linked FV and OKV Failed | 0 / 0 | 0 | 0 | ● |
| 6) No. of PS Incidents with IRAT > 399 | 0 / 0 | 2 | 0 | ● |
| 7) % Isorisk Units Reduction from Previous Year End Total | 2.5 / 13.10 | 25.04 | 12.5 | ● |
| 8) No. of Alarms/console/hour - average | 6 / 8.3 | 9.1 | 6 | ● |
| 9) % Process Safety Training Completion | 8.3 / 2.1 | 47.4 | 66 | ● |
| 10) No. of SH&E Risk Mitigation DCR's Implementation Overdue | 0 / 2 | 29 | 0 | ● |

● Meeting Target ● ≤10% deviation ● >10% deviation

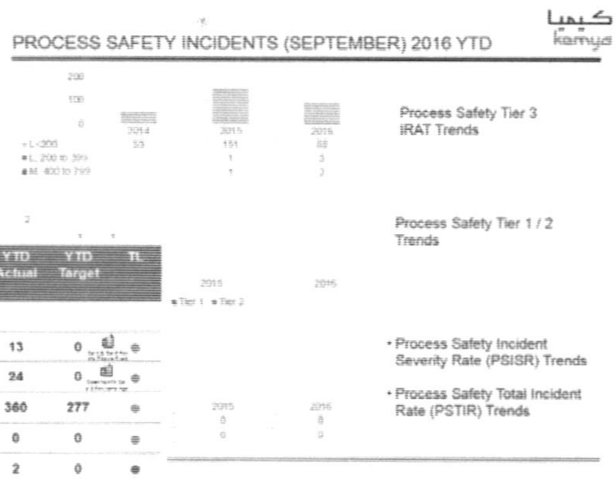


**What gets measured, gets improved**

No. 14

كيميا
kemya

## PROCESS SAFETY ACADEMY – GLIMPSES

No. 15

## PROCESS SAFETY ACADEMY- A STEP TOWARDS BUILDING A PROCESS SAFETY CULTURE

كيميا
kemya

- More companies are now realizing the fact that if they want to have safety as a culture on a sustainable basis, process safety focus on front line supervisors is **a must**.
- Implementing process safety academy (PSA) fine tuned to organization needs based on management system is an effective approach to lead us to safety excellence.
- PSA will be successful if it is supported by management, locally focused and employee-involved. Executive Leadership involvement & support is a key success factor.
- Process safety and personal safety are two sides of the same coin and both need to be part of overall safety program to achieve "complete" safety to sustain safety as a core value.
- Organizations shall enable leaders with adequate process safety skill and resources, to advance safety excellence. Safety leadership development helps achieve other kinds of organizational excellence.

No. 16

### A Step Towards Building A Process Safety Culture

كيميا
kemya

**EFFECTIVE LEADERSHIP**

**INSPIRES** us to

**ENGAGE** stake holders and

CREATE a frame work to

DELIVER a sustainable world-class safety (EHSS) culture,

with

INTEGRITY as the foundation.

THANK YOU

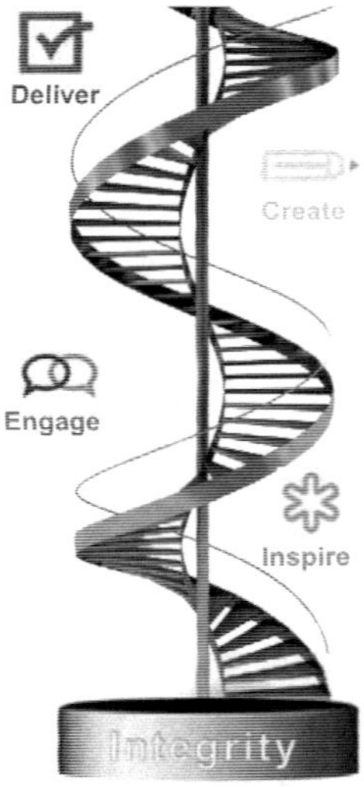

[SABIC Core Values]

No. 17

كيميا
kemya

THANK YOU

كيميا
kemya

# APPENDIX

| LEGEND | | |
|---|---|---|
| FLS | : | Front/ First Line Supervisors |
| HAZAMS | : | HAZard And Machinery Safety |
| HAZOP | : | HAZard & Operability Study |
| IRAT | : | Incident Risk Analysis Tool |
| ISHR | : | Inherent Safety, Health Review |
| OCVM | : | OIMS Critical Verification & Measures |
| OIMS | : | Operation Integrity Management System |
| PHA | : | Process Hazard Analysis |
| PS | : | Process Safety |
| SHEMS | : | Safety, security, Health & Environment Standards |
| TOH | : | Transient Operation Hazard |

| References | |
|---|---|
| 1 | KEMYA PSA Manual |
| 2 | OIMS/SHEMS |
| 3 | KEMYA Performance Statistics |

## "Risk management and Process Safety Leadership"

"Keeping the Balance" for excellence in Safety Performance

Kienbaum

## Process Safety is more than occupational H&S focusing on the long term...

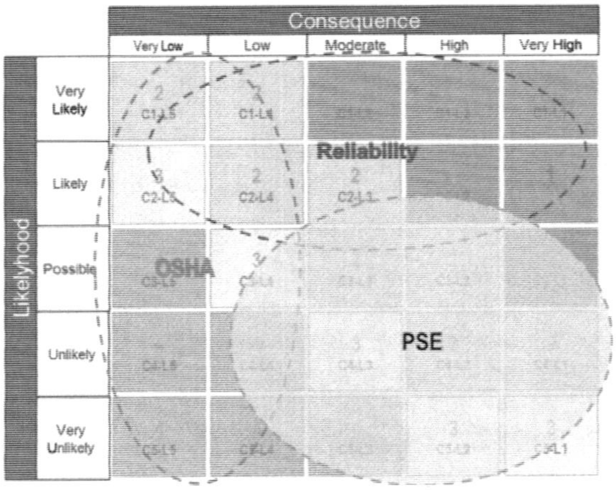


**...and needs leadership with a thorough risk management focus to prevent "Major Accident Hazards" from happening**

Kienbaum
Excellence in People & Organization

## All process safety control systems should be proactively managed...

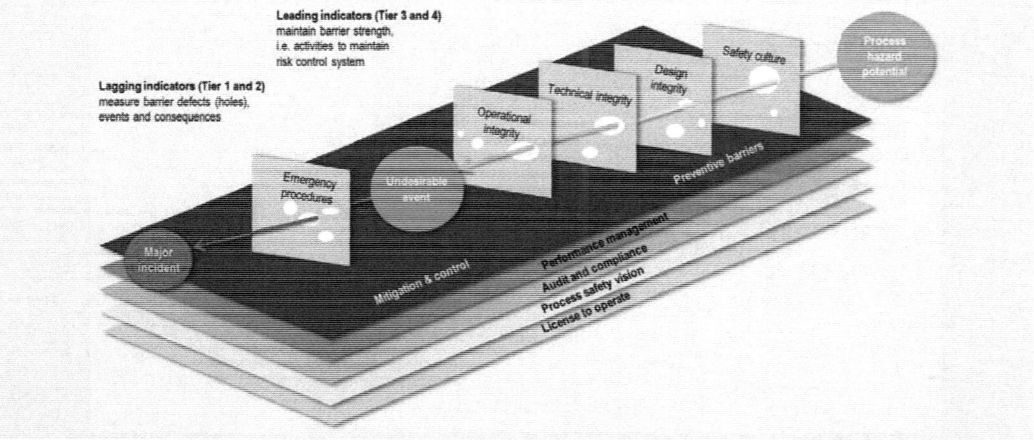


**... as major incidents rarely result from a single cause but rather by multiple failures that coincide and collectively result in an exceptional event with severe consequences**



Kienbaum
Excellence in People & Organizations

## In the core of each of the barriers in an organization is risk management the link to improve and reducing the risk of an incident from happening

Integrated risk process safety management system

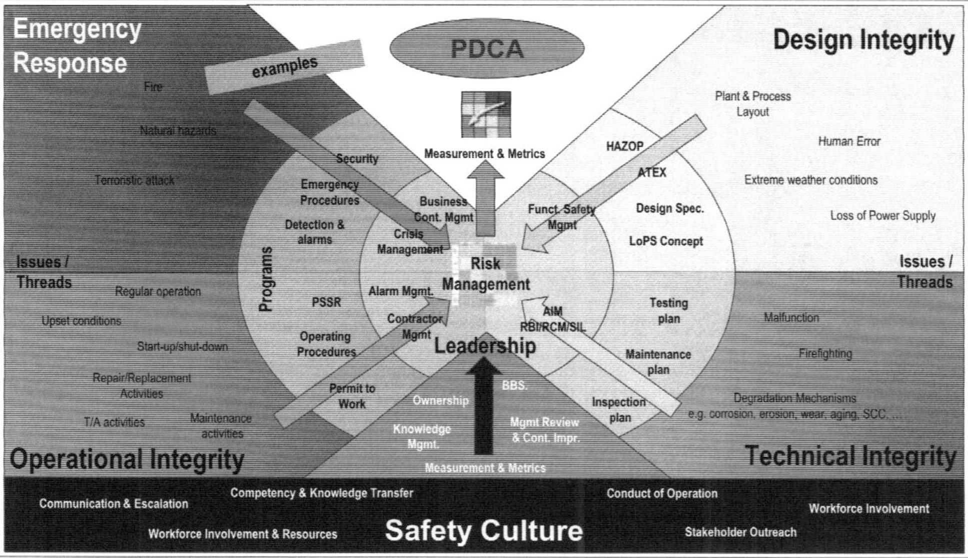




Kienbaum
Excellence in People & Organizations

## In most of the organization risk elements are in place…

Barriers in the PSE management system

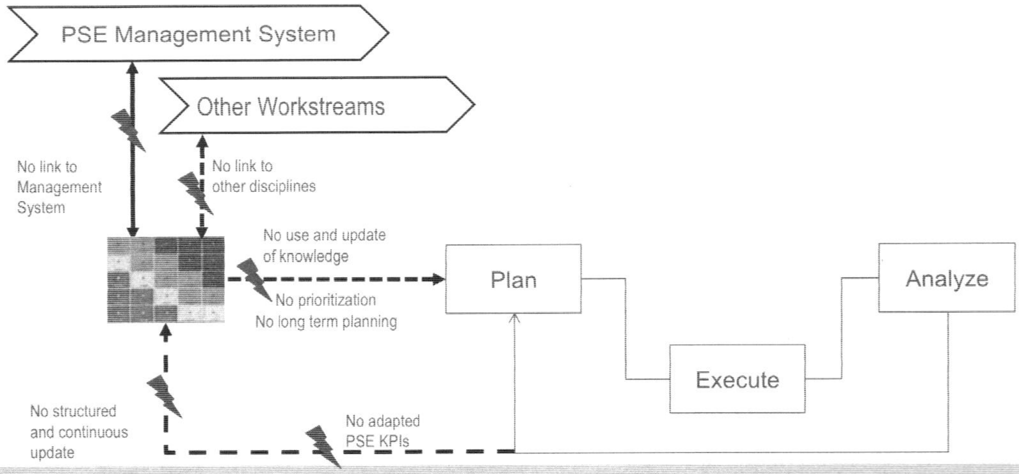


**… but they are not consequently used and utilized to link them to the Process Safety Management System**



## Two examples of the limitations of procedures and systems...

» **Internal & External audits – what is the benefit?**

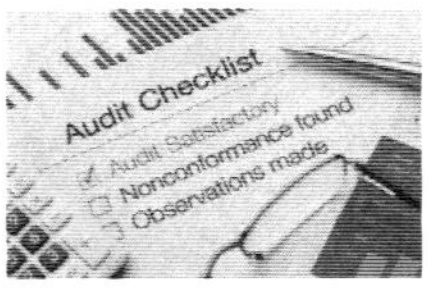


» ***Last Minute Risk Assessment – "tic in the box" list or.....***

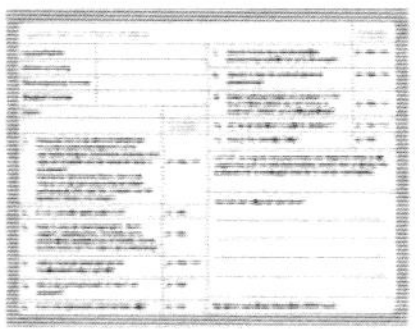

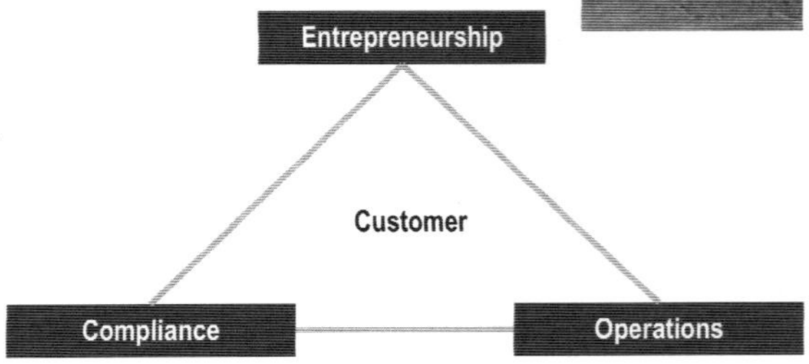


**The consequence of too much focus on the (process safety) management system and compliance driven is that the risk on incidents increases.**

## Experiment with 5 monkeys and the banana – Does it happen in your organization?

"We do lots of stuff that we have no idea why we do it, and we don't even bother to examine it, and in some instances, that does harm"



## An excellent safety culture is depending on a sound balance between a safety management system and safety culture driven by leadership

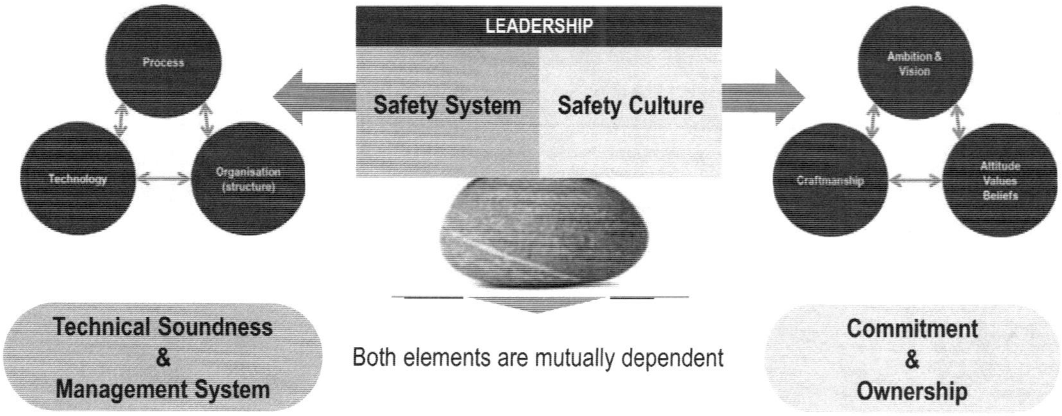


One side doesn't work without the other. If one fails, the other might not be able to compensate.
Leadership has to ensure the right balance.

## The competencies of Safety Leadership in the continuous improvement phase of the journey to safety excellence

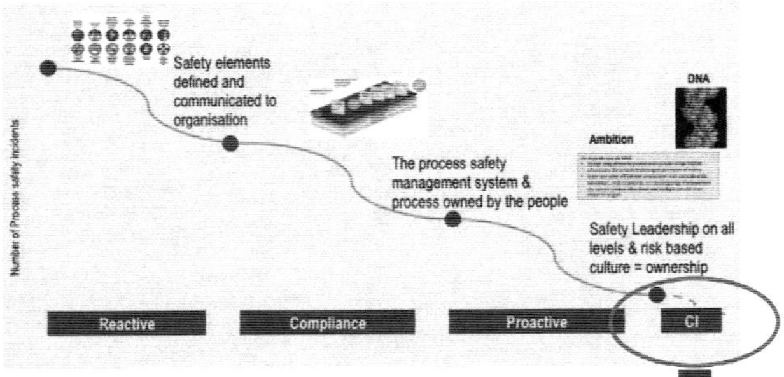


- » **High common ambition** – propagated with great enthusiasm
- » **Clear but not a narrow framework (management system)** – concise picture of the future
- » **Discipline in operations** – continuous improvement with shared quality norms (leading process safety KPI's)
- » **Craftsmanship** – adequate in depth knowledge
- » **Continuous feedback loop** – knowing the unknown



Kienbaum
Excellence in People & Organization

## Linking risk based thinking with Talentleadership as the foundation for continuous improvement

Proactive risk management

| Risk based thinking | | Process Safety KPI performance | |
|---|---|---|---|
| Hazard analysis | • Risk identification through efficient and tailored hazard analysis (PHA/PHR process) | KPI performance Site/Europe | • Aggregation of the performance to develop policy and strategy |
| Select High Impact scenarios | • Highest Consequence Scenarios<br>• Extract key scenarios for each section | KPI performance plant | • Use the KPI's to set priorities and show the developments<br>• Prevention activities & reduce consequences |
| Review with the work teams | • Review in control room and other operational area's<br>• Develop risk awareness with personal | Leadership rounds | • Verification of risk awareness of personal in the organization |
| Integrate in work process | • Verify with management system elements and element owner<br>• Integrate in Shifttour maintenance / production, Inspections, Alarms | Visual board shop floor | • Visualize near misses based on the shift tours, process deviations, etc<br>• Discuss findings in shifts and meeting structure |

Risks need to be clear to define leading KPIs on near miss level of the API pyramid (level 3) and controlling work processes



Kienbaum
Excellence in People & Organization

## Managing risk and improving performance is taking leadership in continuous challenging the current situation

TÜV Kienbaum Excellence in People & Organization

## For questions we are at your disposal

Contact

**Address**

**Kienbaum Consulting International**
Teleport Towers – Kingsfordweg 151
1043 GR Amsterdam
www.kienbaum.com

**TÜV SÜD Industrie Service GmbH**
Westendstr. 199
D-80686 München
http://www.tuev-sued.de/is

**Contact person**

**Kienbaum Consultants International**
**Coen van Driel**
Director
Phone: +31 (20) 6971410
Mobile: +31 6 14328594
coen.vandriel@kienbaum.com

**TÜV SÜD**
**Robert Kauer**
Process safety expert
Phone: +49 89 5791 1277
Mobile: +49 160 3602340
robert.kauer@tuev-sued.de



TÜV Kienbaum Excellence in People & Organization

Kienbaum

Excellence in People & Organization

TÜV SÜD

# Role of Process Safety Leadership to Develop Strong and Vibrant Safety Culture

Nileshkumar Patel
Process Engineering

Creating Value Through Chemistry

"If you always do what you've always done,
you'll always get what you've always got"
-Henry Ford

No One Gets Hurts

# Outline

- Importance of strong and vibrant culture in an organization
- Key leadership attributes and their specific role in developing a strong process safety culture
  - Vision & Just Culture
  - Trustworthiness & reporting Culture
  - Collaboration & learning Culture
  - Feedback and Recognition

**Safety culture "The way we do things around here"**

# Cultures & Its Importance

Culture is defined as the norms of behavior – reflecting an organization's practiced values

- Strong organizational cultures act as a barrier against complacency, omissions, violations, unsafe acts… etc.
- Organization's learning, development and planned change cannot be understood without considering culture as the primary source of resistance to change
- Organization's culture plays a vital role achieving business success and driving the best in class performance & sustainable safety excellence

If leaders do not understand the culture in which they are embedded, those cultures manage them.

## Vision and Just Culture

**Sadara's focus on "Beyond Injury Free"**

Beyond Injury Free is a concept of flawlessly executing our management systems through a culture that embraces safety in everything we do.

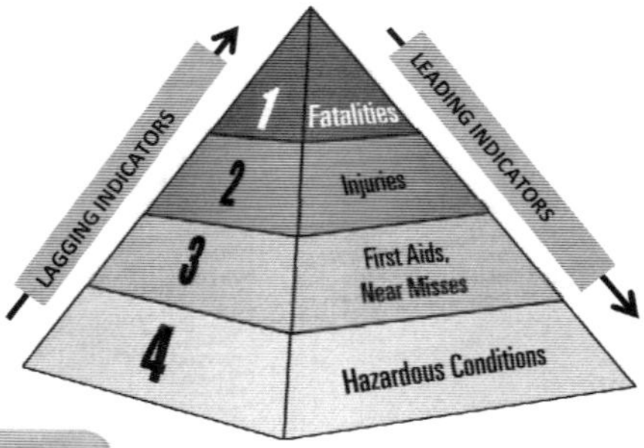


**Management Systems** → Flawless execution of safety management systems through strong safety culture → **Strong Safety culture**

- Leadership at all levels
- Safety is a core value

## Vision and Just Culture

Just Culture, the Sadara Way - embeds 12 **Game Changing Behaviors to create a world-class culture in Sadara**

In Just Culture, people understand the boundary between acceptable and unacceptable behaviors

Unacceptable behaviors are dealt with in a consistent, just and fair manner

- Interdependent culture & structural empowerment through Team Work
- Clear expectations through annual performance goals
- Balance of Consequences (BOC) with emphasis on: Zero tolerance for LCS (Life Critical Standard) violations and significant unsafe acts. Motor vehicle violations with clear escalation requirements.

## Trustworthiness & Reporting Culture

| | |
|---|---|
| Mutual trust environment | ❑ Employees empowered to speak openly and freely<br>❑ Willingness to accept constructive feedback |
| **Enhance value of strong reporting culture** | ❑ Visible demonstration through commitment<br>❑ Timely response and resolution<br>❑ Acting as a role model and ensuring continuous reporting of near misses, incident reporting and observation of unsafe conditions<br>❑ Ensuring maximum independence of investigation & reporting<br>❑ Break negative cycles or under-reporting through confidential reporting system & mutual trust between different layers in the organization. |

## Collaboration & learning culture

| | |
|---|---|
| Behavior related collaboration | ❑ Encouraging cross-departmental collaboration, bringing teams together and encouraging safety discussions<br>❑ Encouraging employees to freely discuss safety concerns amongst themselves<br>❑ Encouraging team work and developing a strong sense of ownership<br>❑ Asking for, listening to, and showing respect for others' views<br>❑ Being open and honest about safety performance<br><br>**"When people are highly motivated, it is easy to accomplish the impossible, and when they are not, it's impossible to accomplish the easy". –Bob Collings** |

## Collaboration & learning culture

Sadara

**Leadership sensitivity to lessons learned**

**Commitment to Resilience**
- Leverage knowledge broadly so the organization benefits from the learning
- Use prior experience and knowledge to minimize impact
- Learn from history and teach what you learn; do not relive it
- Start with a safety moment for each meeting to communicate safety learnings

**Emphasis on high impact Root Cause Investigations (RCI)**
- Training on the accident & swiss cheese model
- Apply swiss cheese model in RCI to identify sharp end, system & strategy holes.

**Leveraging Learning Experience Reports (LER)**
- Knowledge is shared and becomes common knowledge, and any new learning is implemented effectively
- Avoid losing critical knowledge when key people leave

**Study incidents external to organization / looking beyond company**
- Important for the prevention of low frequency-high consequence events, which may have never occurred within the organization, but have occurred within the same industry or plants.

**Deference to Expertise**
- When making decisions and planning activities consult the experts. Experts are those with a better understanding and deeper experience.

## Feedback and Recognition

Sadara

**Influencing through recognition and feedback**

**Why it is important to you?**
- Every employee is a leader in Process Safety
- You can help others in your organization learn to be more effective leaders. Leaders span leaders

**Why start right at the beginning for a new organization like Sadara?**
- People moving into new environments each day
- ZERO Incident culture requires everyone's understanding and commitment

## Feedback and Recognition

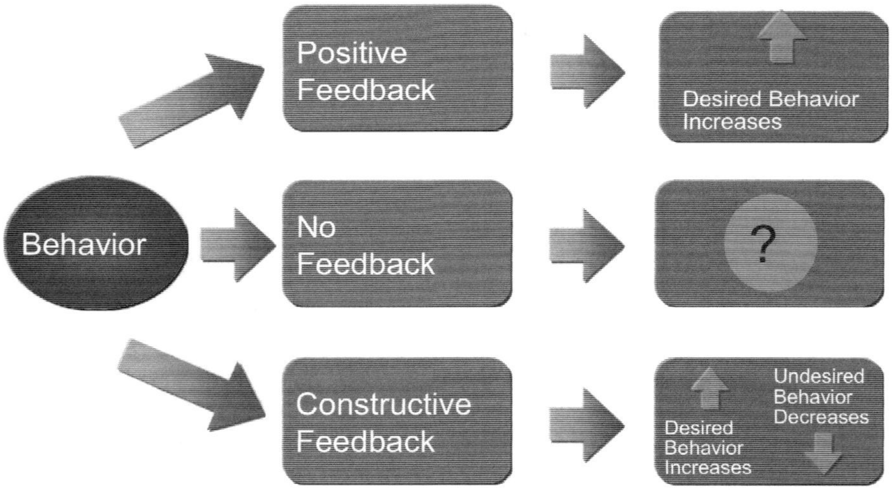


## Feedback and Recognition

**Constructive Feedback vs Positive Feedback**

| What? | So What? | Now What? |
|---|---|---|
| • Describe behavior objectively<br>• Discuss actual behavior you saw, heard, or read about<br>• Don't attempt to guess at "motivation" for the behavior | Describe effect of the behavior on you, on others, and on results | *When Providing Constructive Feedback:*<br>• Ask performer to describe gap in his/her words and identify possible solutions<br>• Mutually agree on next steps, ensuring you describe the behavior you would like to see start or stop using pinpointed language<br>*When Providing Positive Feedback:*<br>• Describe behavior you would like to see continue using pinpointed language |
| Abdullah, I just saw you make that lift without a full barricade in place." | "Not only is this a safety requirement I expect you to follow, but someone could have walked into the area and been exposed to the hazard and potentially injured." | "Why wasn't a barricade used?<br>"What are some ways to prevent it from happening in the future?"<br>"So we agree...in the future, you will make sure the barricade is in place before making any lift?" |

# Feedback and Recognition

Think …. What motivates employees?
- Recognition?
- Feeling of involvement?
- Understanding attitude?

Safety recognition to focus on rewarding positive safety behaviors contributes to a BIF workplace!

Always recognize good safe behaviors.
- Appropriate token gift and / or certificate
- Recognition lunches with management
- A simple "thank you" and recognition in weekly meetings / in presence of peers

Examples
- Excellent housekeeping. Peer Intervention
- Teamwork
- Near Miss Reporting

# Summary: We Get What We Lead !

- **Drive Best in Class Performance**
- **Set the process safety tone for the organization**
- **Role model (influence by leadership)**
- **Encourage collaboration & learning culture**
- **Integrity through interdependent organizational culture**
- **Simply… leverage the best of organization**

**Make it happen !!!!!!!**

# Thank you
# Q & A

January 27, 2016

Sadara

## Summary Finding of the Leadership Forum

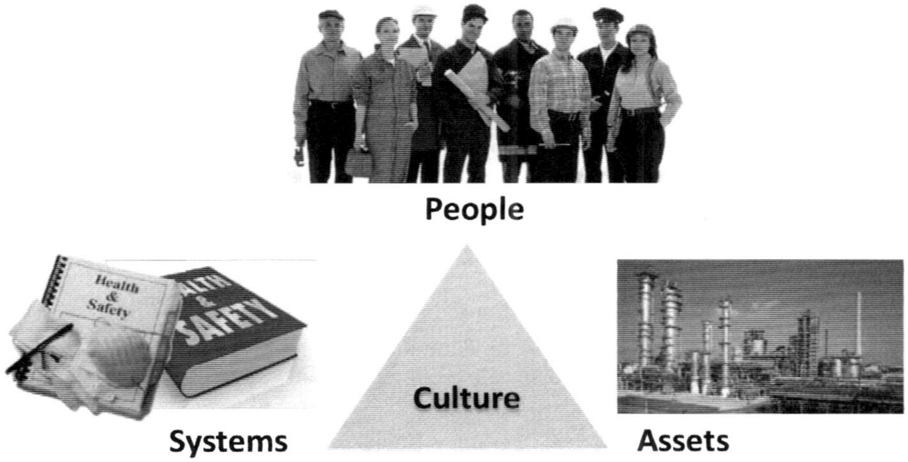




## Summary Finding of the Leadership Forum

**PEOPLE**

- Leadership & Line Management
  - Accountable for performance
  - Visible, engaging & supportive (present in the field)
  - Leadership & management development programs to include safety (PSM)
- Employee Competency & Behavior
  - Development, recognition & retention
  - Design, construction, operation & maintenance
  - Trust & empowerment
- Contractor & subcontractor
  - Recognize challenges & limitations
  - Collaborate & engage with contractors
  - Equal safety expectations

## Summary Finding of the Leadership Forum

### SYSTEMS

- Most companies have PSM system
  - References available, API 754 & OSHA PSM
  - Clear expectations for management, employees & contractors
  - Risk assessment and management
  - Training & competency
  - Communication
  - Design standards
  - Behavior standards
  - Management of change
  - Incident investigation (RCA)
  - Emergency preparedness/response plan
  - Continues improvement
  - **Cyber security, defeats safety systems**
- Meaningful leading & lagging KPIs
- PSM KPIs Board review




## Summary Finding of the Leadership Forum

### ASSETS / FACILITIES

- Design standards
- Material of construction, selection & verification
- Upgrade facilities to new standards, if required
- Knowledge of safe operating envelope
- Inspection programs/Pressure testing
- Management of change
- Alarm rationalization/management
- Asset Integrity Management (book)

## Summary Finding of the Leadership Forum

### Culture

- Safety (PSM) is a value
- Equal to production objectives
- Injuries and incidents are preventable
- Personal responsibility
- Pride and vulnerability
- Together - Leaders, employees & contractors
- Prioritize, what not to work on is just as important as what to work on
- Adjust priorities based on risk and need
- Open reporting, no fear
- Authority to stop work
- Change from compliance to conformance
- Leadership pressure force employees to take shortcuts



## Summary Finding of the Leadership Forum

### CONCLUSION/RECOMMENDATIONS

- Establish regional CCPS chapter
- Develop PSM training programs for regional companies
- Develop means to share learning/best practices
- Collaborate with contractor companies to address common challenges
- Network with other regions, share best practices through CCPS

## Drivers of Critical Chemical Safety Change Program

**3rd CCPS Global Summit on process Safety**
**Dammam, Kingdom of Saudi Arabia**

**Chairperson Vanessa Allen Sutherland**
**U.S. Chemical Safety and Hazard Investigation Board**

**December 5, 2016**

CSB U.S. Chemical Safety and Hazard Investigation Board

# About the CSB

- Independent federal agency charged with investigating industrial chemical accidents.
- Board members are appointed by the President and confirmed by the Senate.
- CSB conducts root cause investigations of chemical accidents at fixed industrial facilities.
- The agency does not issue fines or citations, but does make recommendations to plants, regulatory agencies such as to OSHA, EPA, industry organizations, and labor groups.

www.csb.gov

CSB U.S. Chemical Safety and Hazard Investigation Board

# Agenda

- **Overview of the CSB and 2017 – 2021 Strategic Plan**
- **Drivers of Critical Chemical Safety Change**
- **Critical to Chemical Safety Change: Preventive Maintenance**

www.csb.gov

CSB U.S. Chemical Safety and Hazard Investigation Board

# 2017-2021 Strategic Plan

- **Mission and Vision**
- **Strategic Goals**

www.csb.gov

CSB U.S. Chemical Safety and Hazard Investigation Board

# Mission and Vision

**MISSION**

Drive chemical safety change through independent investigations to protect people and the environment.

**VISION**

A nation safe from chemical disasters.

www.csb.gov

CSB U.S. Chemical Safety and Hazard Investigation Board

# Strategic Goals

**GOAL 1:** Prevent recurrence of significant chemical incidents through independent investigations.

**GOAL 2:** Advocate safety and achieve change through recommendations, outreach, and education.

**GOAL 3:** Create and maintain an engaged, high-performing workforce.

www.csb.gov

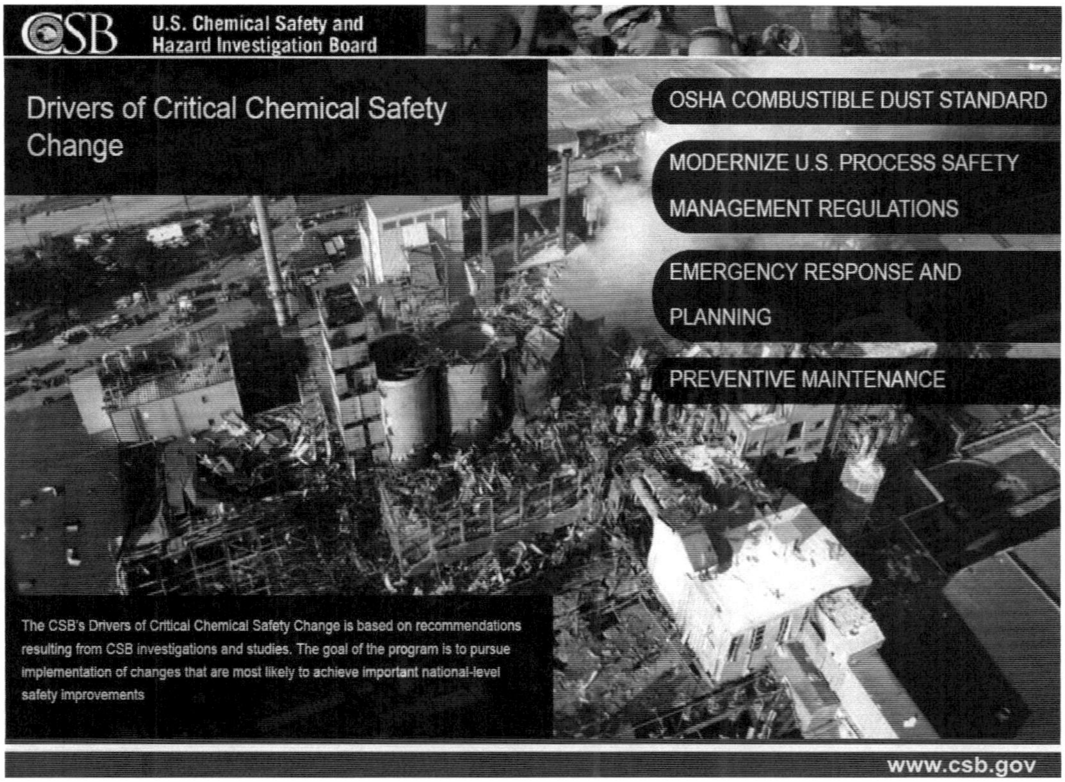


CSB U.S. Chemical Safety and Hazard Investigation Board

## Drivers of Critical Chemical Safety Change

- Identify the most important chemical safety improvement goals of the CSB.
- Pursue implementation of changes most likely to advance chemical safety.
- Focus special advocacy efforts by Board Members and staff.
- Inform deployment decisions and allocation of resources.

www.csb.gov

## Drivers of Critical Safety Change Considerations

- The nature of the risk and estimated extent of exposure to workers and offsite populations;
- Previous loss and potential for future loss of life or property, illnesses, and environmental damage;
- Very high risks disproportionately affecting discrete but highly identifiable groups of individuals
- Strong concern of important sectors of a community, civic leaders or the like; and
- The possibility that advocacy will help bring about change.



CSB U.S. Chemical Safety and Hazard Investigation Board

## Critical to Chemical Safety Change: Preventive Maintenance

***A lack of and poor preventive maintenance programs has been a recurring root cause in CSB investigations.***

**Common root causes:**

- Inadequate mechanical integrity programs
- Delayed or deferred preventive maintenance
- Ageing equipment – condition of equipment, not just age in years

www.csb.gov

## CSB Recommendations

- 11 investigations with 19 open recommendations.
- Recommendations to address gaps in facility, corporate, regulatory & industry programs.
- Several of the 11 investigations occurred at oil refineries.



CSB U.S. Chemical Safety and Hazard Investigation Board

## Tesoro Refinery Fatal Explosion and Fire Anacortes, Washington - 2010

www.csb.gov

## Tesoro Anacortes

- Explosion and fire led to the fatal injury of seven employees.
- Heat exchanger that failed during a maintenance operation was nearly 40 years old a the time of the incident.
- Immediate cause was long-term, undetected High Temperature Hydrogen Attack (HTHA) of steel equipment, which led to the pressure vessel rupture.

CSB U.S. Chemical Safety and Hazard Investigation Board

## Tesoro inspection strategy

- Tesoro inspection strategy relied on design operating conditions rather than verifying actual operating parameters.
- Inspection for HTHA is unreliable – microscopic cracks are difficult to identify.
- High-chromium steel is resistant to HTHA, not in use by Tesoro.

www.csb.gov

U.S. Chemical Safety and
Hazard Investigation Board

## Preventing HTHA

- Tesoro, like others in the industry, used published data from API, called Nelson Curves, to predict HTHA.
- CSB found these curves unreliable – historical data for HTHA that does not reflect actual operating conditions.
- API does not require users to verify operating conditions in establishing operation limits for equipment.

www.csb.gov

U.S. Chemical Safety and
Hazard Investigation Board

## Preventing HTHA

- CSB recommended that API prohibit use of carbon steel in processes above 400F and greater than 50 psia hydrogen partial pressure.
- API updated API 941: *Steels for Hydrogen Service at Elevated Temperatures and Pressures in Petroleum Refineries and Petrochemical Plants*

Steels for Hydrogen Service at Elevated Temperatures and Pressures in Petroleum Refineries and Petrochemical Plants

Downstream Segment

API RECOMMENDED PRACTICE 941
SEVENTH EDITION, AUGUST 2008

www.csb.gov

U.S. Chemical Safety and Hazard Investigation Board

# CSB assessment of API 941 revisions

- The updated version of API 941 provides incremental improvements, but it fails to address important elements of the CSB's recommendation.
- API's updated carbon steel Nelson Curves do not take into account all of the estimated process conditions where the catastrophic failure occurred due to HTHA at the Tesoro Anacortes refinery.
- Because of these deficiencies, the Board voted to make this recommendation "Closed – Unacceptable Action."

www.csb.gov

U.S. Chemical Safety and Hazard Investigation Board

# CSB Safety Alert: HTHA

- In the absence of industry guidance that incorporates findings from the Tesoro Anacortes failure, the CSB issued a safety alert to provide additional guidance for industry.

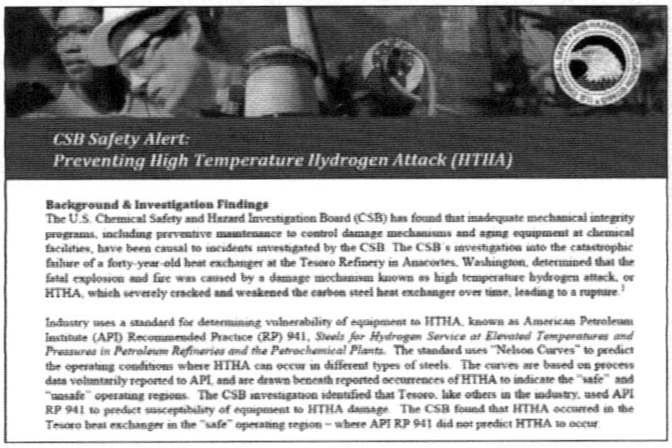

*CSB Safety Alert:*
*Preventing High Temperature Hydrogen Attack (HTHA)*

**Background & Investigation Findings**
The U.S. Chemical Safety and Hazard Investigation Board (CSB) has found that inadequate mechanical integrity programs, including preventive maintenance to control damage mechanisms and aging equipment at chemical facilities, have been causal to incidents investigated by the CSB. The CSB's investigation into the catastrophic failure of a forty-year-old heat exchanger at the Tesoro Refinery in Anacortes, Washington, determined that the fatal explosion and fire was caused by a damage mechanism known as high temperature hydrogen attack, or HTHA, which severely cracked and weakened the carbon steel heat exchanger over time, leading to a rupture.[1]

Industry uses a standard for determining vulnerability of equipment to HTHA, known as American Petroleum Institute (API) Recommended Practice (RP) 941, *Steels for Hydrogen Service at Elevated Temperatures and Pressures in Petroleum Refineries and the Petrochemical Plants.* The standard uses "Nelson Curves" to predict the operating conditions where HTHA can occur in different types of steels. The curves are based on process data voluntarily reported to API, and are drawn beneath reported occurrences of HTHA to indicate the "safe" and "unsafe" operating regions. The CSB investigation identified that Tesoro, like others in the industry, used API RP 941 to predict susceptibility of equipment to HTHA damage. The CSB found that HTHA occurred in the Tesoro heat exchanger in the "safe" operating region – where API RP 941 did not predict HTHA to occur.

www.csb.gov

CSB U.S. Chemical Safety and Hazard Investigation Board

## CSB Safety Alert: HTHA

- Identify all carbon steel equipment in hydrogen service that has the potential to harm workers or communities due to catastrophic failure;
- Verify actual operating conditions (hydrogen partial pressure and temperature) for the identified carbon steel equipment;
- Replace carbon steel process equipment that operates above 400 ° F and greater than 50 psia hydrogen partial pressure; and
- Use inherently safer materials, such as steels with higher chromium and molybdenum content.

www.csb.gov

U.S. Chemical Safety Board
1750 Pennsylvania Ave, NW, Suite 910
Washington, DC 20006
202-261-7600

Visit our website: www.csb.gov

YouTube USCSB

@chemsafetyboard

U.S. Chemical Safety Board

# The Engineers' Role in Process Safety

Khalid Al Hamid,
Director, Engineering
Sadara Chemical Company

December 5, 2016

Creating Value Through Chemistry

## Outline

- Why Process Safety?
- When to Engage in Process Safety?
- The Engineers' Role in Process Safety
- Sadara's Journey in Process Safety
- Summary

## Why Process Safety?

In absence of Process Safety, bad things happen:

- Loss of Lives
- Loss of Property
- Environmental Impact
- Business Interruption
- Legal Issues
- Reputation Impact (loss of future business)



## When to Engage in Process Safety?

It is a total lifecycle engagement

- Initial Design Development
- Construction and Commissioning
- Start Up, Normal Operations and Beyond

## The Engineers' Role in Process Safety

It is a total lifecycle engagement

- Initial Design Development
  - Technology Selection
  - Standards Application
  - Engineering Firm Competency
  - Credible Scenario Identification
  - Sound Design Basis
  - Owners Involvement
  - Process Safety Reviews



## The Engineers' Role in Process Safety

It is a total lifecycle engagement, cont.

- Initial Design Development
- Construction and Commissioning
  - Construction Verification
  - Appropriate Hydro-test and Preservation
  - Temporary Equipment Deployment
  - Documentation Handover: Equipment Registration, etc.
  - Interface Management
- Start Up, Normal Operations and Beyond

## The Engineers' Role in Process Safety

It is a total lifecycle engagement, cont.

- Initial Design Development
- Construction and Commissioning
- Start Up, Normal Operations and Beyond
  - Pre-startup Safety Review
  - Procedure Knowledge
  - Corrosion Management
  - MOC Control
  - Safety Audits
  - Fitness for Service



## Agenda

- Why Process Safety?
- When to Engage in Process Safety?
- The Engineers' Role in Process Safety
- Sadara's Journey in Process Safety
- Summary

# Sadara Overview

A strategic partnership developed by **Saudi Aramco** & **Dow Chemical Company** formed in **October 2011**

- US $**20** Billion investment;
- **Only** complex to crack Naphtha in GCC;
- **Largest** chemical complex ever constructed in a single phase: **26** world scale plants with over **3 million** metric tons of capacity;
- **Differentiated Product Slate: 14** out of the **26** plants are new to KSA, enabling new value chains & downstream opportunities.

December 5, 2016 9

# Sadara Value Chain

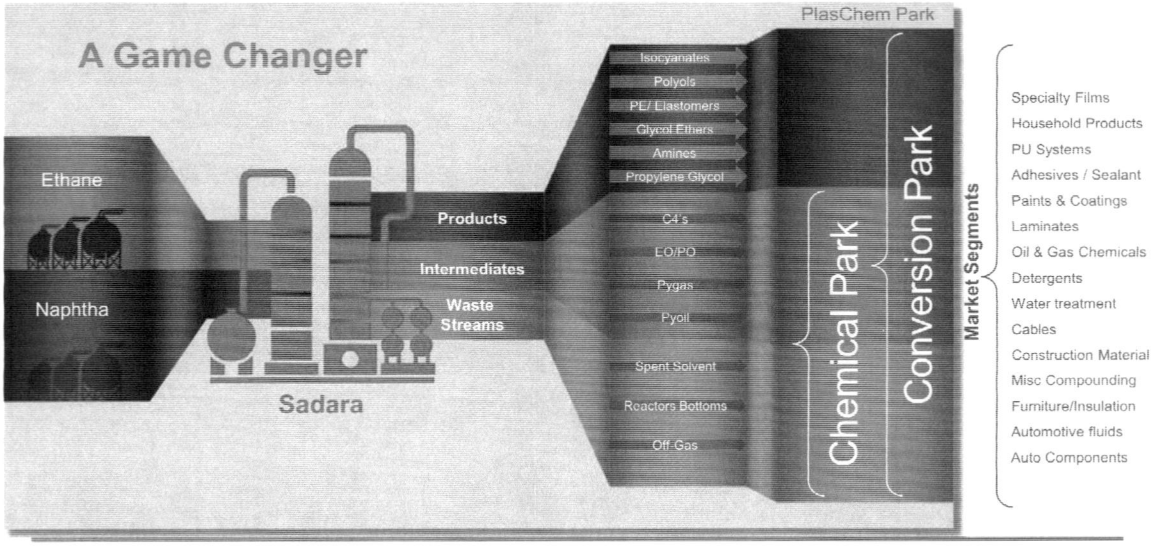


December 5, 2016 10

# Collaborating at Unprecedented Levels

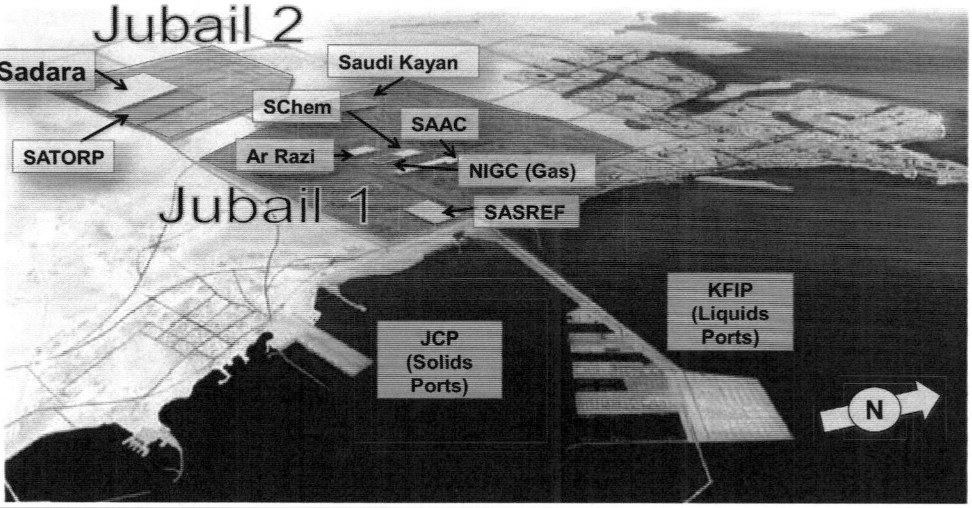


*Note: Map is illustrative and not to scale*

December 5, 2016

11

Sadara's Journey in Process Safety through
# Conceptual Design

- Size & Technology Selection
  - State of the art proven technologies
  - Inherently safe design, a prime consideration
- Integration
  - Extensive modeling to integrate to minimize storage of hazardous materials
- Selection of Site Location
  - Away from populated areas
- Developed well defined & realistic Execution Plan

December 5, 2016

## Sadara's Journey in Process Safety through
# Engineering

ﺻﺪﺍﺭﺓ Sadara

- Resources
  - Staffed with Dow & Saudi Aramco experts
- Standards and Specifications
  - Used over 6,000 Dow and Saudi Aramco Design Standards
  - Updated more than 2,000 Dow Engineering Standards and Specifications to reflect local Safety directives
- Layout
  - Applied Chemical Exposure Index (CEI) and Fire and Explosion Index (F&EI) for safe design and layout

December 5, 2016

## Sadara's Journey in Process Safety through
# Engineering (Cont'd)

ﺻﺪﺍﺭﺓ Sadara

- Process Safety in Design
  - Reactive Chemicals Program implemented to minimize risk – includes testing to help establish safe operating parameters
  - Chemical Reactivity Matrix developed to assess compatibility
  - Eliminated overpressure scenarios where practical (inherently safe design) & designed appropriate safety systems for rest of systems
  - Developed Electrical Area Classification for safe design of electrical & instrument systems
  - Full compliance with Dow's Loss Prevention Principles (LPPs) and Saudi Aramco Safety & Security Directives

December 5, 2016

## Sadara's Journey in Process Safety through
# Engineering (Cont'd)

- Process Safety in Design, cont'd
  - Essentially used fail safe design of instruments & controls
  - Conducted extensive safety design reviews with EPC contractors
    - ✓ PHA, HAZOP, FMEA & 3D-model reviews
  - Extensively reviewed critical EPC deliverables
    - ✓ PFD's, P&ID's, Equipment, Piping & Instrument design
  - Conducted Layers of Protection Analysis to close all gaps

December 5, 2016

## Sadara's Journey in Process Safety through
# Commissioning & Startup

- "Beyond Injury Free" vision at Enterprise level to achieve zero incidents
  - Identified & registered critical service equipment & piping for routine inspection
  - Developed offline simulator for all plants & trained all operators
  - Developed comprehensive operating & maintenance procedures
  - Trained 4,000+ new employees the "Sadara Way," to develop new culture with safety mindset for all activities
  - Trained 1,000+ Sadara new hires at existing Dow, Saudi Aramco & other JV facilities worldwide under experienced supervision
  - Developed & implemented comprehensive site-wide incident management & disaster response plan

December 5, 2016

Sadara's Journey in Process Safety through
## Commissioning & Startup (Cont'd)

- Identified and addressed many design improvements:
  - Leak isolation for key headers
  - Additional scenarios with a potential to introduce cryogenic gases in carbon steel headers
  - Column internal design & installation verification
  - Extensive end-to-end System Reviews identified to ensure adequacy of hydro test, thermal relief & other design parameters

December 5, 2016

Sadara's Journey in Process Safety through
## Use of Well-Established Work Processes

Enhance safety in Design & Operation through use of:

- Operating Discipline Management System (ODMS)
- Global Project Methodology (GPM)
- Management of Change (MOC)
- Pre-Startup Safety Review (PSSR)
- Many other well proven Dow & Saudi Aramco Work Processes

December 5, 2016

Sadara's Journey in Process Safety through
## Normal Operation

- Strictly follow **ALL** safe work practices
- Maintain critical documents evergreen
  - PFDs, P&IDs, Operating Procedures
- Report all environmental or process safety incidents to appropriate authorities
- Conduct Root Cause Investigations for process safety & environmental incidents
- Share & leverage learning from internal & external incidents
- Full engagement of all employees in Sadara's "Beyond Injury Free" vision for zero incidents

December 5, 2016

## Summary

- Process Safety is a total lifecycle engagement
- Engineers play a major role in this Process Safety Journey
- Sadara is committed to Safety

December 5, 2016

# Thank you

# Q & A

Sadara

# Practical Implementation Challenges for Process Safety Framework – Executive Decision Making

3rd CCPS Global Summit on Process Safety
Dammam, Saudi Arabia

Mahesh Murthy
Process Safety Management
+ 372 5432 9002
mahesh.murthy@outlook.com

## In this presentation ....

- Project background
- Organization challenges
- Implementation strategy
- Lessons learnt

*If you think safety is expensive……try having an accident – Trevor Kletz – Father of modern Process Safety*

## Project background

- Mid-tier oil company
- Daily production of 80,000 barrels of crude
- Operating under Production Sharing Agreement and joint partnership
- Designed to local safety standards
- Recent spate of 4 process safety incidents
- Corporate audit – damning report
- Readjusting priorities – business sustainability and continuity challenges
- Declining production, increasing maintenance costs, potential souring reservoir and low oil prices

# Implementation challenges

Risk visibility

Organization structure

# Risk visibility

- HAZOP, LOPA and SIL baseline activities
  - ➢ Safety Critical Element (SCE) register
- Risk Register development
- Process Safety Management Framework ( CCPS – RBPS)
- Business Case and ALARP demonstrations

## Risk visibility – outcomes

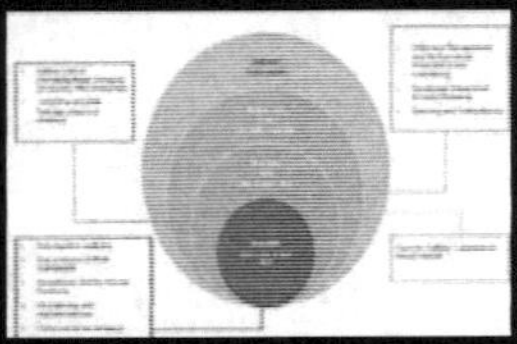

- Prioritized risks (Safety, Asset, Reputation and Environment)
  - ➢ 12% of total registered asset tags designated as SCE
  - ➢ Deferral management Program initiated
- Risk register – risk review committee
  - ➢ Enterprise risk management and risk register integration

- Process Safety 5-year implementation plan
- Interface with capital expense projections

## Organization structure

- Reporting structure
- Leadership annual performance targets
- Integrated Activity Plan – accountability model
- Budget and Resources planning

# Organization structure - outcomes

- Re-organization with direct report to Managing Director
- Executive Leadership had the following Business Performance Contract;
  - ➢ Sponsoring one of the CCPS- RBPS elements
  - ➢ Chairing /Participation in at-least 2 risk committee meetings in a year
  - ➢ 2 site visits with full Personal Protection Equipment (PPE) – frontline engagement

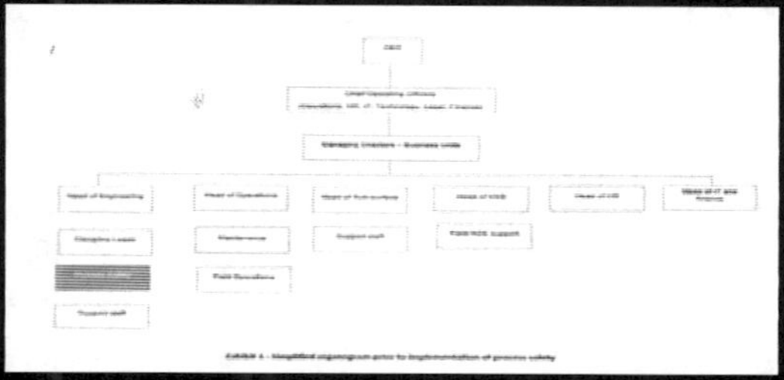

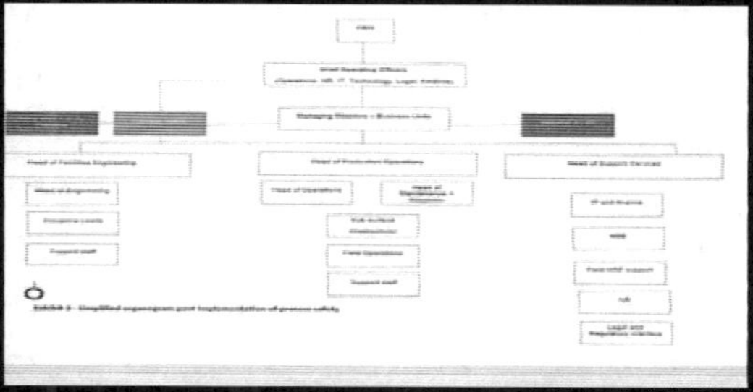

# Organization structure - outcomes

- Integrated Activity Plan – accountability model
  - ➢ HAZOP, LOPA and SIL studies on revalidation 5 year plan
  - ➢ Close out actions tracking through IAP – Primevera and Synergi – HSE software
- Budget and Resources planning
  - ➢ Cost codes for all safety, optimization and integrity study recommendations
  - ➢ Focal points and accountability established

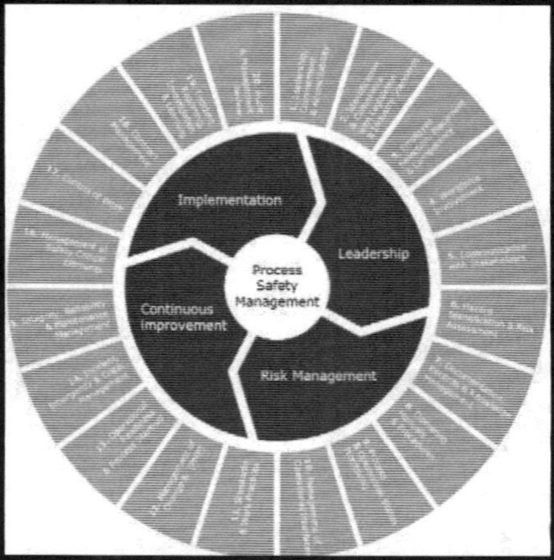

## Lessons learnt

- HAZOP followed by LOPA gave improved risk prioritization
- CCPS – RBPS framework helped to create "Accountability model"
- Integrated Activity Planning using Primavera helped prioritize, identify resource gaps and strategize capital improvement implementation projects
- Leadership performance targets helped set right expectations and mindset

*Engineers do not fail in engineering......*
*but fail in imagination*

*Rex Tiller ...CEO, Exxon Mobil ,*
*Congressional Committee hearings*

## Some companies implement Process Safety with better results than others. What was the difference between these two companies?

DUPONT

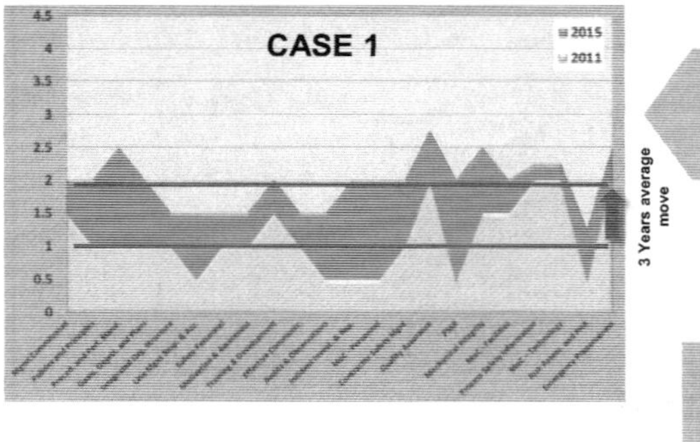


Average rate of increase in PSM Performance ~ 70% in 3 years

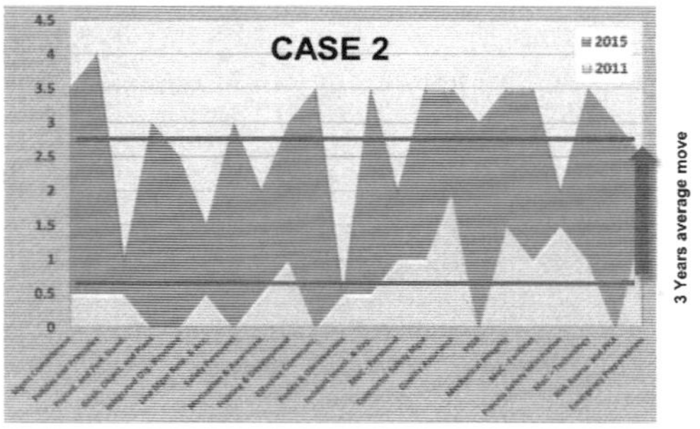


Average rate of increase in PSM Performance ~ 300% in 3 years

Source: DuPont Benchmark Analysis

DuPont Sustainable Solutions

 2

## There are 4 necessary ingredients for implementing a process safety culture… the four main pillars

*A PSM and Behavioral Safety management system that identifies, governs and protects against risks*

*Leadership that sets the PSM and Behavioral Safety improvement vision and drive the efforts forward*

*An environment that encourages employees' engagement in the improvement, implementation, and sustain processes*

*A performance management process that monitors performance proactively and identifies actions to ensure continuous improvement and benefits realization*

A SUCCESSFUL PROCESS SAFETY CULTURE ESTABLISHED

COMPREHENSIVE SYSTEM + LEADERSHIP COMMITMENT + EMPLOYEES ENGAGEMENT + PERFORMANCE MANAGEMENT =

DuPont Sustainable Solutions

 3

## DuPont's 22 Safety Element to Safety Excellence – The system component of Safety and Process Safety Culture

LEADERSHIP
- Management Commitment
- Policies and Principles
- Goals, Objectives, and Plans
- Procedures and Performance Standards

ORGANIZATION
- Line Management Accountability and Responsibility
- Safety Personnel
- Integrated Org. Structure
- Motivation and Awareness

OPERATION
- Observations and Audits
- Incident Investigation
- Training and Development
- Effective Communication

PERSONNEL
- Emergency Preparedness and Contingency planning
- Management of Change – Personnel
- Contractor Safety Management

FACILITIES
- Quality Assurance
- Pre-startup Safety Reviews
- Mechanical Integrity
- Management of Change – Facilities

TECHNOLOGY
- Process Safety Information
- Management of Change – Technology
- Risk Assessment and Process Hazard Analysis

DuPont Sustainable Solutions

 4

## Operational discipline – the soft side of 'process safety culture' – is essential for operational effectiveness

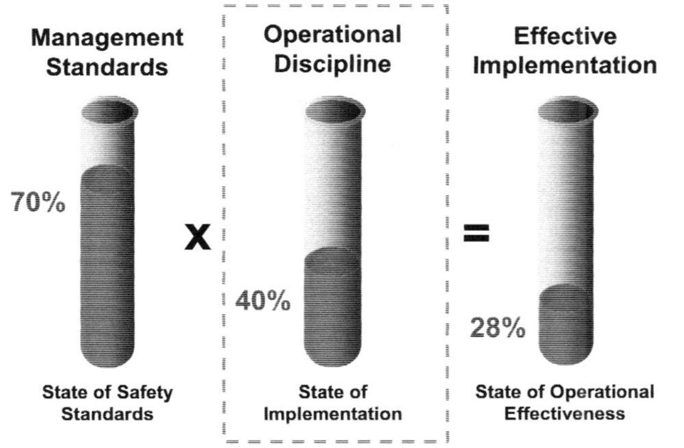


Operational effectiveness can be summarized as *"everyone...do it right...every time"*

**Characteristics of Operational Discipline**

- Leadership by example
- Practice consistent with procedures
- Sufficient resources
- Employee involvement
- Active lines of communication
- Strong teamwork
- Common shared values
- Up-to-date documentation
- Absence of shortcuts
- Excellent housekeeping
- Pride in the organization



 

## Senior leaders' visible and effective commitment has the highest level of influence on organizational behavior – culture change starts at the top

*Not Exhaustive*

**If a leader does this…** → **What will be the cultural impact?**

- Devotes 95% of their time to operations activities
- Relies only on KPIs, performance and anecdotes without going into the field to get the full story
- Always blames their people for incidents
- Never holds or attends safety meetings
- Delays critical work activities or turnarounds without analyzing the risks
- Declines or delays all safety related capex projects
- Struggles to hold their line managers accountable for safety performance
- Never communicates safety messages
- Does not invest in training and capabilities of their people

New Culture???

*"Your actions are so loud that I cannot hear your words…"*



 

## Leadership Commitment – Establish a process safety culture by building capability, influencing, role-modeling, and aligning systems

"I will change my mindsets and behaviors, if ..."

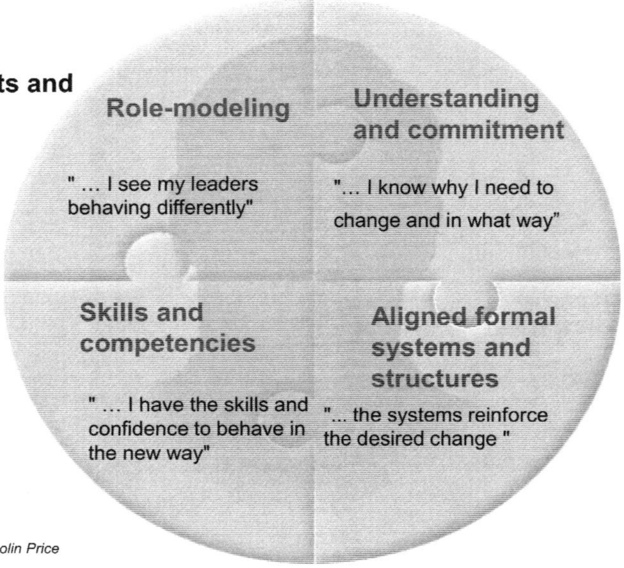


*Source: Beyond Performance - Scott Keller and Colin Price (The Influence Model )*

DuPont Sustainable Solutions

 7

## Employee Engagement – Employees at all levels need to help establish the safety culture

**Employee Engagement through the Integrated Safety Organization**

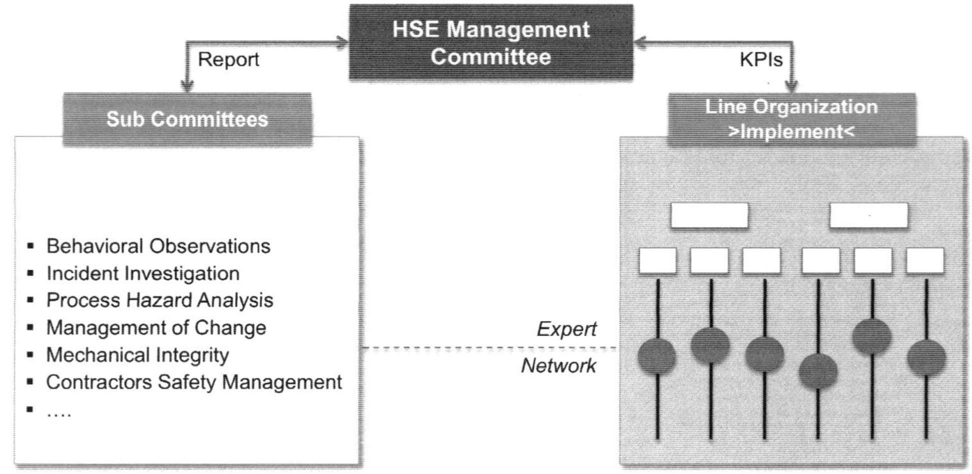


- Behavioral Observations
- Incident Investigation
- Process Hazard Analysis
- Management of Change
- Mechanical Integrity
- Contractors Safety Management
- ....

DuPont Sustainable Solutions

 8

## What does Employee Engagement really look like in the context of safety and process safety?

*Not Exhaustive*

- Employees are empowered to stop work or shut down equipment if it is unsafe
- Employees have mechanisms for making suggestions to enhance process safety and the ability to correct issues in their areas
- There is regular two-way communication across white collar and blue collar ranks with cascading safety messages
- Per Operating Area, everyone understands the top risks that must be controlled, the safeguards in place to control them, and how to respect these controls during work activities
- Employees have a set of shared safety values and beliefs that can be felt across the organization
- Process Safety Management Systems are owned by safety and line organization members who regularly drive system enhancements and thought leadership
- Blue Collar employees regularly participate in process safety activities such as procedure revisions, PHA/HAZOP's, MOC and PSSR field activities, etc.
- Employees receive training and education on process safety topics that are relevant for their roles

*As leaders, we must empower all levels of our organizations to drive strong process safety performance.*



 

## Performance Management – Align Strategy to Objectives and Area/Individual KPI's

# Use KPIs – Monitor Leading and Lagging indicators of safety culture

*Illustrative*

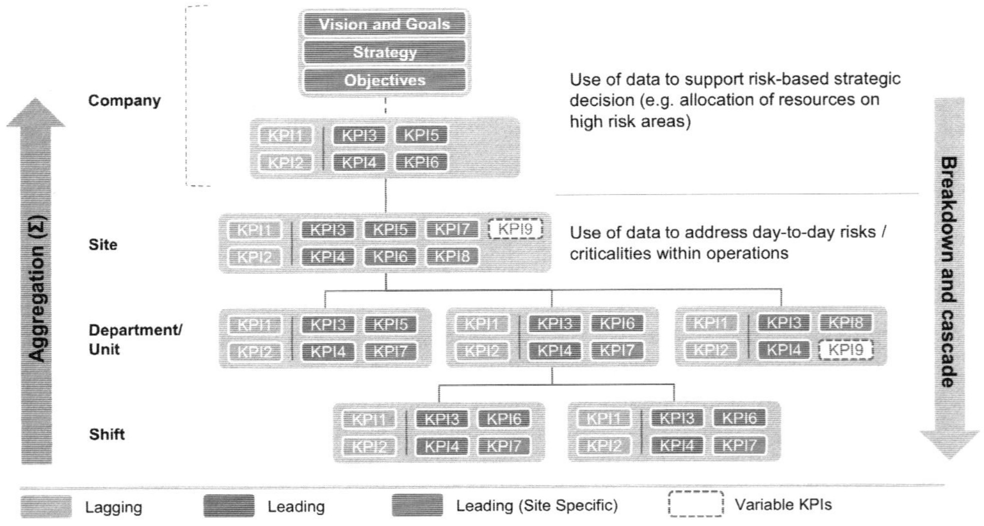




 

# Monitor Performance frequently and Manage deviations pro-actively to establish and maintain the right safety culture

*ILLUSTRATIVE*

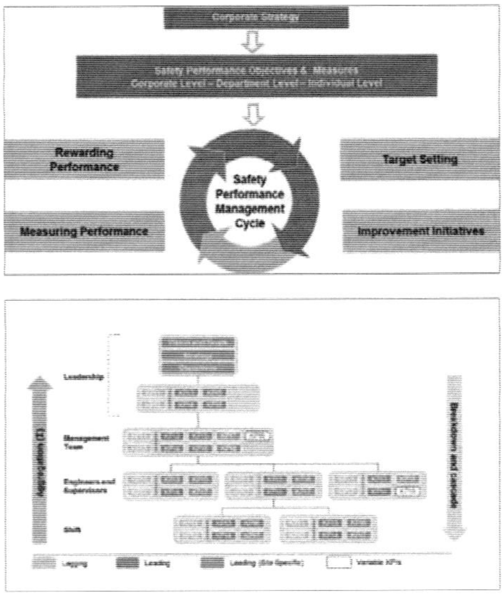


Meet or Exceed expectations → **Reward and Recognition** (e.g. Bonus, merit increase, career advancement)

Does not meet expectations → **Training & Development Needs** (e.g. Training program, coaching program, target review, counseling)



 

## The necessary ingredients for establishing and maintaining a process safety culture… the four main pillars

*A PSM and Behavioral Safety management system that identifies, governs and protects against risks*

*Leadership that sets the PSM and Behavioral Safety improvement vision and drive the efforts forward*

*An environment that encourages employees' engagement in the improvement, implementation, and sustain processes*

*A performance management process that monitors performance proactively and identifies actions to ensure continuous improvement and benefits realization*

A SUCCESSFUL PROCESS SAFETY CULTURE ESTABLISHED

COMPREHENSIVE SYSTEM + LEADERSHIP COMMITMENT + EMPLOYEES ENGAGEMENT + PERFORMANCE MANAGEMENT =



## In addition to the 4 Pillars, there are basic change methodologies that must be embedded into the implementation approach.

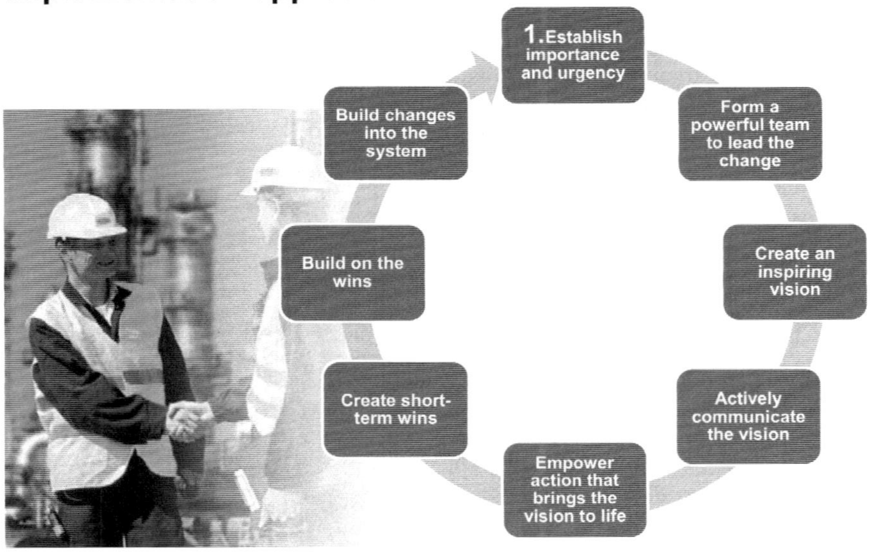


*(Source: Dr. John Kotter's 8-step change management model)*

## Illustrative Examples
## Comparable Sites PSM Journey over 3 years

AVERAGE RATE OF INCREASE IN PSM PERFORMANCE
**Case 1: ~ 70% in 3 years**

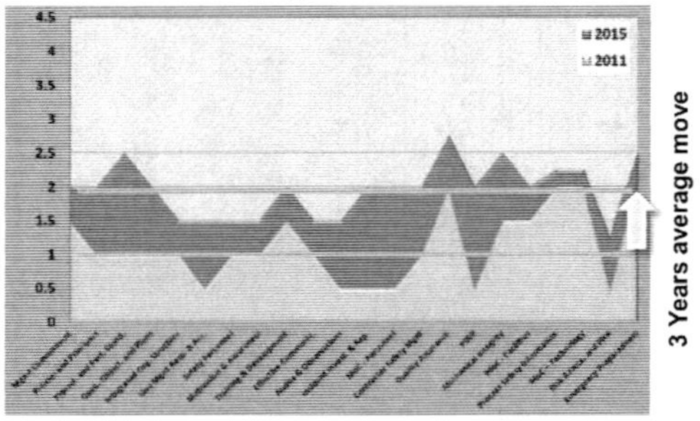


AVERAGE RATE OF INCREASE IN PSM PERFORMANCE
**Case 2: ~ 300% in 3 years**

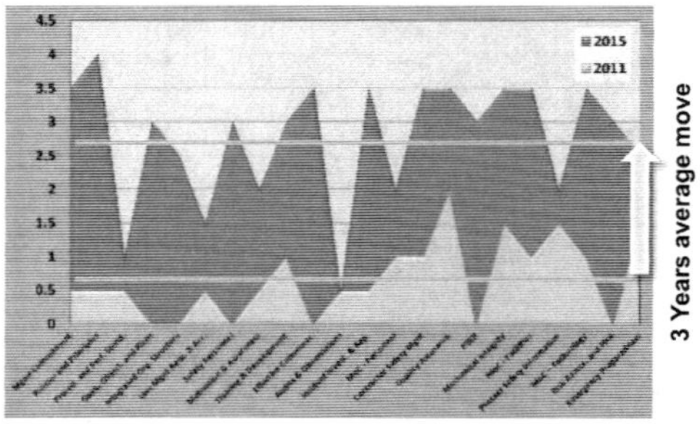


**KEY SUCCESS FACTORS FOUND DURING THE IMPLEMENTATION**

- Accountability driven by top management through the line
- Strong Field Presence of Top Management
- Strong emphasis on follow up and action management with team leaders
- Direct feedback to blockers/resistors
- Ownership and Execution of Key Activities by all organizational levels
- Positive Recognition for early adopters and key successes
- Advancement of good performers
- Focused performance appraisals and reward schemes
- Strong Communication with total organization on awareness of topics and status of implementation

DuPont Sustainable Solutions

 15

DuPont Sustainable Solutions

DUPONT
From Operational Risk to
Operational Excellence:
Identifying Common Ground
June 16th 2016
Srinivasan
Ramabhadran

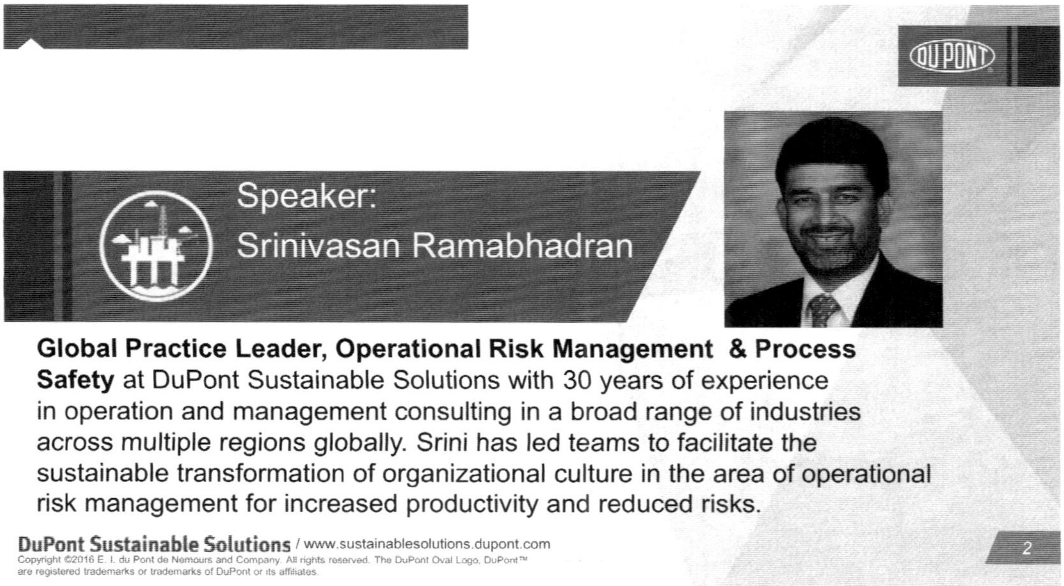
DUPONT
Speaker:
Srinivasan Ramabhadran
Global Practice Leader, Operational Risk Management & Process Safety at DuPont Sustainable Solutions with 30 years of experience in operation and management consulting in a broad range of industries across multiple regions globally. Srini has led teams to facilitate the sustainable transformation of organizational culture in the area of operational risk management for increased productivity and reduced risks.
DuPont Sustainable Solutions / www.sustainablesolutions.dupont.com
Copyright ©2016 E. I. du Pont de Nemours and Company. All rights reserved. The DuPont Oval Logo, DuPont™ are registered trademarks or trademarks of DuPont or its affiliates.
2

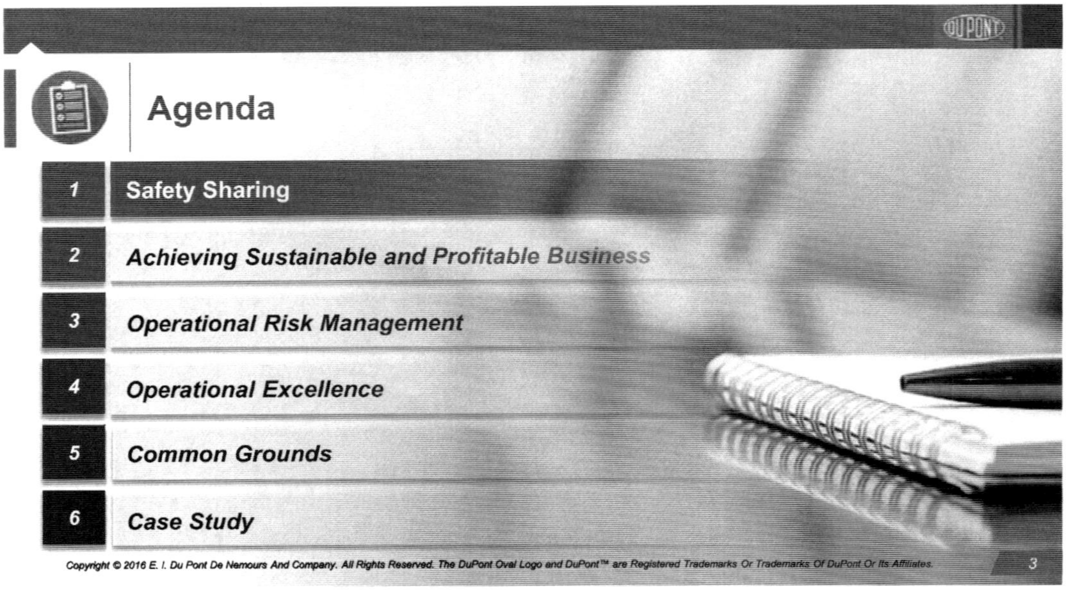
Agenda
1 Safety Sharing
2 Achieving Sustainable and Profitable Business
3 Operational Risk Management
4 Operational Excellence
5 Common Grounds
6 Case Study
Copyright © 2016 E. I. Du Pont De Nemours And Company. All Rights Reserved. The DuPont Oval Logo and DuPont™ are Registered Trademarks Or Trademarks Of DuPont Or Its Affiliates.
3

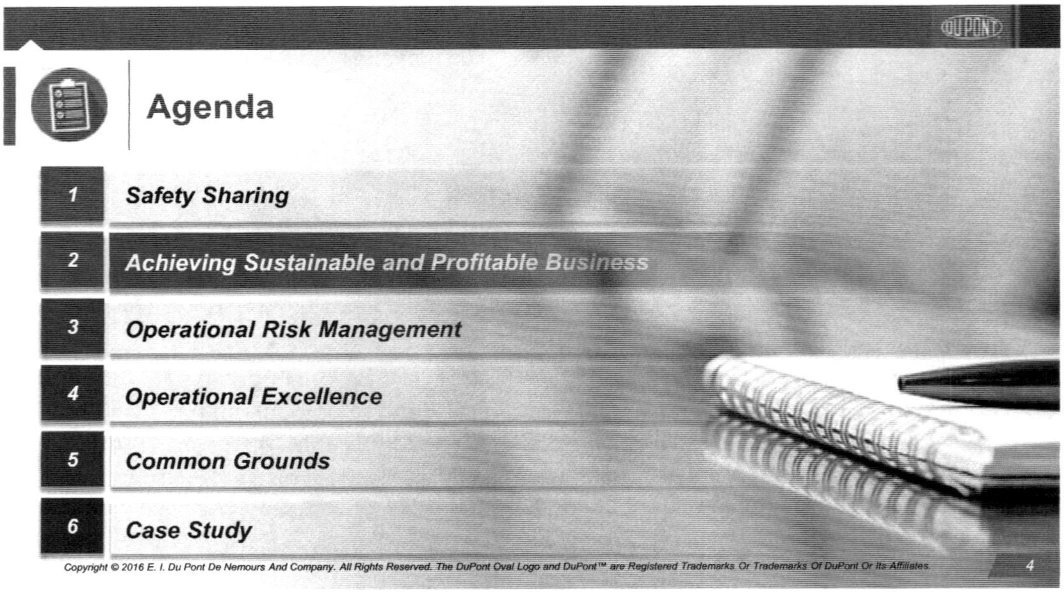
Agenda
1 Safety Sharing
2 Achieving Sustainable and Profitable Business
3 Operational Risk Management
4 Operational Excellence
5 Common Grounds
6 Case Study
Copyright © 2016 E. I. Du Pont De Nemours And Company. All Rights Reserved. The DuPont Oval Logo and DuPont™ are Registered Trademarks Or Trademarks Of DuPont Or Its Affiliates.
4

# Journey to Sustainable and Profitable Business

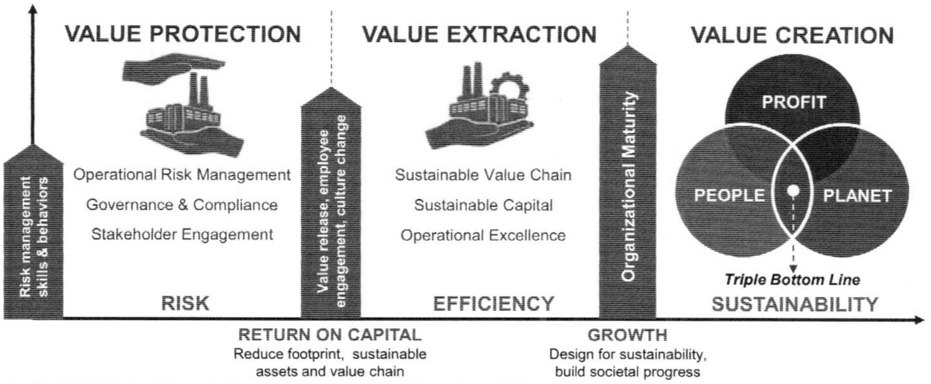






# Agenda

1. *Safety Sharing*
2. *Achieving Sustainable and Profitable Business*
3. *Operational Risk Management*
4. *Operational Excellence*
5. *Common Grounds*
6. *Case Study*

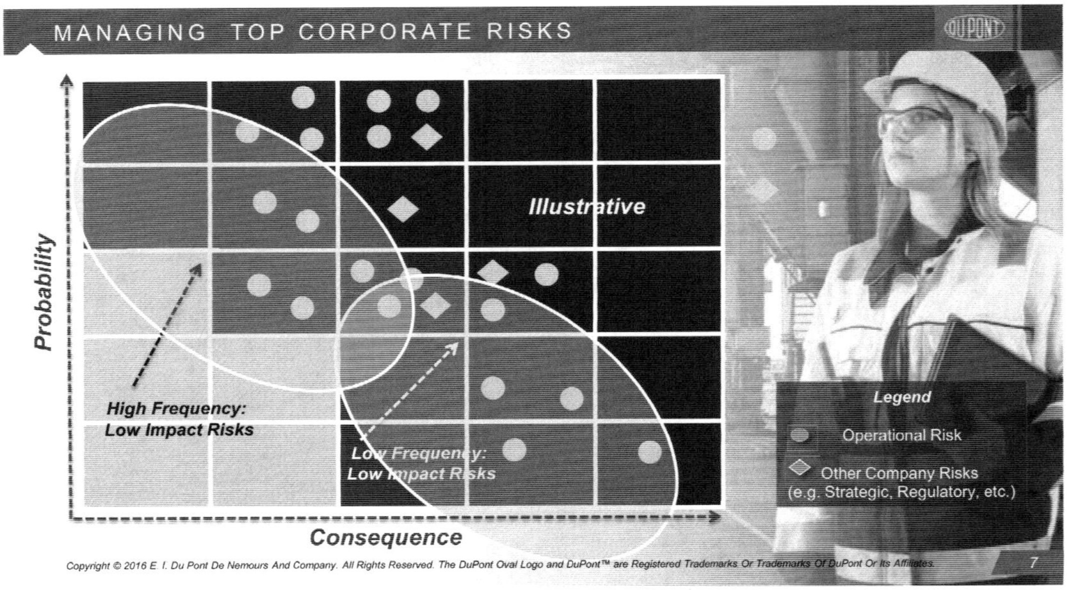


## OPERATIONAL RISK MANAGEMENT

**Identifying, evaluating and controlling losses by differentially managing associated risks.**

HOW TO ASSESS AND MITIGATE RISKS
HIGH
PROBABIILITY
Treat?
Terminate Unsustainable
Competitive Advantage
Transfer?
LOW
HIGH
CONSEQUENCE
MEASURE OF RISK
Copyright © 2016 E. I. Du Pont De Nemours And Company. All Rights Reserved. The DuPont Oval Logo and DuPont™ are Registered Trademarks Or Trademarks Of DuPont Or Its Affiliates.
9

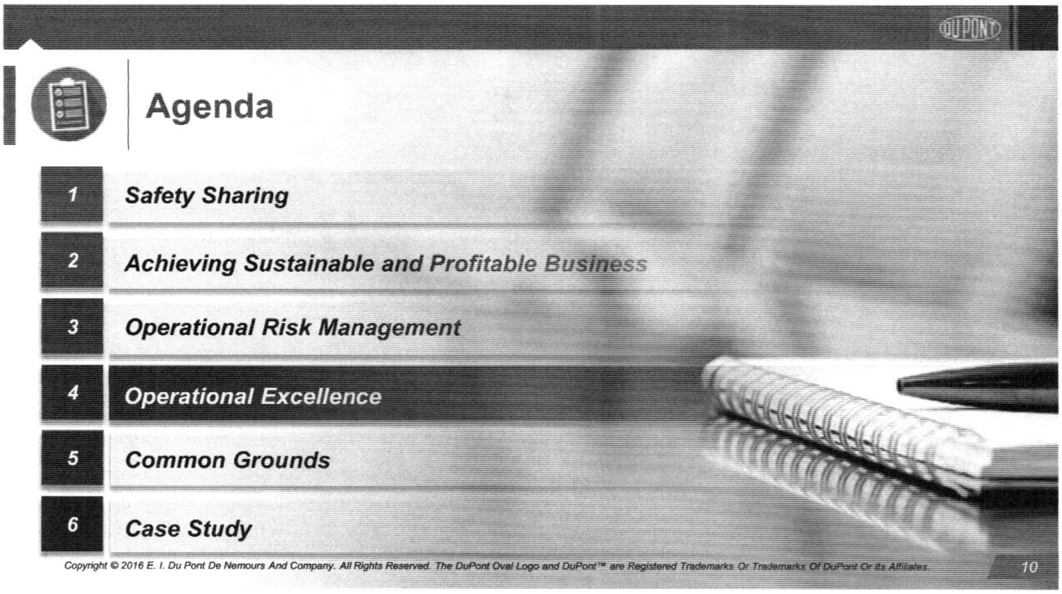
Agenda
1 Safety Sharing
2 Achieving Sustainable and Profitable Business
3 Operational Risk Management
4 Operational Excellence
5 Common Grounds
6 Case Study
Copyright © 2016 E. I. Du Pont De Nemours And Company. All Rights Reserved. The DuPont Oval Logo and DuPont™ are Registered Trademarks Or Trademarks Of DuPont Or Its Affiliates.
10

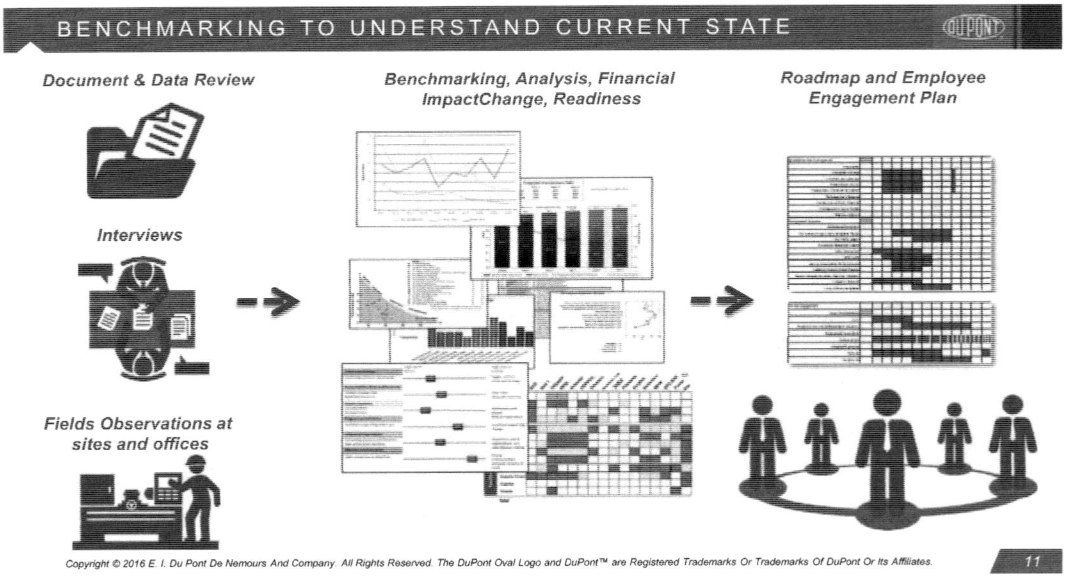


## SYNTHESISING TO FACILITATE PRIORITISATION

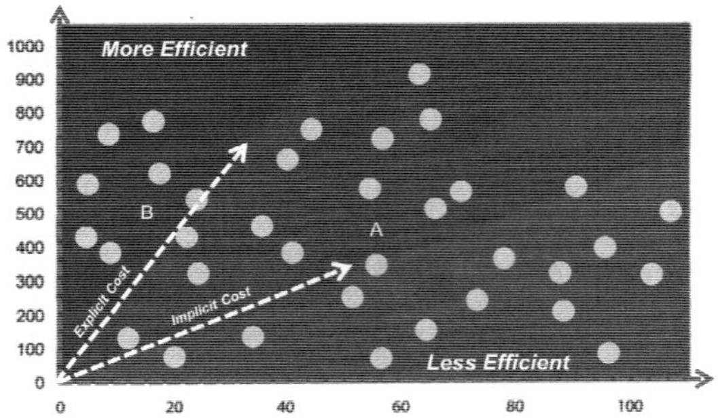


*More Efficient*
*Implicit Cost*
*Less Efficient*

- Integrates with the Business Planning process
- Explicitly defines costs and value
- Sets priorities for project categories and specific projects.
- Requires Leadership Credibility & Interdependence



12

BEST PRACTICES FOR IMPROVEMENT
DUPONT
Value
Increase revenue
Reduce operational costs
Reduce CAPEX
Improve HSE performance
Yield Improvement
Optimize mix
Increase Availability
Cheaper feedstock
Energy efficiency
Materials & Services use
Labor efficiency
Rigid Investment planning
Working capital Reduction
Process Safety & Asset Integrity
People safety
OPERATIONS EXCELLENCE
LEAN & SIX SIGMA
ENERGY MANAGEMENT
MAINTENANCE & RELIABILITY
TURNAROUND MANAGEMENT
CAPITAL PROJECT EFFECTIVENESS
OPERATIONAL RISK MANAGEMENT
Copyright © 2016 E. I. Du Pont De Nemours And Company. All Rights Reserved. The DuPont Oval Logo and DuPont™ are Registered Trademarks Or Trademarks Of DuPont Or Its Affiliates.
13

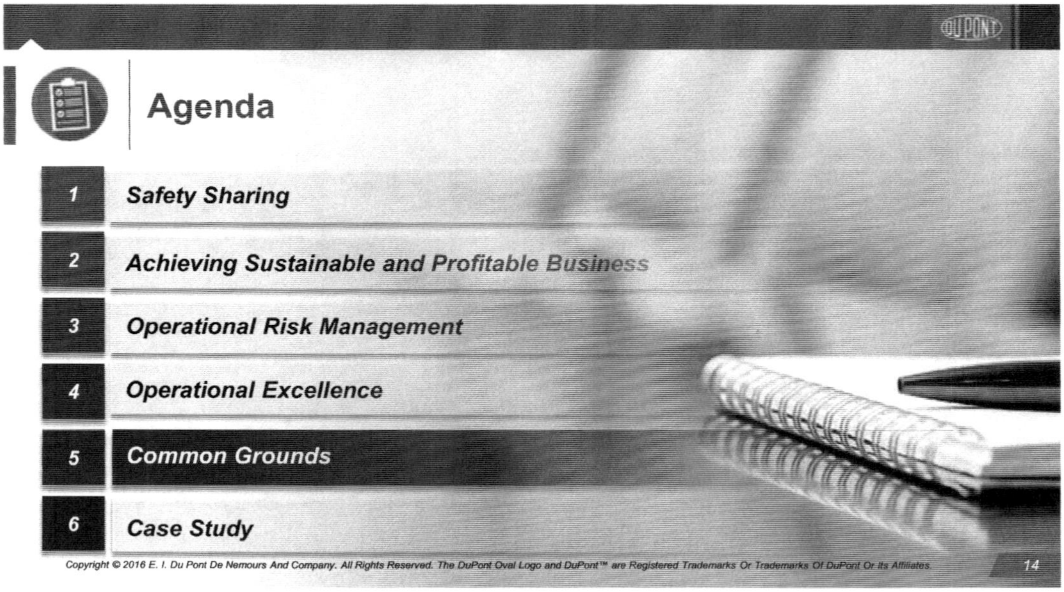
DUPONT
Agenda
1 Safety Sharing
2 Achieving Sustainable and Profitable Business
3 Operational Risk Management
4 Operational Excellence
5 Common Grounds
6 Case Study
Copyright © 2016 E. I. Du Pont De Nemours And Company. All Rights Reserved. The DuPont Oval Logo and DuPont™ are Registered Trademarks Or Trademarks Of DuPont Or Its Affiliates.
14

## Common Ground

1. Corporate Push to Site Pull
2. Integrated Approach that focuses on Outcomes
3. Operational Discipline
4. Need for Transformation

## PUSH TO PULL APPROACH

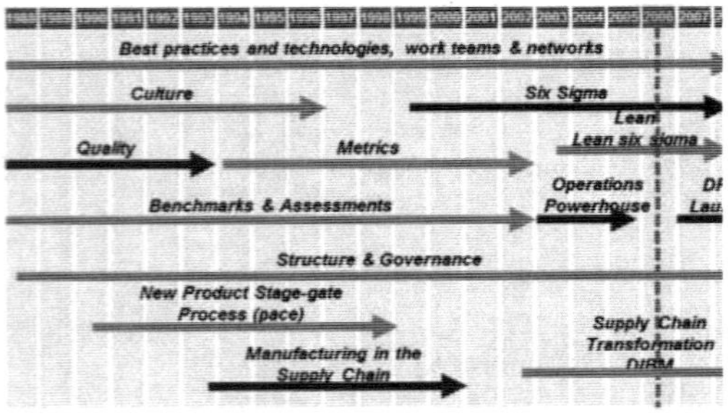


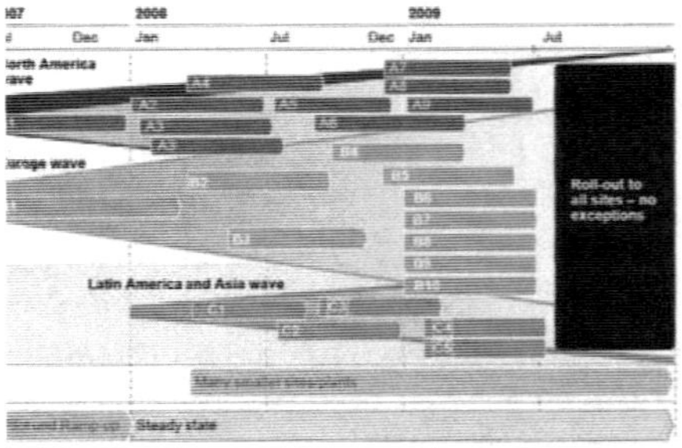

## INTEGRATED APPROACH

## FOCUS ON OUTCOMES

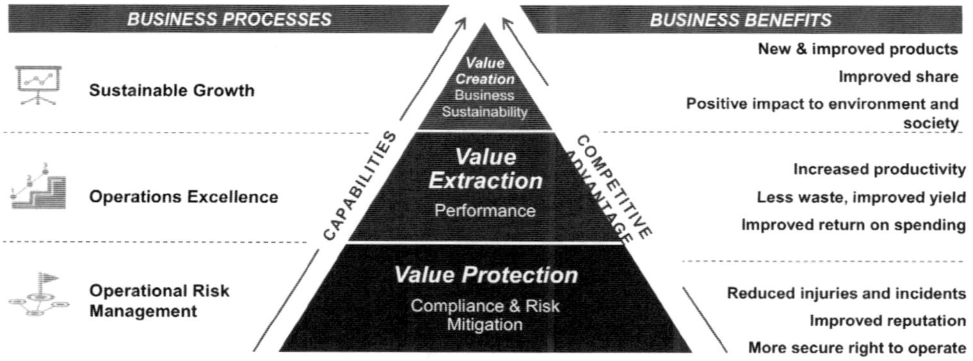

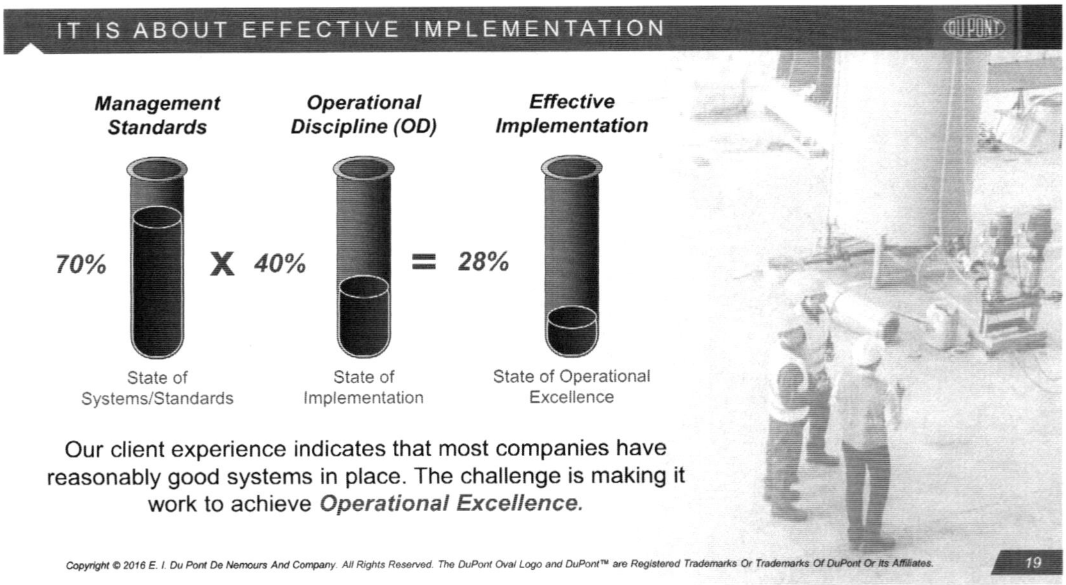
IT IS ABOUT EFFECTIVE IMPLEMENTATION
Management Standards
Operational Discipline (OD)
Effective Implementation
70%
X
40%
=
28%
State of Systems/Standards
State of Implementation
State of Operational Excellence
Our client experience indicates that most companies have reasonably good systems in place. The challenge is making it work to achieve Operational Excellence.
Copyright © 2016 E. I. Du Pont De Nemours And Company. All Rights Reserved. The DuPont Oval Logo and DuPont™ are Registered Trademarks Or Trademarks Of DuPont Or Its Affiliates.
19

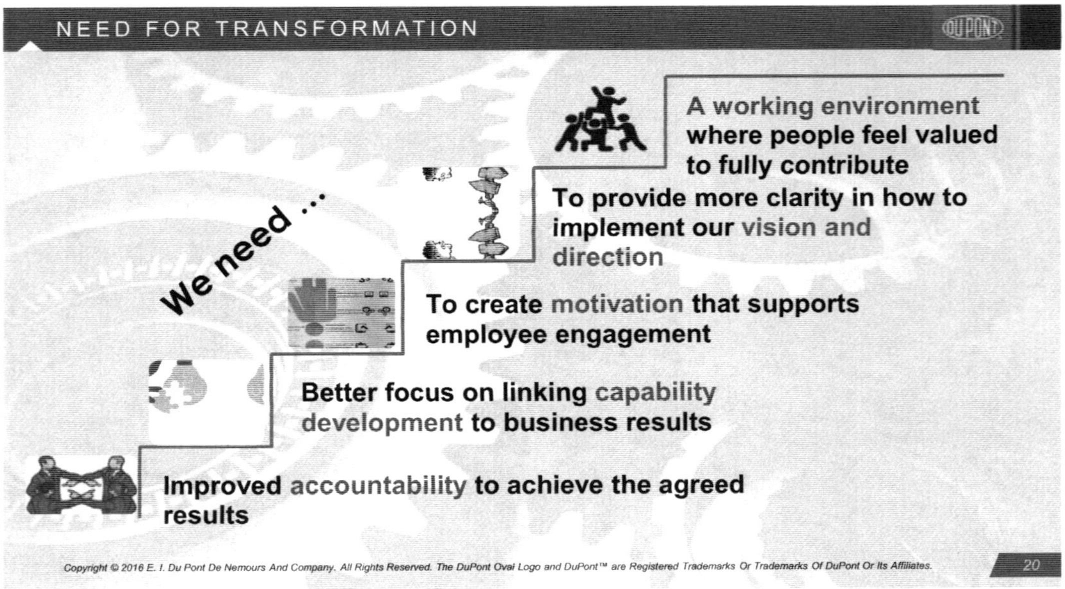
NEED FOR TRANSFORMATION
We need …
A working environment where people feel valued to fully contribute
To provide more clarity in how to implement our vision and direction
To create motivation that supports employee engagement
Better focus on linking capability development to business results
Improved accountability to achieve the agreed results
Copyright © 2016 E. I. Du Pont De Nemours And Company. All Rights Reserved. The DuPont Oval Logo and DuPont™ are Registered Trademarks Or Trademarks Of DuPont Or Its Affiliates.
20

# Agenda

1. *Safety Sharing*
2. *Achieving Sustainable and Profitable Business*
3. *Operational Risk Management*
4. *Operational Excellence*
5. *Common Grounds*
6. *Case Study*





## PRIORITIZING FOCUS AREAS FOR MAXIMUM GAIN

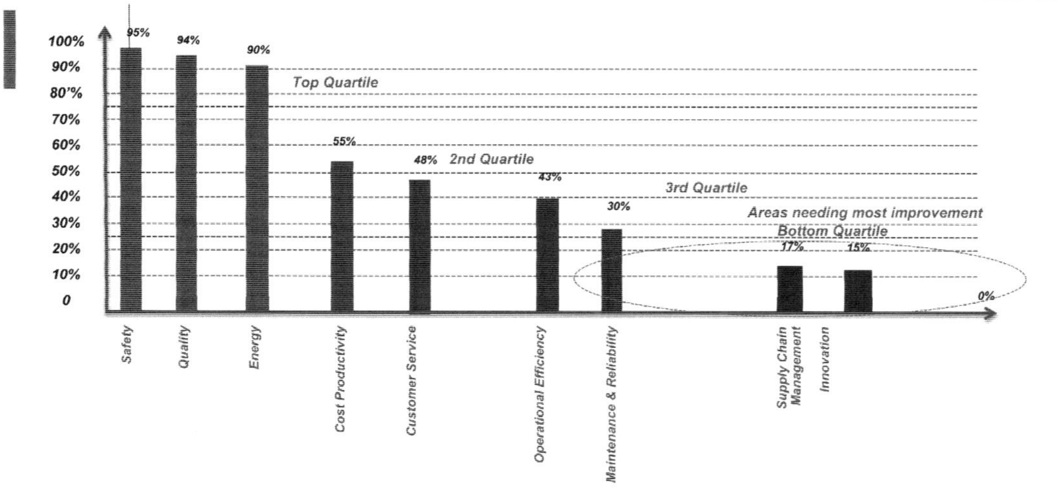

KEY LEARNINGS

- Making choices
- Integrated Business Approach
- Leadership and Governance Mechanisms
- ….and more importantly **PEOPLE**



23

- Join our LinkedIn Group to continue the discussion:
- DuPont Sustainable Solutions LinkedIn Group
- Contact Us
- Ramabhadran.Srinivasan@dupont.com
- EHSAN.AKHAVAN@dupont.com

Copyright © 2016 E. I. Du Pont De Nemours And Company. All Rights Reserved. The DuPont Oval Logo and DuPont™ are Registered Trademarks Or Trademarks Of DuPont Or Its Affiliates.

24

DuPont Sustainable Solutions
Copyright ©2016 E. I. du Pont de Nemours and Company. All rights reserved. The DuPont Oval Logo, DuPont™ are registered trademarks or trademarks of DuPont or its affiliates.
Questions?

# SUCCESSFUL IMPLEMENTATION OF PHA'S @HADEED

DEBOTTLENECK PROJECT - DR MODULE D

CHEMISTRY THAT MATTERS™

## OUTLINE

- HADEED - AN INTRODUCTION
- PHA IMPLEMENTATION @ HADEED
- DR MODULE D - INTRODUCTION.
- SCOPE OF DBN PROJECT & MAJOR CHANGES.
- PHA STUDY IN DEBOTTLENECK PROJECT.

No. 1

# HADEED OVERVIEW

In 1983 , the SAUDI IRON & STEEL COMPANY (HADEED) started its journey of success. Today, it produces up to 6 MILLION TONS of superior quality long and flat products. It houses approximately 4518 employees, out of which 3950 DIRECT HIRE and 568 CONTRACTORS.

HADEED has three divisions:

- STEEL PLANT AND DIRECT REDUCTION (SPDR)
- ROLLING MILLS (RM)
- MANUFACTURING SUPPORT (MS)

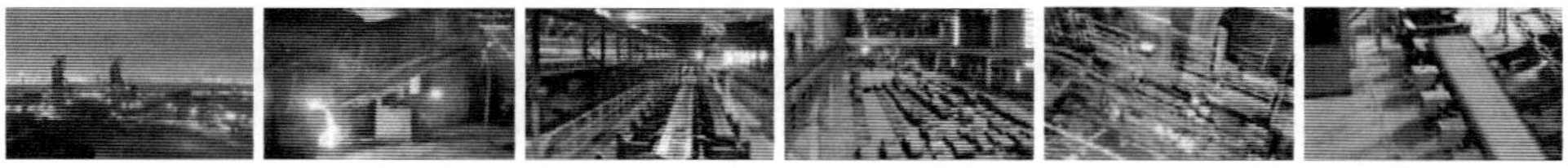

No. 2

# HADEED PROCESS PLANTS/UNITS

| DIRECT REDUCTION PLANT | STEEL PLANT MELT SHOPS | ROLLING MILLS | UTILITIES |
|---|---|---|---|
| 5 | 6 | 15 | 26 |

No. 3

## PROCESS HAZARD ANALYSIS IMPLEMENTATION IN HADEED

| | |
|---|---|
| INITIATION | • Initial PHA Study for all CAPEX Projects<br>• Lead by Third Party "Certified Vendor" with contribution of HADEED, SABIC Project & Contractor.<br>• Conducting on 3 Phases:<br>➢ Phase 1 "Pre-feasibility/feasibility phase reviews"<br>➢ Phase 2 "Basic design phase reviews"<br>➢ Phase 3 "Detail design phase reviews" |
| REVALIDATION | • Annually review of HADEED PHA Plan.<br>• Prioritization Study conducted to determine the interval of Revalidation for each System/Unit. "Maximum interval of five years<br>• After 2 cycle of Revalidation, 3rd one will treat as Initial. |
| SOFTWARE SOLUTION FOR STUDY/ACTION TRACKING | • Using Hazard Review software for studies.<br>• IT solution System use for maintain PHA Report. Workflow based solution for Initiation, level of Management Approval & assignment as well as actions tracking & records. |
| PHA LEADERS/ TEAM | • PHA Leaders requirements:<br>➢ Training on General Risk Assessment & PHA Methodologies for a nominate employees (Engineers, Plant Operations, Maintenance & Safety).<br>➢ Operation Years of Experience required. (engineer with at least 3 years, plant operations staff with at least 8 years experience)<br>➢ Participated as member in at least (2) studies. |

No. 4

## INTRODUCTION OF DR-D PLANT (HYL PROCESS).

**PROCESS IN BRIEF:**

The HYL process uses Steam reformer to reform natural gas to produce the reducing gases the H2 and Co.

After the reformer, the reducing gas Heated to 950 – 1000 deg.C and introduced at the bottom of the reduction Zone to reduce iron ore while descending by gravity and produce DRI , prior the DRI discharge carburization is take placed by means of Natural Gas at the cooling zone.

No. 6

## SCOPE OF DE-BOTTLENECKING (DBN) PROJECT AND MAJOR CHANGE

The Module D of HYL in HADEED was started up in 1997, but actually the plant is showing some problems and limitations; showing that the production level is below 140 ton/H DRI.

PURPOSE :
1. Increase plant availability above 95%.
2. Increase plant productivity to 155 ton of DRI/hr.

**TO ACHIEVE THE TARGETED PRODUCTION, THE MAJOR MODIFICATIONS CAN BE DIVIDED INTO TWO PARTS :**

- Modification on Reactor Charge System from existing one bin charging system to 4 (Four) bins charging system.
- Incorporation of new steam package boiler ,new Oxygen injection System, new High Pressure Nitrogen Skid, new Cooling Gas Ejector System.

No. 7

# PHA HISTORY OF DIRECT REDUCTION MODEL D

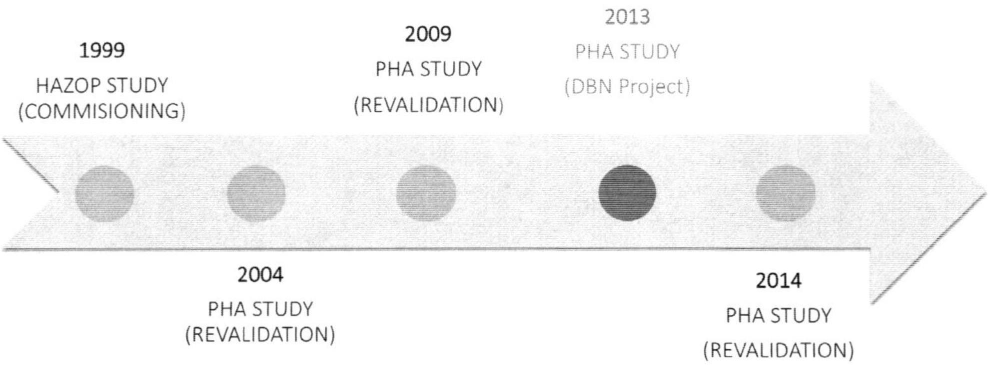




# PROCESS HAZARD ANALYSIS METHODOLOGY

**HADEED's safety objectives for the DR-D DBN Project included:**

1. Identifying process hazards.
2. Reducing or eliminating the risk that the hazards pose to employees, the public, and/or the environment.

**ANALYSIS SCOPE:**

The PHA covered all new major equipment for the DR-D DBN Project. The scope for the studies conducted in all PHAs phases included evaluation of the drawings and manuals for potential equipment failures, human errors, and process problems that could result in consequences of interest

## PROCESS HAZARD ANALYSIS METHODOLOGY

**For this analysis, consequences of interest were limited to events (e.g., hazardous material releases) that could present:**

1. serious injury to employees and/or.
2. imminent and substantial endangerment to public health and the environment.

**The analysis also included minor consequences such as:**

1. Those resulting from routine occupational hazards (e.g., employee slips or falls),
2. less significant or less widespread environmental effects (e.g., a small reportable spill).
3. Operability problems that could lead to economic consequences (business interruptions, off-specification product, etc.).
4. Extreme external events such as natural disasters and sabotage were not considered as causes; however, the effects of equipment damage resulting from typical external impact causes (e.g., vehicle collision, crane failure) were considered.

No. 10

## COMPOSITION OF THE TEAM

**To perform the analysis, HADEED assembled a team that meets the requirements of SABIC SHEM 02.01 "Process Safety Risk Assessment" element:**

1. PHA study team leader.
2. Scribe/secretary.
3. Personnel shall be included in the PHA study:

➔Process Engineer.
➔Operations Specialist/Engineer.
➔Safety, E&IH Engineer.
➔Instrumentation, Electrical & Control Systems Engineer
➔For initial PHA include Project Engineer/Mechanical Engineer (Design).

The PHA team members, included experienced operations and engineering personnel possessing a thorough understanding of previous DR-D plants as well as the DR-D DBN Project's design and intended operations.

No. 11

# TEAM MEMBERS

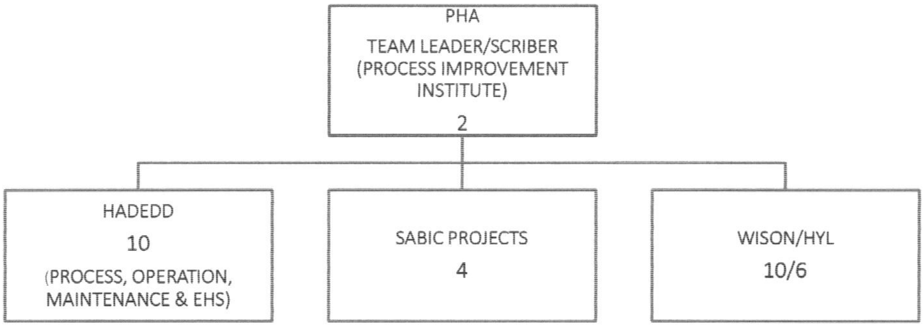




# ANALYSIS TECHNIQUES

To ensure adequate and efficient coverage of process hazards, the PHA team chose appropriate analysis techniques as specified in **SABIC's SHEM 002** PHA standard:

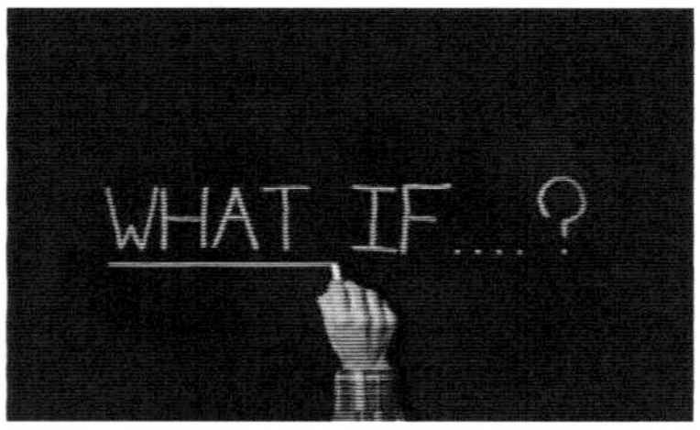

The team used approved methodology: **What-If /Checklist** in reviewing and updating the PHA:

- The *What-If* analysis technique was used to review DR-D process equipment.
- The *Checklist* analysis technique was used to review general human factors issues and general facility siting issues.

## ANALYSIS TECHNIQUES

- *What-If Analysis:*

The PHA team reviewed selected nodes for iron ore handling, the reduction furnace and dust handling processes. For accident scenarios that could result in a consequence of interest, the team reviewed the safeguards that either:

i. Help reduce the likelihood of the scenario.
ii. Or help mitigate the consequence.

- *Checklist Analyses:*

The PHA team reviewed a checklist related to general human factors issues and general facility siting issues.

*i.* <u>*Facility siting*</u> checklist was used to review the locations of process equipment and buildings relative to other equipment around and within the DRD process, as well as to offsite surroundings.
*ii.* <u>*Human factors*</u> checklist was used to evaluate the general working conditions that may adversely impact the ability of facility personnel to safely perform required tasks.



## RISK RANKING

HADEED, the process owner, provided a semi-quantitative system for determining the overall risk based on:

- Consequence category (Environmental, Personnel Safety, Economics).
- Estimated consequence severity ranking by category.
- Estimated frequency of occurrence of the consequence.

The frequency rating applied equally to the different consequence areas

- Personnel Appearance. Used only when personnel safety was a consequence
- Safeguard Effectiveness.

After the ratings for frequency, personnel appearance, and safeguard effectiveness were determined, the likelihood of the current design was determined using the following equation:

***Likelihood = -1 x (Frequency + Safeguard Effectiveness + Personnel Appearance)***

## DBN PHA STUDY

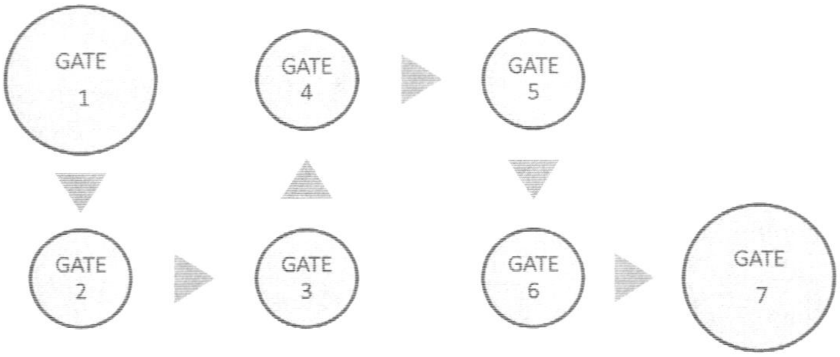


GATE 4: PHA **phase 1** was conducted during Basic engineering in china July 2 - 13, 2012.

GATE 5: PHA **phase 2** was conducted during detail engineering in china Feb 11 - 15, 2013.

GATE 6: PHA **phase 3** was conducted during detail engineering in china May 2013.



## DBN PHA STUDY, PREPARATION

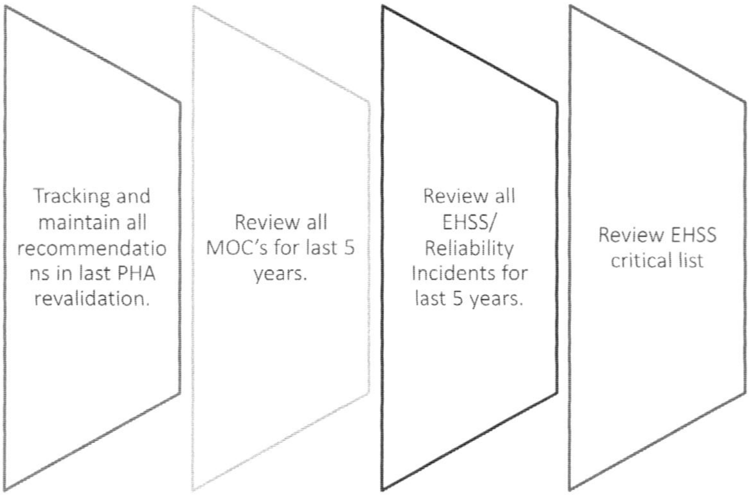

## DBN PHA STUDY, RECOMMENDATIONS

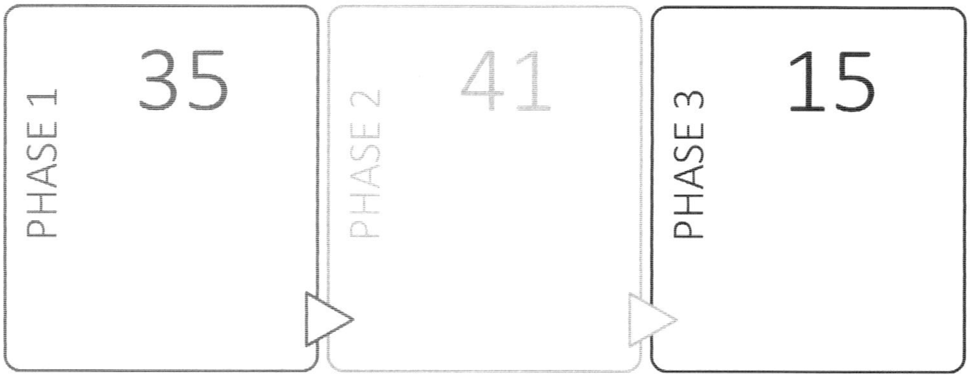


No. 18

## CONCLUSION

DBN Project one of the critical enhancement project conducted in HADEED, Where PHA study was contributed to achieving the main project scope by achieving many milestones where:

1. Achieving production by reaching to 160 ton/hr instate of 155 ton/hr.
2. Project was awarded from SABIC as successful project.
3. Mechanical completion was done ahead of target date.

Excellency of PHA Implementation was highly contributed & demonstrated well in Safe commissioning as criticality of integrated the process of DBN with existing DR-D as challenge during 3 phases of PHA Study.

No. 19

## AWARDING FOR BEST PROJECT PERFORMANCE AMONG SABIC MEGA PROJECTS, 2013

سابك
sabic

No. 20

# THANK YOU

# Reactive Chemicals Management Processes

Jamal AlGhamdi
Process Safety Technology Leader
5th December 2016

## Purpose of RC Management Processes

Sadara

# Purpose of RC Management Processes

**To prevent uncontrolled chemical reactions that have the potential to result in**

- injury
- property damage, or
- environmental harm

**An example of an uncontrolled chemical reaction: Explosion of railcar containing ethylene oxide and wash water heel, inadvertently "stored" for 23 days →**

Sadara

# Reactive Chemical Events

A Reactive Chemical Event is defined as an uncontrolled or unexpected chemical reaction that has the potential for injury, property loss, or environmental release.

## Reactive Chemical Reporting

- All RC Incidents must be reported and investigated
- The Goal is PREVENTION of Reactive Chemical events
- Reporting of all Near Misses and Accidents is intended to improve awareness and learning across the entire company

## Preventing UNCONTROLLED Chemical Reactions can only be achieved by...

- Understanding of the chemistry and processes that we work with
- A disciplined approach to operations and process changes

صدارة
Sadara

**Napp Technologies Incident:**
*jacket leaked into blender!!*

April 21, 1995

- 5 worker fatalities
- ~400 evacuated
- Facility destroyed
- Surrounding businesses damaged
- Fish killed by runoff

**Water leaked into a chemical mixing tank containing aluminum powder, sodium hydrosulfite, potassium carbonate and benzaldehyde**

# Three Key Principles of successful RC Management Processes

صدارة
Sadara

1. The RCMP plays a key role in **avoiding circumstances** that put people, environment, equipment or businesses at risk

1. The focus of the RCMP is **understanding the energy** of our systems and conditions under which it can be released

2. The **"Keystone"** of the RCMP is **"Owner Responsibility"**

## The Owners Responsibility

After all the testing, consultation, and reviews…

*The owner of the project, process, or facility must make the appropriate decisions concerning Reactive Chemical issues!*

*We are all "Owners" of the materials we handle!*

## The Owner's Responsibility is to Ensure:

- All appropriate operations personnel have a fundamental understanding of the reactivity of all the chemicals
- Reactive chemical risks are identified
- Controls and operating discipline are in place formalized documented and periodic reviewed
- All Reactive Chemical events are appropriately investigated and reported
- Root causes are determined for all events and communicated to prevent future events

## The types of RC reviews

## Why Do We Have Reviews?

- Ensures that nothing is missed/overlooked
- Provides many "sets of eyes" with broad and varied expertise to contribute
  - Multifunctional disciplines
  - External (from other technologies or plants)
- All items discussed that need resolution are captured and must be addressed

## Types of RC Reviews

- Management of Change (MOC)
- Reactive Chemicals/Process Hazard Analysis (RC/PHA's) :
  - New Project
  - Facility – Recurring (every 3-5 yrs)
  - New Leader reviews (within 90 days)

## Management of Change (MOC) Review

Small changes sometimes have major effects on the potential reaction hazards...

Examples of changes:

- New supplier of raw materials or method of delivery
- Equipment changes
- Procedure changes
- Changes in shutdown or startup plans or procedures
- New computer control systems or program changes

## RC/PHA* Reviews

*Reactive Chemicals/Process Hazard Analysis

- Reviews are a key part of the RCMP
- These are formal meetings of key stakeholders with external representation
- Scope is adjusted to reflect process scale and risk
- Formal and standardized agenda is used for the Site

## Key Components of RC/PHA Review

- Review process chemistry and hazards
- Review Reactive Chemicals test data
- Examine the processes in detail for worst case scenarios (and other ways of potentially having a reactive chemicals event)
- Analyze <u>each</u> phase of operations for any place where a Reactive Chemical event could occur.

**Consider the following:**

- Equipment or utilities failure
- Not following procedures, including startup and shutdown
- Time delay in any step of the process
- Extended hold or shut down period on plant operation
- Other conditions or actions that can lead to a severe RC incident (many items to consider are triggered by Questionnaire)
- Identify lines of defense that will be employed to avoid Reactive Chemical events and worst case scenarios at each point
- Complete or update the RC/PHA Questionnaire

# RC/PHA Questionnaire Example Section

## EXISTING FACILITY REVIEW

Use "X" to mark the answers

| C# | Q# | QUESTION | YES | NO | NA | RESPONSE (provide written response and include a copy or a path to the e-document if requested) |
|---|---|---|---|---|---|---|
| F. | 0 | **Reactive Chemicals Data** | | | | |
| F. | 0 | The objective of these questions is to determine if the appropriate test data have been obtained. It is preferred that the test data be consolidated into a folder for easy review. (If in a folder please provide a copy or a path in the Response box.) | | | | |
| F. | 1a | Have Reactive Chemicals test data been gathered and evaluated versus worst case scenarios for the following? Provide a link to summarized info. | | | | |
| F. | 1b | -- Process materials and mixtures that are normal to the operation? | | | | |
| F. | 1c | -- Mixtures that may result from abnormal conditions? (This would include upsets like feeding too fast or too slow or adding the components in the wrong order.) | | | | |
| F. | 1d | -- Materials of construction throughout the process? | | | | |
| F. | 1e | -- Evaluation of Dust Explosion Hazards under normal and possible abnormal conditions as per LPP 11.1 Combustible Dusts (e.g. particle size distribution (PSD), minimum ignition energy (MIE), rate of pressure rise (Kst), maximum explosion pressure (Pmax))? | | | | |
| F. | 1f | -- Appropriate thermal stability data been generated on reactive mixtures, formulations and end products; regardless whether handled in drum or larger volumes? | | | | |
| F. | 1g | -- Auxiliary process materials, e.g., cooling tower chemicals, process or instrument seal fluids, that are normal to the operation? | | | | |
| F. | 1h | -- Maintenance materials including oils, lubricants and solvents ? | | | | |
| F. | 1i | -- Materials or residues that might be or might form shock sensitive materials? | | | | |

RC-PHA Menu | RCPHA Introduction & Protocol | RCLHA Introduction & Protocol | PHA Questionnaire | Appendix A | Appendix B1, B2 | Appendix C | Action Items | MOC | Scenarios

## How We Identify Worst Case Scenarios

***Ask yourself:*** *If I lost control and my process ran-away with no heat removal, how high could the temperature go?*

- ***This is probably a worst case scenario.***
- *What are our lines of defense?*

## Worst Case Scenario Strategy

- The worst case scenarios are identified
- Lines of defense are developed for each scenario
- The Lines of defense are:
  - Standardized across the technology
  - Part of the in-plant training
  - Included in the operating discipline of the facility

## Layers of Protection Analysis

- LOPA (Layers of Protection Analysis) should be done for worst case scenarios where additional, more in-depth analysis is warranted.
- LOPA is a systematic, qualitative, in-depth approach for hazard evaluation that uses a structured method.

## Kaltech Explosion, New York City

- <u>Common Waste materials</u> (nitric acid and lacquer thinner) were mixed without being identified or characterized
  - **<u>No Effort</u>** was made to determine <u>compatibility</u> of the materials
- **Root Cause: Kaltech did not manage its hazardous waste**

31 Injured!

# The RC Committee

## The Reactive Chem Committee (RCC)

- Reactive Chemical Committee (RCC) to be established
- The RCC serves mainly as a site multi-discipline resource for conducting reviews
- The RCC is composed of experienced representatives from many technology disciplines:
  - Manufacturing
  - Engineering
  - Other Technology Centers
  - Process Safety, EH&S
  - the Reactive Chemicals Discipline
  - and other appropriate functions

## RCC Responsibilities

- Provide "outside-in" evaluation by RCC members from many different functions for RC/PHA Reviews
- Train employees on Reactive Chemicals, using tools and resources provided by the Reactive Chemicals Discipline
- Implement ways to enhance awareness of Reactive Chemical issues
- Help investigate Reactive Chemicals events
  - Establish "Corporate Memory" of Reactive Chemical events, i.e. a strong focus on the reporting and prevention of repetitive Reactive Chemical accidents

## Santa Paula, CA, USA
## Santa Clara Waste Water Facility

**Nov 18, 2014**

Truck was delivering industrial waste - Probable reaction of sulfuric acid and organic peroxide waste

Dozens sent to hospital!
Community affected, highway, schools and businesses closed!

## The Role of the Process Safety & Reactive Chemicals Groups in the RC Process

- Process Safety **team is responsible for:**
  - Facilitating RC/PHA Reviews
  - Loss Prevention Principles and audits
  - LOPA
- Reactive Chemicals **team is responsible for:**
  - Supporting RC/PHA reviews
  - Thermal stability, flammability & reactive hazard testing, data interpretation, and consultation

## Reactive Chemicals Team

- Reactive Chemicals Subject Matter expert is part of the Process Safety Organization
- Reactive Chemicals testing expertise is aligned with the Central Analytical Laboratory

The ***Team*** of RC-S and Lab analysts provide Process Owners with information to help understand the reactive chemicals risk

# Plant Compatibility Chart

صدارة
Sadara

SIP Training

**Sadara Chemical Co. Compatibility Chart**

**Kevin Darst**

Chart compiled to provide a chart for Sadara Chemical Co.

Materials produced and used by the facility

**Compatibility is scenario dependent**

Version: 18-Nov-2014 (v1.00) [Current date: 26-Nov-2014]

☐ Display chart in local language (English)

☑ Display compatibility numbers inside cells

100% complete

Select a Site

Update Workprocess

Additional RC-data List

Chemicals List

| Health | Flammability | Instability / Reactivity | PRODUCTS / STREAMS | Air Flammability | Aniline | Nitric Acid (conc.) | n-Propane | Polyethylene | Propylene glycol | Propylene oxide | Sodium hydroxide (20 % aq | TDI | Water |
|---|---|---|---|---|---|---|---|---|---|---|---|---|---|
| 3 | 2 | 0 | Aniline | 0 | I | | | | | | | | |
| 4 | 0 | 0 | Nitric Acid (conc.) | 0 | 2 | I | | | | | | | |
| 1 | 4 | 0 | n-Propane | 2 | 0 | 1 | I | | | | | | |
| | | | Polyethylene | 0 | 0 | 1 | 0 | I | | | | | |
| 0 | 1 | 0 | Propylene glycol | 1 | 0 | 2 | 0 | 0 | I | | | | |
| 3 | 4 | 2 | Propylene oxide | 2 | 2 | 2 | 0 | 0 | 2 | I | | | |
| 3 | 0 | 1 | Sodium hydroxide (20 % aqu.) | 0 | 0 | 2 | 0 | 0 | 1 | 2 | I | | |
| 3 | 1 | 1 | TDI | 0 | 1 | 2 | 0 | 0 | 2 | 1 | 2 | I | |
| 0 | 0 | 0 | Water | 0 | 0 | 2 | 0 | 0 | 0 | 2 | 1 | 2 | I |

صدارة
Sadara

# THANK YOU

Petrochemical Corporation of Singapore (Private) Limited

# Major Hazard Installations must have Competent People for Safe and Reliable Operations

**Lucas NG H K**
**General Manager of Plant**

3rd CCPS Global Summit on Process Safety
December 4-5, 2016, Dammam, Kingdom of Saudi Arabia

3RD CCPS GLOBAL SUMMIT ON PROCESS SAFETY

## Major Hazard Installations must have Competent People for Safe and Reliable Operations

1. Introduction – PCS
2. Major Hazard Installations
3. Must have Competent People
4. Structured Competency Development Programs
5. Competency in Management of Change
6. Leaders must Lead in Process Safety
7. Conclusion

2

## Major Hazard Installations must have Competent People for Safe and Reliable Operations

1. **Introduction – PCS**
2. Major Hazard Installations
3. Must have Competent People
4. Structured Competency Development Programs
5. Competency in Management of Change
6. Leaders must Lead in Process Safety
7. Conclusion

3

## 1. Introduction – PCS

- Upstream company of Singapore Petrochemical Complex on Jurong Island
  - supply high quality ethylene, propylene, acetylene, butadiene, 1-butene, MTBE and benzene to companies within Island
- 1977 – incorporation
- 1984 – first ethylene plant (cracker) in Southeast Asia on stream
- 1997 – second cracker unit on stream
- Present shareholders
  - Japan-Singapore Petrochemicals Co., Ltd (50%) (JSPC led by Sumitomo Chemical)
  - QPI and Shell Petrochemicals (Singapore) Pte Ltd (50%) [*QPI: Qatar Petroleum International]

Source: PCS

4

## 1. Introduction to PCS – Integration with Jurong Island

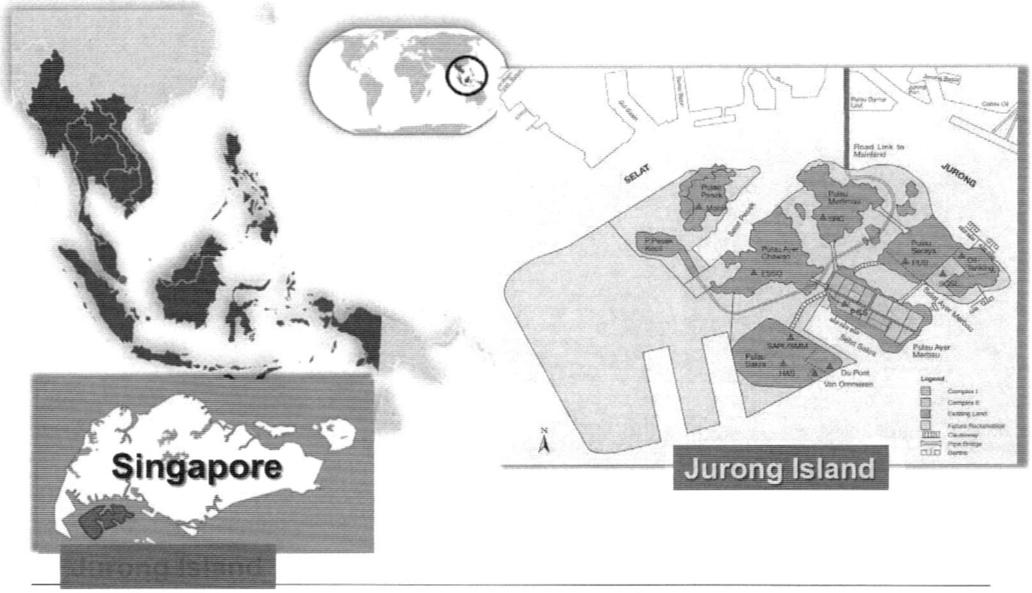


Source: websites 5

## 1. Introduction to PCS – Integration with Singapore Petrochemical Complex (SPC)

Source: PCS 6

## Major Hazard Installations must have Competent People for Safe and Reliable Operations

1. Introduction – PCS
2. **Major Hazard Installations**
3. Must have Competent People
4. Structured Competency Development Programs
5. Competency in Management of Change
6. Leaders must Lead in Process Safety
7. Conclusion

7

## 2. Major Hazard Installations

Crude oil production, petroleum refining, petrochemical manufacturing, chemical processing

↓

**Concerns**
many catastrophic past incidents occurred

↓

**Complexity of MHIs**
design, engineering, construction, operations, facilities integrity & reliability

↓

**Crucial to fully understand installations**
hazards and mitigation measures

↓

**Emphasize**
process safety management system

↓

**Competent in managing process safety**
management of change

8

## Major Hazard Installations must have Competent People for Safe and Reliable Operations

1. Introduction – PCS
2. Major Hazard Installations
3. **Must have Competent People**
4. Structured Competency Development Programs
5. Competency in Management of Change
6. Leaders must Lead in Process Safety
7. Conclusion

9

## 3. Must have Competent People

**Why?**

**Extended turnaround intervals**
– from 2 to 5 years intervals

| Operators | Maintenance crew | Challenges |
|---|---|---|
| • Lack opportunity to "practice" critical operations<br>• Experience diluted due to retirement, attrition | • Non-routine facilities maintenance<br>• Work on off-days, weekends, public holidays<br>• Handing and taking over | • Hire fresh / experienced people<br>• Speed up internal competency development<br>• Reduce attrition rate |

10

## 3. Must have Competent People (cont'd)

### Holistic approach competency development

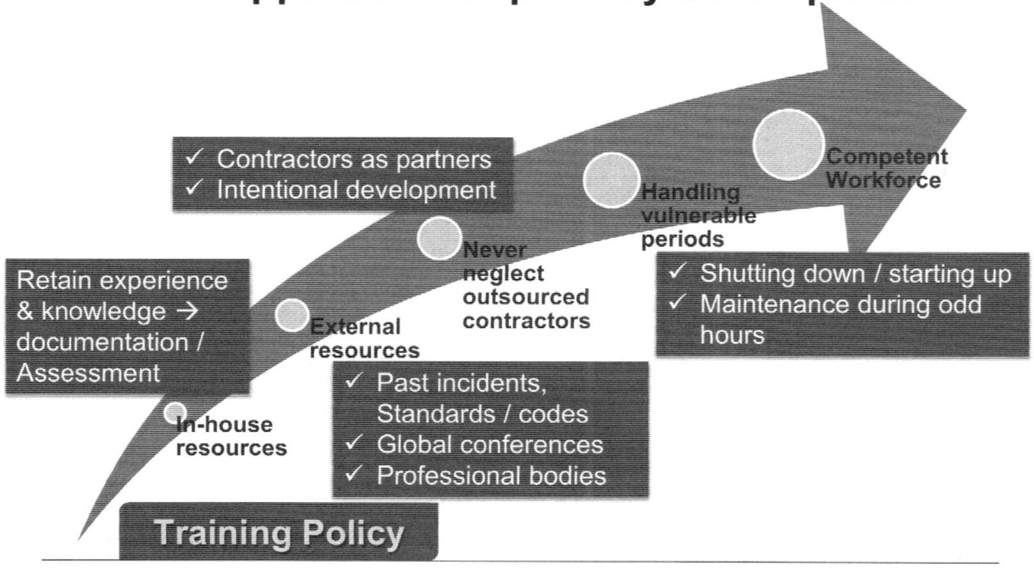


11

## Major Hazard Installations must have Competent People for Safe and Reliable Operations

1. Introduction – PCS
2. Major Hazard Installations
3. Must have Competent People
4. **Structured Competency Development Programs**
5. Competency in Management of Change
6. Leaders must Lead in Process Safety
7. Conclusion

12

## 4. Structure Competency Development Programs

- How to be a learning organization?

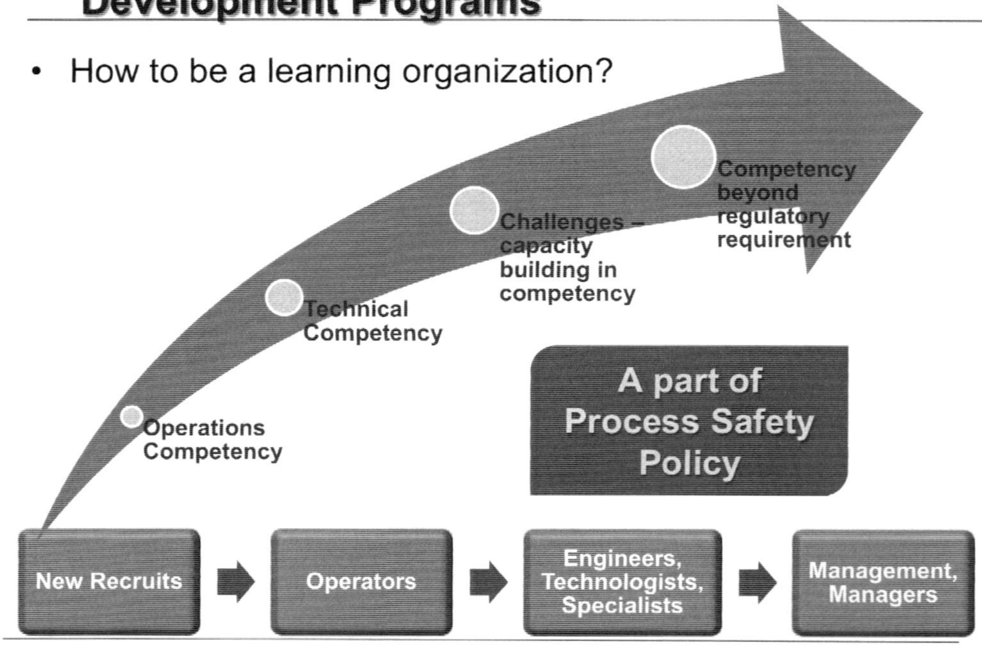




## 4. Structure Competency Development Programs (cont'd)

### Operations competency

- Technicians, supervisors, superintendents, managers – know "a bit of everything"
  - Process, equipment, instrument, electrical
- Process know-how, know-why briefing / discussion
  - Plant upset / emergency response, safe equipment / instrument preparation for maintenance
- Training simulator training
  - Full simulator / PC unit operation simulation training
  - In shift training – own time

### Technical competency

- Engineers, technologists, maintenance specialists – know "a bit more of everything"
  - Process design, equipment troubleshooting etc
- Process design bases
- Process hazards analysis – HAZOP, LOPA, Bow-tie etc training by expert
  - PHA participation OJT
- Participate in local and international professional bodies – IES, AIChE, CCPS

## 4. Structure Competency Development Programs (cont'd)

### Challenges – capacity building in competency

- Process safety continuing education
  - CCPS book review and team presentation, discussion
- Taskforce review other's incident investigation report
  - Disseminate lessons learnt
- External specialized training providers
  - Example: **Process Safety Boot Camp by CCPS**, LOPA training by PSRG

### Competency beyond regulatory requirement

- No regulations can exhaustively cover vast varieties of facilities
- Management to recognize its unique workplace – back to basic principles
- Develop competent people within organization – hazards, mitigate risks, communicate to all levels
- Top-down and bottom-up

15

## Capacity Building – Process Safety Competency

2nd session – July 2014
3rd session – June 2015
4th session – Sept 2016

http://www.aiche.org/ccps/news/institute/04-09-2013/ccps-conducts-process-safety-bootcamp-singapore-first-time **Source: PCS** 16

## 4. Structure Competency Development Programs (cont'd)

- Back to Basic Principles – Asset Integrity & Reliability

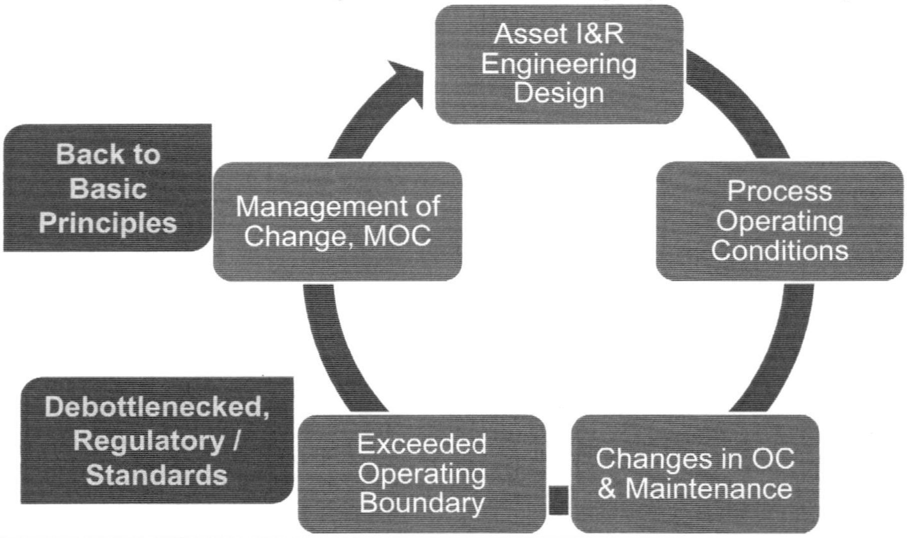


17

## 4. Structure Competency Development Programs (cont'd)

- "To know a bit of everything" – examples: Metallurgy, Insulation, Lubrication

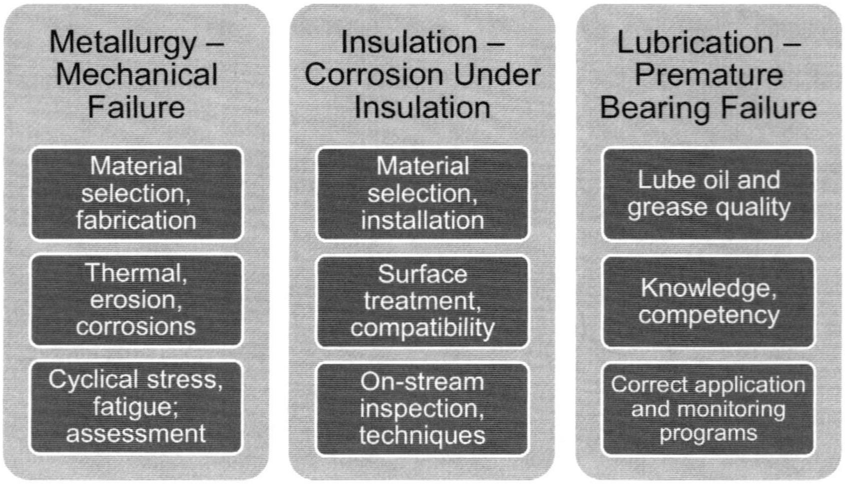


18

## Major Hazard Installations must have Competent People for Safe and Reliable Operations

1. Introduction – PCS
2. Major Hazard Installations
3. Must have Competent People
4. Structured Competency Development Programs
5. **Competency in Management of Change**
6. Leaders must Lead in Process Safety
7. Conclusion

19

## 5. Competency in Management of Change

**Good knowledge of Management of Change (MOC)**

**AIChE Discussion Central**
**April, 2016**
**"MOC System Challenges"**

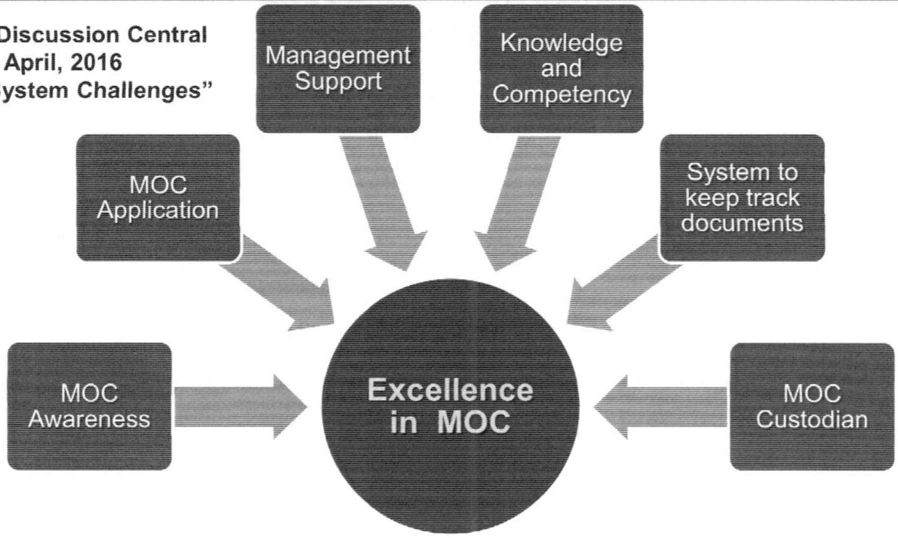


https://engage.aiche.org/viewdocument/re-moc-system-challenges?ssopc=1

**Source: PCS**

20

## 5. Competency in Management of Change (cont'd)

| MOC Awareness | MOC Application | Management Support |
|---|---|---|
| • Policy, protocols<br>• Technical, organizational<br>• Know-how, know-why | • Flexible, nimble, appropriate; guidelines<br>• MOC review: big or small – process hazards analysis | • Collaboration of all levels in organization<br>• Two way communication<br>• Review past MOC performance for any gap |

21

## 5. Competency in Management of Change (cont'd)

| Knowledge and Competency | System to keep track documents | MOC Custodian |
|---|---|---|
| • Staff to be competent<br>• Process safety knowledge, appropriate tools<br>• MOC team good knowledge similar to HAZOP team<br>• Good teamwork culture | • Gate-keeper tracking all documents<br>• Special procedures for temporary change<br>• Tracks PSSR | • Another "gate-keeper"<br>• Track and prompt owner for another action should temporary MOC becomes "permanent temporary"<br>• A competent successor to custodian |

22

## Major Hazard Installations must have Competent People for Safe and Reliable Operations

1. Introduction – PCS
2. Major Hazard Installations
3. Must have Competent People
4. Structured Competency Development Programs
5. Competency in Management of Change
6. **Leaders must Lead in Process Safety**
7. Conclusion

23

## 6. Leaders must Lead in Process Safety

- Cultivating corporate culture of process safety

24

## 6. Leaders must Lead in Process Safety
**(cont'd)**

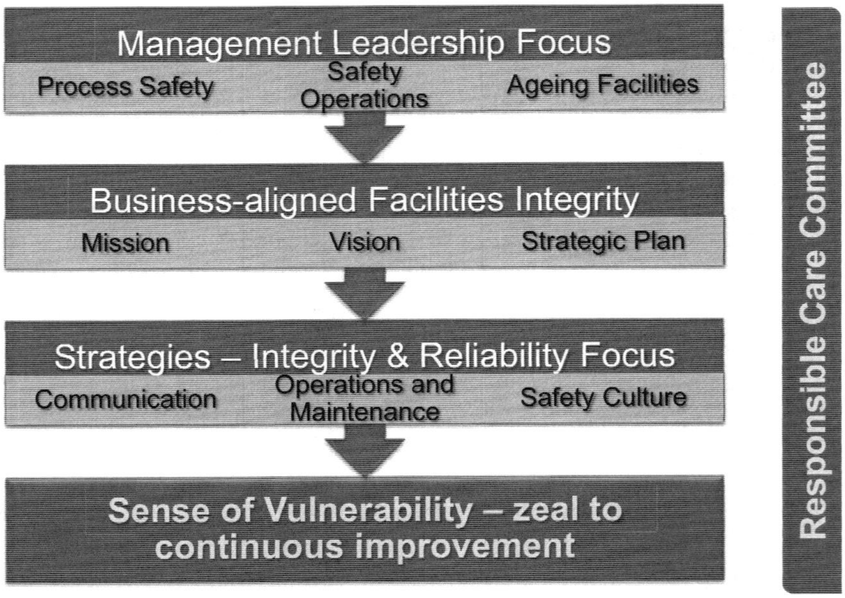


25

## 6. Leaders must Lead in Process Safety
**(cont'd)**

Source: PCS

26

## 6. Leaders must Lead in Process Safety
(cont'd)

**Sense of Vulnerability – zeal to continuous improvement**

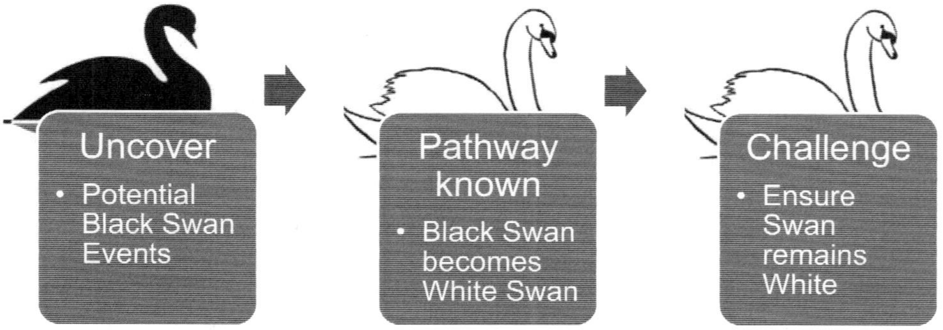


**Eliminate Black Swan Events**

[**Black swan events** are typically random and are unexpected]
["Surviving the black swan, strategies for process safety...", John F. Murphy]

**Source: The Internet**



## Major Hazard Installations must have Competent People for Safe and Reliable Operations

1. Introduction – PCS
2. Major Hazard Installations
3. Must have Competent People
4. Structured Competency Development Programs
5. Competency in Management of Change
6. Leaders must Lead in Process Safety
7. **Conclusion**

# Conclusion – Takeaways

- Ensure Knowledge and Competency for MHI
- PS Competency Development for everyone
- Management commitment and visibility
- Tap industry's best practices
- Constant Competency and Resources Review for continuous improvement

*"Competent people in MHI not just limited to own staff but also include outsourced contractors as well"*

29

3RD CCPS GLOBAL SUMMIT ON PROCESS SAFETY

December 4-5, 2016, Dammam, Kingdom of Saudi Arabia

# THANK YOU

**Lucas NG H K**
**General Manager of Plant**

Petrochemical Corporation of Singapore (Private) Limited

bizSAFE Mentor

Responsible Care

30

THE IMPORTANCE OF SHIFT HANDOVER (SHO) IN PREVENTING PROCESS SAFETY INCIDENTS

December, 2016

SAMI ALMAIYAD
MOHAMMED AL-KANFARI

A SABIC and CTE Joint Venture

## OUTLINES

- SHO Risks
- What can go wrong on SHO?
- Major Incidents and SHO
- SHO Risk management
- Recommended Practices in SHO
- Conclusion

No. 1

## SHIFT HANDOVER (SHO)

- Shift handover is a critical aspect of day-to-day operation management

✓ **The goal of effective shift handover** is the accurate, reliable communication of task-relevant information across shift changes, thereby ensuring continuity of **safe** and **effective** working.

No. 2

## SHO RISKS

Risk is always introduced on every shift change.

The incoming shift does not understand operational threats especially:

- During plant maintenance, particularly when this work continues over a shift change.
- When safety systems have been over-ridden.
- Plant abnormal operation.
- Following a lengthy absence from work.
- When handovers are between experienced and inexperienced staff.

No. 3

ابن سينا
ibn sina

## WHAT CAN GO WRONG IN SHO?

Unreliable communications can result from a variety of problems

including:

- Missing information
- Unnecessary information
- Inaccurate information
- Poor or variable quality of information.
- Misunderstandings
- Failing to carry forward information over successive shifts.

Different understanding of the current state of the process lead to miscommunications.

No. 4

ابن سينا
ibn sina

## WHAT CAN GO WRONG IN SHO?

50% of all operating mistakes happen in the first 30 minutes after a handover. *

Every 2nd incident in process industry is related to communication error during shift handover.**

Because the person coming on doesn't get the handover he or she needs.

* Larkin Communication Consulting

** Bradley Ford of KBC Advanced Technologies

No. 5

# MAJOR INCIDENTS AND SHO

ابن سينا
ibn sina

## 2014 DUPONT LAPORTE TOXIC CHEMICAL LEAK

A U.S. Chemical Safety Board investigation found a series of **shift communication mistakes** that began five days before the incident.

No. 7

ابن سينا
ibn sina

## 2005 BP TEXAS CITY

BP own report- "there was no written expectations with explicit requirements for shift handover."

CSB report- "the condition of the unit – specifically, the degree to which the unit was filled with liquid raffinate- **was not clearly communicated from night shift to day shift"**

No. 8

ابن سينا
ibn sina

## 2005 BUNCEFIELD

Standards Group- "effective **shifts/crew handover communication** arrangement must be in Place to ensure the safe continuation of operation."

No. 9

## NOT A NEW DISCOVERY

Windscale Vitrification 1991- Plant Shield Door Incident

- Temporary override was not properly logged
- Over-reliance on one-way written communication.

Piper Alpha 1988- Status of condensate pumps not known

- Status of pressure safety valve failed to be communicated during shift handover.

Sellafield 1983- Presence of radio active material in tank pumped to sea.

- Failure to accurately describe the tank's contents, coupled with transcription errors made in the written logbook.





## COMMON FAILURE MODE

The inaccurate or incomplete communication of information from shift to shift.

Failures to effectively communicate information about maintenance actions in progress, but incomplete, at the time of a shift handover.

Poor logbook design and poorly-conducted shift handovers are the two most common causes of such lapses in communication across shifts.

The two problems are closely intertwined, and can have significant consequences.

ابن سينا
ibn sina

## NOT A NEW DISCOVERY BUT!

80 % of the world's industrial operations still lack of structured approach to shift handover*

* A Critical Shift: How Adding Structure Can Make Shift Handover More Effective. March 2012

No. 12

ابن سينا
ibn sina

## THE PROBLEM

- Shift handover is a complex, error prone activity, performed frequently "High Risk"
- It can't 'engineered out'
- Partly driven by systems and procedures
- Highly dependent on behaviors of people involved
- Rarely cited as a Root Cause of accidents.

No. 13

ابن سينا
ibn sina

## SHO RISK MANAGEMENT

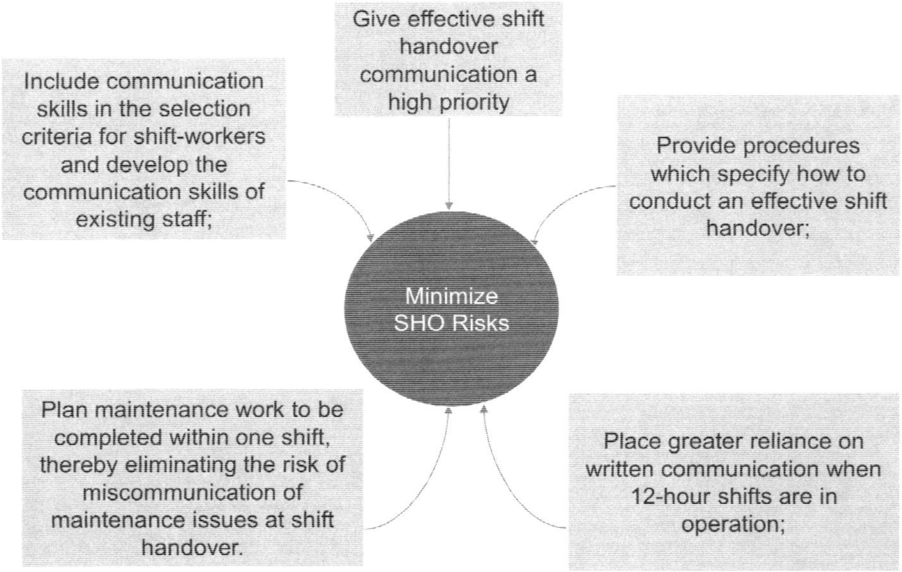


No. 14

# RECOMMENDED PRACTICES IN SHO

ابن سينا
ibn sina

## RECOMMENDED PRACTICES IN SHO

1. Allow sufficient time for the handover;

Note: More time should be given when plant is under maintenance, following a lengthy absence and between experienced and inexperienced staff .

2. Conduct the handover Face-To-Face.

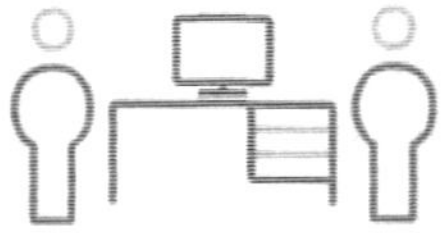

3. Eliminate distractions.

4. Ensure the person leading the handover:

- Gives an overview of the handover content,
- Makes positive statements about safety issues,
- Talks through the log items, and
- Summarizes the handover at the end.

No. 16

ابن سينا
ibn sina

## RECOMMENDED PRACTICES IN SHO

5. Ensure feedback takes place to seek clarification and confirmation; and

6. Ensure the person receiving the handover takes notes.

7. Use structured shift logs that clearly indicate what should be reported in the handover presentation.

8. Train operators in the skill of conducting effective shift handovers and in effective two-way communication.

9. Always have system in place to assess the effectiveness of the handover process.

No. 17

ابن سينا
ibn sina

## EXAMPLE OF STRUCTURED SHIFT LOGS

1. EHSS (permit to work, safety logs, etc)
2. Process Safety Issues (ESD override, Frequent alarms, exceeding safe operating limits, etc.)
3. Highlights (Operational logs)
4. Maintenance items
5. Instructions
6. Shift manning

No. 18

ابن سينا
ibn sina

## EXAMPLE OF STRUCTURED SHIFT LOGS

| Date______________________ Time__________________ | |
|---|---|
| **The following points are to be discussed at shift handover** | **Discuss & Tick** |
| **Health, Safety, Environmental** | |
| Safety or Environmental limits exceeded or issues. Reported? | ☐ |
| SOL exceedinaces | ☐ |
| Disabled and/or nuisance alarms | ☐ |
| MOC/s implemented or initiated during shift. Review for understanding | ☐ |
| **Process Safety** | |
| Loss of primary containment (LOPC) | ☐ |
| Emergency Shutdown Systems or Relief Systems bypassed | ☐ |
| New risk assessment actions | ☐ |
| **Operations** | |
| Operating plan status (rates, specs, yields, energy, special sample, etc.) | ☐ |
| Operating instruction Status (special instructions, orders, etc.) | ☐ |
| Charge increases, decreases, composition changes | ☐ |
| Process lineup changes (off spec product corrective action, tank transfers. Etc.) | ☐ |
| Process controls or instrument problems or issues (vessel & tank levels) | ☐ |
| Chemicals or catalyst inventories and /or deliveries | ☐ |
| Inventory trends, issues threats | ☐ |
| Equipment out of service, backup equipment unavailable (pumps, fans, et | ☐ |
| Equipment/control valve bypass open | ☐ |
| **Maintenance** | |
| Maintenance work in progress/scheduled. Status | ☐ |
| Preparation of equipment ofr maintenance (PM's regular maintenance. Etc.) | ☐ |
| Permit equip locked out; instrument work, DCS system work | ☐ |
| Emergency Notifications | ☐ |
| **Any new information since logbook submission** | ☐ |
| | |
| **Other information not covered above** | ☐ |
| | |

| Signature: Incoming Shift Person | Signature: Outgoing Shift Person |
|---|---|
| | |

No. 19

ابن سينا
ibn sina

## ROLES AND RESPONSIBILITIES

**Outgoing personnel** to ensure the incoming personnel understand the status of the business, process and systems before leaving the workplace.

**Incoming personnel** to ensure they are in a position to carry out actions and make decisions before they allow the outgoing personnel to leave the workplace.

No. 20

ابن سينا
ibn sina

## CONCLUSION

50% of all operating mistakes happen in the first 30 minutes after a handover.

Shift handover is a complex, error prone activity, performed frequently "High Risk"

Poor handover could lead to major process safety incidents

It is vital that all personnel involved in the shift-handover process are fully aware of their roles and responsibilities

Include communication skills in the selection criteria for shift-workers and develop the communication skills of existing staff.

The best practice in effective shift handover is face-to-face, two way communication. Use structured shift logs make it even better.

More time will be needed to communicate when differing 'mental pictures' exist.

No. 21

ابن سينا
ibn sina

# Q&A

No. 22

ابن سينا
ibn sina

# Thank You

No. 23

أرامكو السعودية
saudi aramco

# Proactive Analysis of Human Performance Issues in Process Safety

3rd CCPS Global Summit on Process Safety

December 4-5

Dammam, Saudi Arabia

where energy is opportunity™

أرامكو السعودية
saudi aramco

# System Overview

where energy is opportunity™

## Reactive Investigation

- Plan the investigation
- Determine events/conditions
- Select causal factors
- Select root causes
- Select generic causes
- Select corrective actions
- Implement

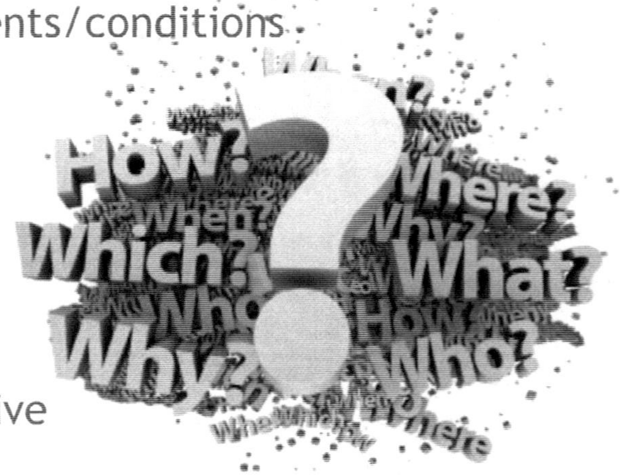




أرامكو السعودية
saudi aramco

## Proactive Process Review

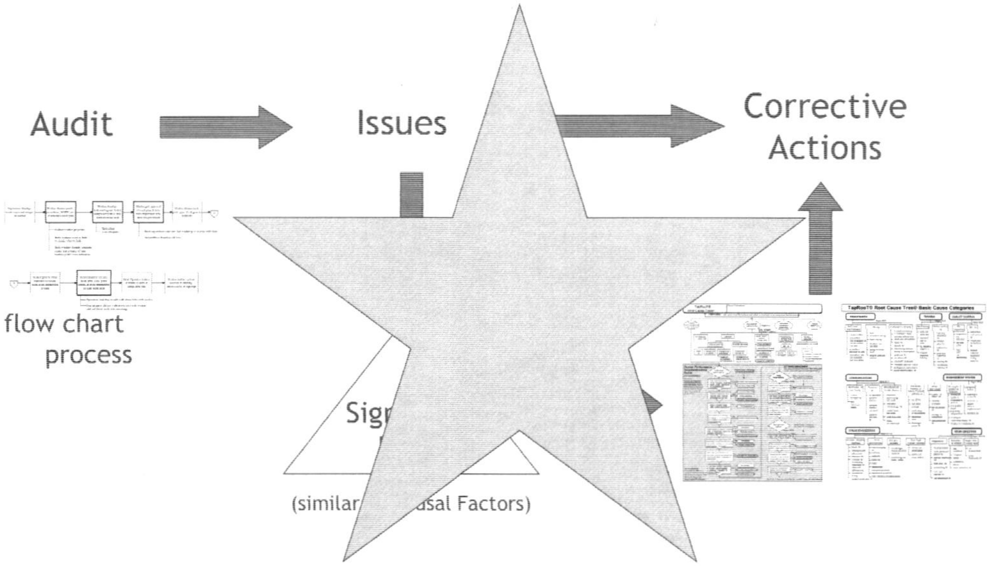






## Proactive Process Review

- Plan the investigation / Chart process/plan audit
- Determine events/conditions / Perform audit/note discrepancies
- Identify causal factors / Select significant issues
- Select root causes
- Select generic causes
- Select corrective actions
- Implement

أرامكو السعودية
saudi aramco

# The Need for an Expert System & System Characteristics

where energy is opportunity™

## The Need for an Expert System

- Not everyone is an expert
- We tend to focus on what we know
- We tend to have operational biases
- We tend towards confirmation bias

## Characteristics of a Good Root Cause Analysis Expert System

- Has a sound scientific basis
- Has a history of rigorous development and field implementation
- Allows transparency
- Can be used with existing systems (e.g. Six Sigma, BBS, etc.)



## Characteristics of a Good Root Cause Analysis Expert System

- Is well structured (reproducible)
- Produces actionable root causes
- Identifies generic causes
- Allows for advanced trending

## Characteristics of a Good Root Cause Analysis Expert System

- Produces actionable root causes
- Identifies generic causes

11

أرامكو السعودية
saudi aramco

أرامكو السعودية
saudi aramco

# Case Study

where energy is opportunity™

Bypassing an Emergency Shutdown Device

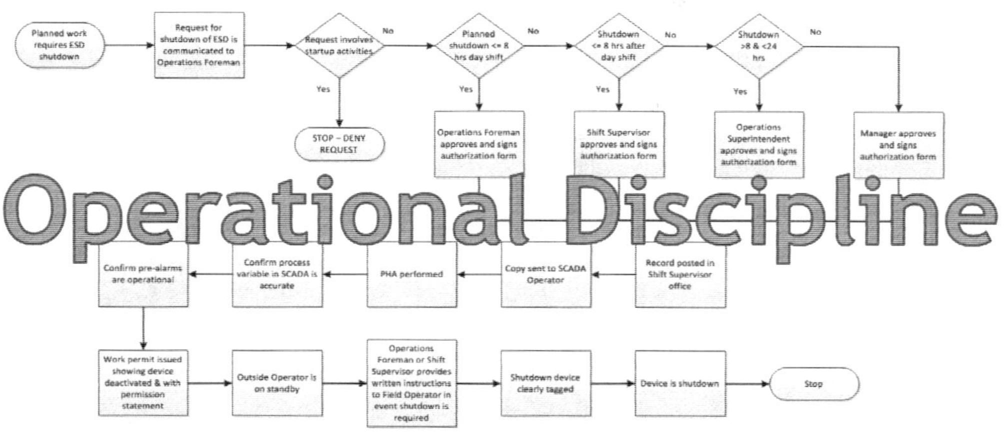


Step 1 – Flow Chart the Process





Bypassing an Emergency Shutdown Device

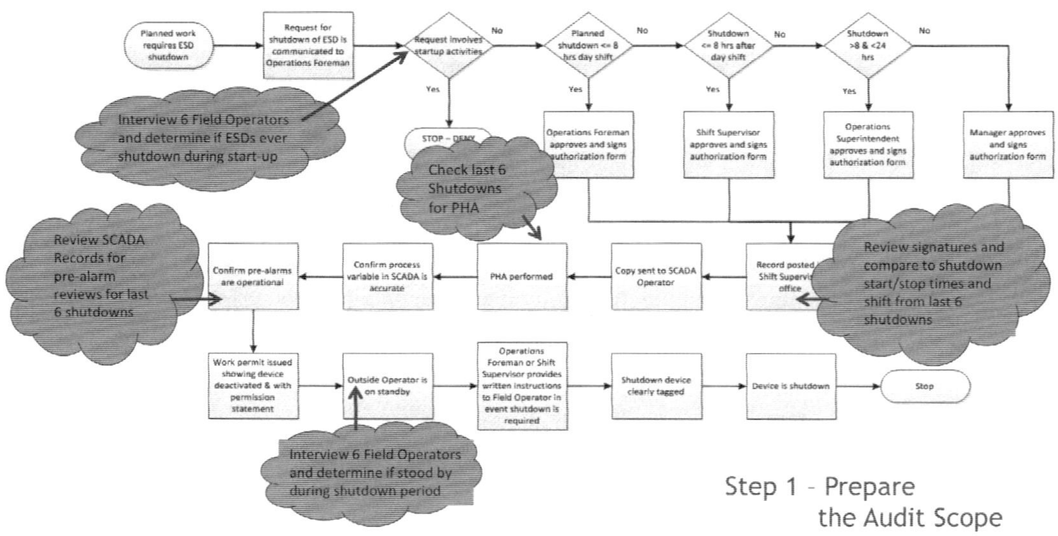


Step 1 – Prepare the Audit Scope

Bypassing an Emergency Shutdown Device

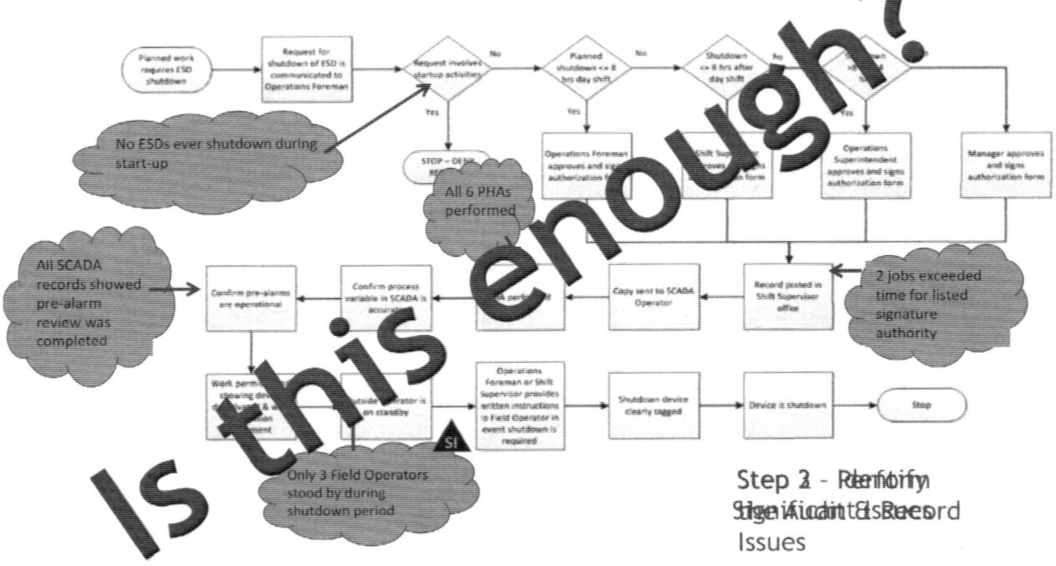




أرامكو السعودية
saudi aramco

# Information from Field Operators & Ops Foremen

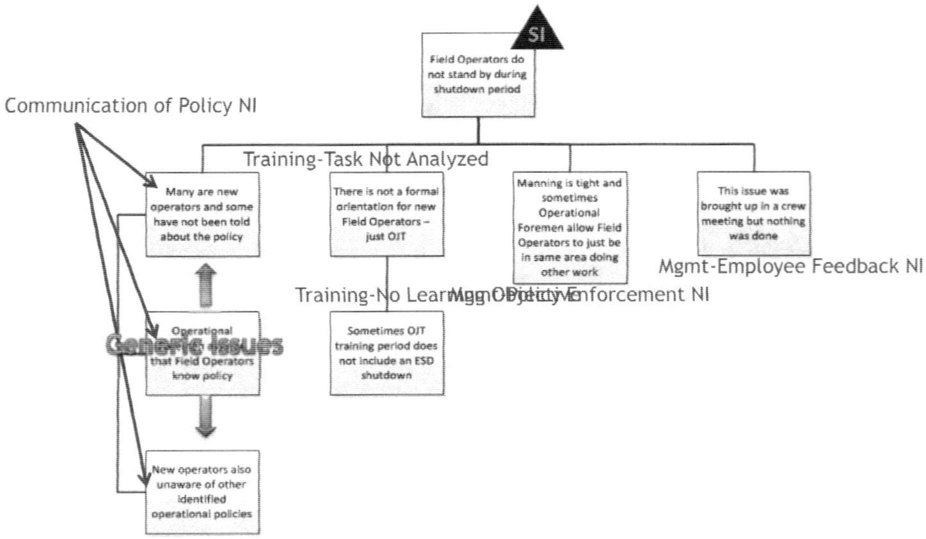




أرامكو السعودية
saudi aramco

# RCA

**Significant Issue Root Cause Analysis**

**Significant Issue #1:** Field Operators do not stand by during ESD shutdown period

| Root Cause | Basis for Specific Root Cause | Evidence[†] | CA # | Corrective Action Recommendation |
|---|---|---|---|---|
| Human Performance Error – Management System – SPAC Not Used – Comm. Of SPAC NI | Were standards or policies not communicated from management down through the organization?<br><br>Many new Field Operators have not been told about the policy. They were also unaware of several other key operational policies. | Interviews with Field Operators and Operational Foremen | 1-1 | The Operational Superintendent will identify all key operational policies important for new Field Operators to know. Training sessions will be held for all Field Operators to cover the important elements of these policies. Completion by 6/30/2013. |
| | | | 1-2 | The Compliance Group will develop a new Field Operator orientation that includes the communication of the identified policies. Training in these policies will be added to the Field Operator training matrix and tracked in the training system. The process flow for new policy development will be routed by the Compliance Group Leader for determining if the policy or revisions impact safety of operations. If so the new policy or revisions will be communicated in the orientation. Completion by 6/30/2013. |
| Human Performance Error – Management System – SPAC Not Used – Enforcement NI | Was enforcement of the SPAC seen as inconsistent by the employees?<br><br>Operations Foremen allowed Field Operators to just be "in the area" of the ESD device when shutdown, not monitoring the device. | Interviews with Field Operators and Operational Foremen | 1-3 | The Compliance Group Leader will develop an audit, both of orientation and the policy communication and monitoring of ESD devices during shutdown. The audits will be unannounced. Results will be communicated to the Department Manager on a monthly basis. Completion by 7/15/2013, first audit results by 8/1/2013. |
| Human Performance Error – Work Direction – Scheduling NI | Was it clear that too few workers were assigned to successfully accomplish the work?<br><br>Having too few operators led to unmonitored ESDs during shutdown. | Interviews with Field Operators and Operational Foremen | 1-4 | The Department Manager will have a review performed by OCD within 60 days. The recommendations of the review will be implemented with 90 days of their issue. |

Corrective Action to

Reason for

Question from Expert

Identified Root Cause

Reason from Review Why the

Yes Answer

Question Was Answered Yes

[†]Evidence includes interview statements, documents, analytical reports, equipment positions, etc. It shows why the basis of the root cause is a fact, rather than opinion

[‡]NI – Needs Improvement



# Effectiveness of Corrective Actions

Field operators should stand by during ESD shutdown.

**OR**

The Compliance Group will develop a new Field Operator orientation that includes the communication of the identified policies. Training in these policies will be added to the Field Operator training matrix and tracked in the training system. The process flow for new policy development will be routed by the Compliance Group Leader for determining if new policies or revisions impact the safety of operations. If so the new policy or revisions will be communicated in the orientation.

## Summary

An expert system meeting the criteria outlined can be used to audit the human performance side of process safety proactively, identify significant issues, provide the structure to uncover both specific and generic root causes, and assist with the development of corrective actions to prevent and/or minimize process safety-related incidents. At Saudi Aramco Upstream (and other Business Lines) TapRooT® is being used for this purpose.

19

أرامكو السعودية
saudi aramco

أرامكو السعودية
saudi aramco

# Questions?

where energy is opportunity™

# Event & Conditions Chart

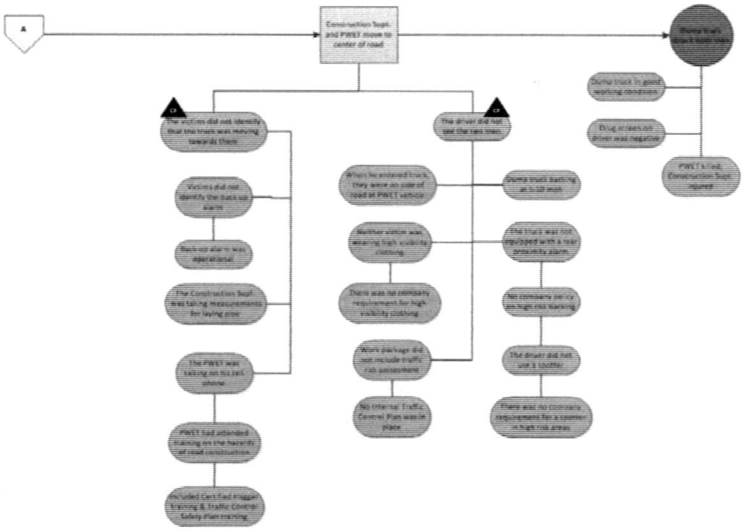

21

أرامكو السعودية
saudi aramco

# Causal Factor

An error or equipment failure that, if corrected, could have prevented the incident from occurring or would have significantly reduced its consequences

22

أرامكو السعودية
saudi aramco

# Root Cause

A root cause is the absence of best practices or the failure to apply knowledge that would have prevented the problem or significantly reduced the consequences.

# Generic Cause

A systemic problem that allows root causes to exist.

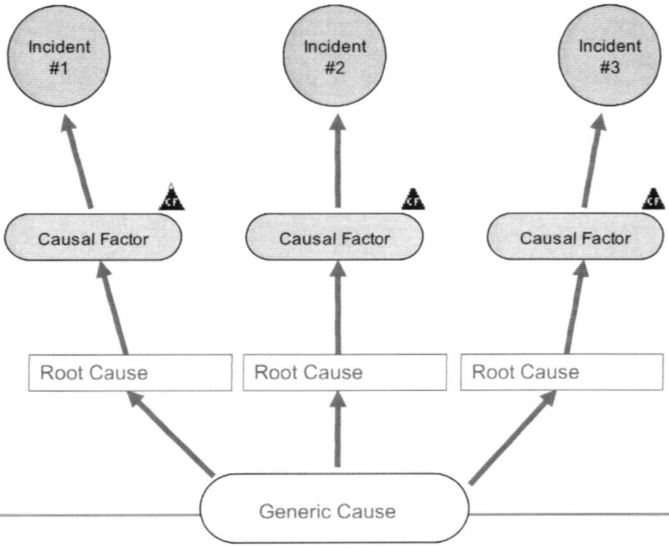

# Word with "F"...

FINISHED FILES ARE THE
RESULT OF YEARS OF
SCIENTIFIC STUDY COMBINED
WITH THE
EXPERIENCE OF YEARS.

Answer: SIX (6)

أرامكو السعودية
saudi aramco

# The Impact of Uncontrolled Human Factors on Process Safety

Abdulaziz Alajlan
05/12/2016

where energy is opportunity™

## Outline

- Definition of Human Factors
- Objectives of Addressing Human Factors
- Human Failure Types
- Human Factors Key Elements
- Physical Ergonomics
- Impact of human failures to Organizations
- Lines of Defence Against Human Failure
- The Role of Management of Change (MOC)

## What is Human Factors ?

- Human Factors is the integration and application of the scientific knowledge about:
  - Management Systems
  - People
  - Facilities

  to improve their interaction in the workplace
- Integration of the design achieved through applying knowledge of the capabilities, limitations, and needs of people to the:
  - Facilities layout
  - Equipment and tools used
  - Tasks performed (manual handling, info processing, required skills)
  - Work environment (temp, noise)

## Human Factors Contribution

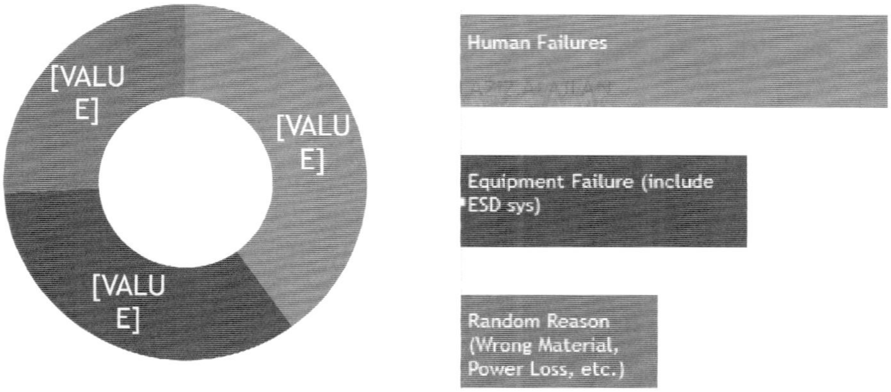


Source: TNO investigations of 216 accident

## Benefits of Addressing Human Factors

- Improved safety & health
  - Reduced both probability and consequences of incidents
- Anticipate and respond to internal and external expectations
  - Government Regulations
  - Litigation
- Improved job effectiveness / efficiency
  - Minimize errors, less downtime & higher quality
- Maximize Profit by Preventing Incidents



## Human Failure Types

| Type of Error | Control Measures |
|---|---|
| • Action Error (associated with familiar tasks)<br>- Slip (little attention required)<br>- Lapse (short term memory required) | ▪ Robust Design and sufficient time<br>▪ Checklist and cross check<br>▪ Distraction Removal |
| • Thinking Error (involving decision making failures)<br>- Mistake (Rule-Based)<br>- Mistake (Knowledge-Based) | ▪ Proper planning and exercise drills<br>▪ Clear Overview<br>▪ Competency and training |
| • Non-Compliance Error (deviation from rules)<br>- Routine<br>- Situational<br>- Exceptional | ▪ Risk Perception<br>▪ Organization Culture<br>▪ Effective Supervision |

## Human Factors Key Elements

- Workplace Design
  - Facility Layout
  - Accessibility
- Equipment Design
  - Control Panels
  - Hand tools
- Work Environment
  - Vibration - Lighting
  - Noise - Temperature
- Physical Activities
  - Force - Repetition
  - Posture
- Job Design
  - Behavior Based Safety
  - Workload - Work Schedule
- Information Transfer
  - Procedures - Instructions
  - Labels
- Personal Factors
  - Fatigue - Stress
  - Motivation



## Physical Ergonomics

- Ergonomics is a field that Study both the person and the job/task. It aims to provide comfort and safety of the human body by adjusting tools, work stations and work processes to fit the worker.

- Major Consequences
  - Musculoskeletal Disorders
    - Carpel Tunnel Disorder
    - Back Pain
- Evaluation and Analysis
  - Task Analysis
    - RULA
    - REBA

# Impact of Human failures

- There are direct and indirect impact of human errors
  - Financial Impact

    "If you want to maximize profit, stop losses"

Contribution of Human Errors in Major Disasters

Example: PIPER ALPHA platform in North Sea. Explosion killed 167

| | | |
|---|---|---|
| • Improper location of the valve | → | • Slip action error |
| • Routes to lifeboats blocked | → | • No person to take control |
| • Fire resistant wall not blast resistant | → | • Routine violation |



# Lines of defense

- ➢ Elimination
- ➢ Substitution
- ➢ Engineering Approach
- ➢ Administrative or Work Practice Control
- ➢ Personal Protective Equipment

## The Role of Management of Change (MOC)

- MOC play a major role in closing the gap and ensure smooth transfer of information.
- Management of Change defined as a process to ensure that changes that would take operating conditions outside the safe range are properly evaluated, and that appropriate safeguards are implemented.
- In every case, the fundamental reason for the accident is that someone, somewhere lost control of the operation, *i.e.*, they allowed operating conditions to deviate beyond their safe range.
- If changes are properly controlled, incidents will be minimized.



## Conclusion

- Human Factors is the integration of Management System, Facility and people to improve the interaction in the workplace.
- Human Factors tend to improve health & safety and increase job efficiency and employee effectiveness.
- Facility, job and equipment design as well as personal factors and physical activities are key elements playing a major role in human performance.
- Incidents are multi causal and human factors are one of the common causal factors.
- Incidents can be prevented if procedures are properly set and effectively followed.

## Questions

THE KAITEKI COMPANY
Mitsubishi Chemical Holdings Group

# Human Factors in Chemical Process Safety


**Masaki Nakagawa**

**Mitsubishi Chemical Corporation**
**Process Safety Technology Department**
**Environment, Safety and Quality Department**
**1-1, Marunouchi 1-chome, Chiyoda-ku**
**Tokyo 100-8251, JAPAN**


1. Introduction
2. Objective
3. Variation Tree Analysis (VTA) and Why-Why analysis
4. Education program
5. Conclusion

MITSUBISHI CHEMICAL

## Outline

THE KAITEKI COMPANY
Mitsubishi Chemical Holdings Group

1. Introduction
2. Objective
3. Variation Tree Analysis (VTA) and Why-Why analysis
4. Education program
5. Conclusion

MITSUBISHI CHEMICAL

THE KAITEKI COMPANY

# Organization

2015.10.13

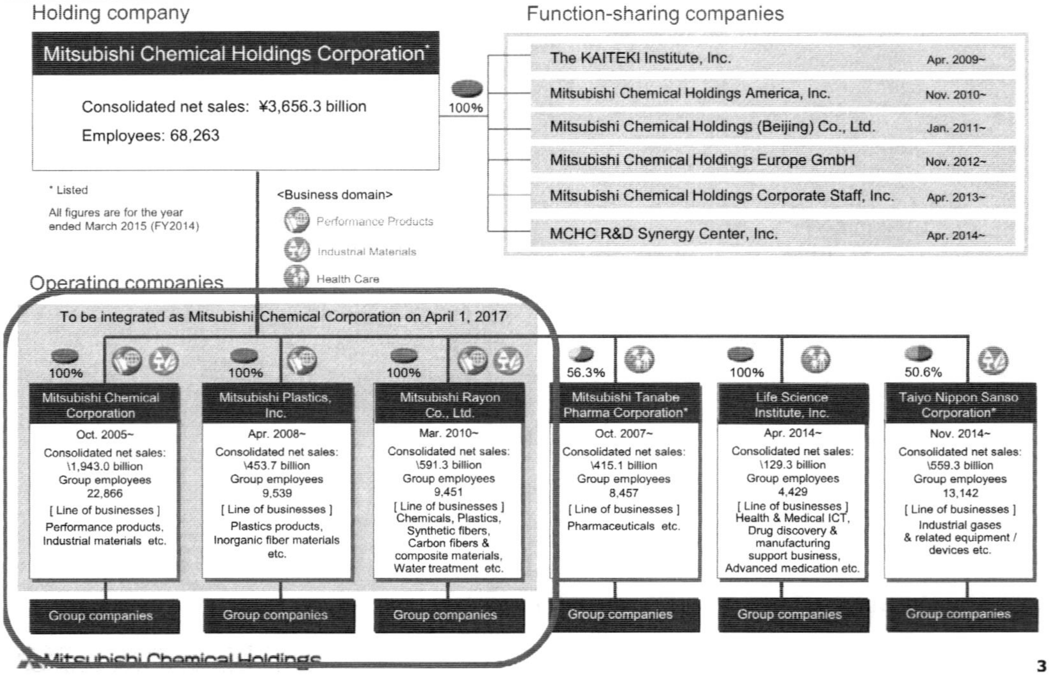


Mitsubishi Chemical Holdings

3

# Outline

THE KAITEKI COMPANY
Mitsubishi Chemical Holdings Group

1. Introduction
2. Objective
3. Variation Tree Analysis (VTA) and Why-Why analysis
4. Education program
5. Conclusion

MITSUBISHI CHEMICAL

# Objective

THE KAITEKI COMPANY
Mitsubishi Chemical Holdings Group

Recently, the number of accidents in the industry is increasing in Japan. One of the causes is human factors. It is important to analyze root causes of accidents when they occurred. Variation Tree Analysis (VTA) and Why-Why analysis are often used to analyze an accident due to human factors but there are some problems that it takes much time and manpower to analyze an accident. Therefore, these methods need to be improved.

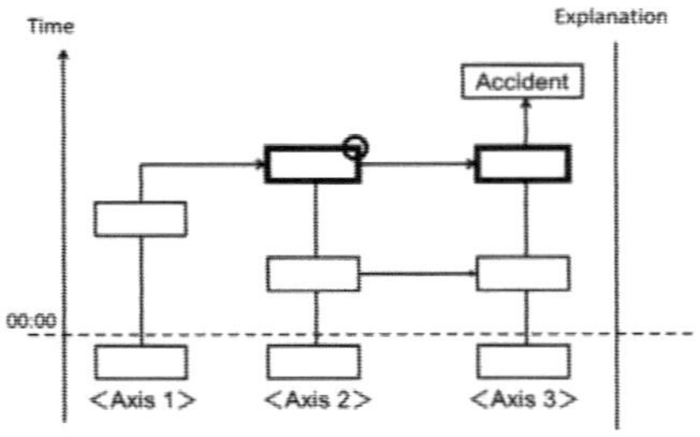


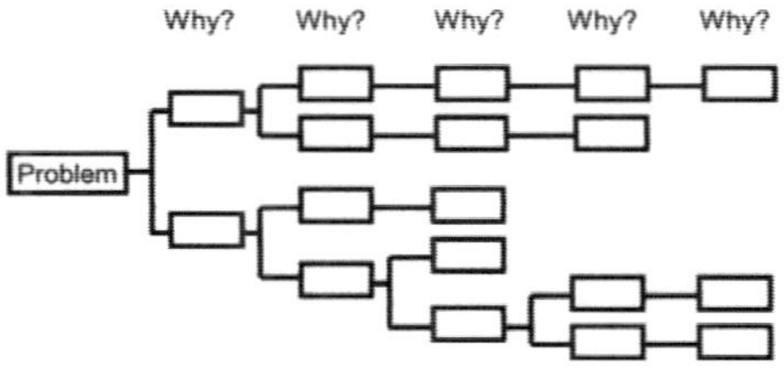


MITSUBISHI CHEMICAL

# Outline

THE KAITEKI COMPANY
Mitsubishi Chemical Holdings Group

1. Introduction
2. Objective
3. Variation Tree Analysis (VTA) and Why-Why analysis
   - 3-1. What is VTA and Why-Why analysis?
   - 3-2. Investigation Result
   - 3-3. Case Study
4. Education program
5. Conclusion

MITSUBISHI CHEMICAL

# What is VTA ?

J.Rasmussen

VTA ( Variation Tree Analysis) is the method to employ time series analysis of the circumstances at the time of an accident, with particular emphasis on the actions and judgments of human. The progress of the accident along the time series is illustrated in a VTA. A situation, work, judgement or behaviour that diverges from the usual condition in the VTA is picked up as an elimination node. After the key-points of an accidents are identified through VTA, the causes of the accident are analyzed by Why-Why analysis.

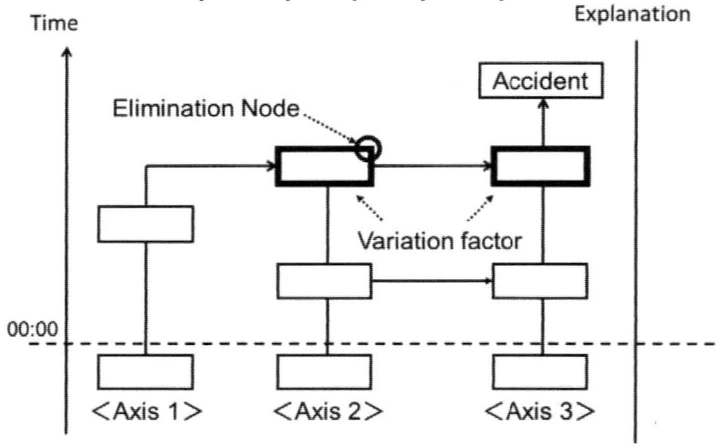


# What is Why-Why analysis ?

Why-Why analysis is the method to identify the root causes of accidents and develop countermeasures by continuously questioning the causes of accidents logically.

< Why-Why analysis procedure >

1) Starting with a problem, asking why it occurred.
2) Then, turning the answer to the first question into a second why question.
3) After that, next answer becomes the third why question and so on. Asking more or less whys until identifying the root cause for the appropriate countermeasures.

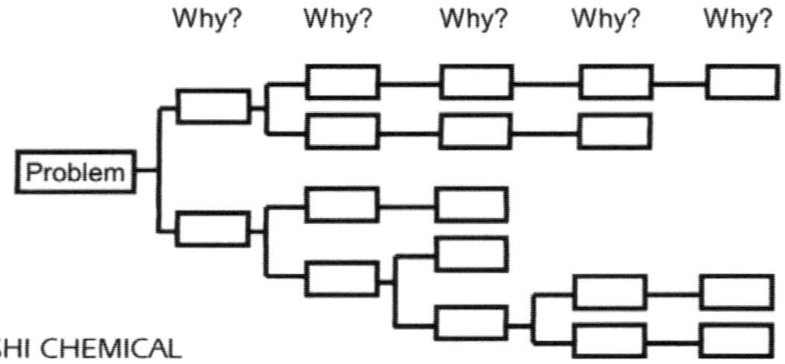


MITSUBISHI CHEMICAL

## VTA and Why-Why analysis

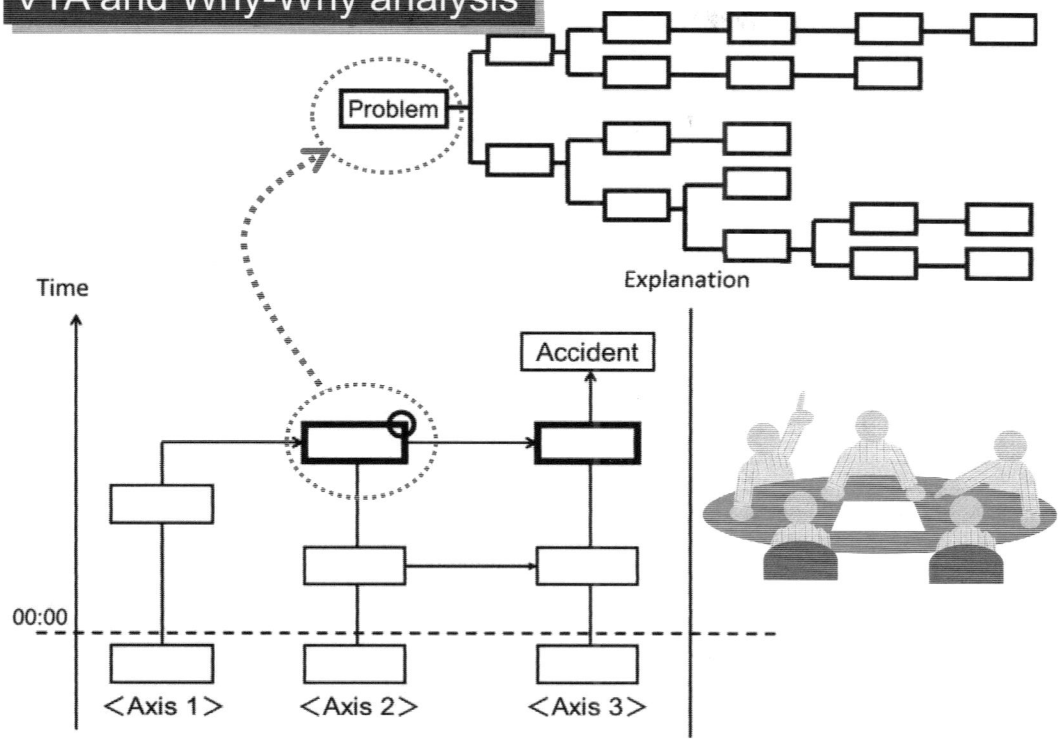


## Outline

THE KAITEKI COMPANY
Mitsubishi Chemical Holdings Group

1. Introduction
2. Objective
3. Variation Tree Analysis (VTA) and Why-Why analysis
   - 3-1. What is VTA and Why-Why analysis?
   - ➡ 3-2. Investigation Result
   - 3-3. Case Study
4. Education program
5. Conclusion

MITSUBISHI CHEMICAL

# Problem of Why-Why analysis

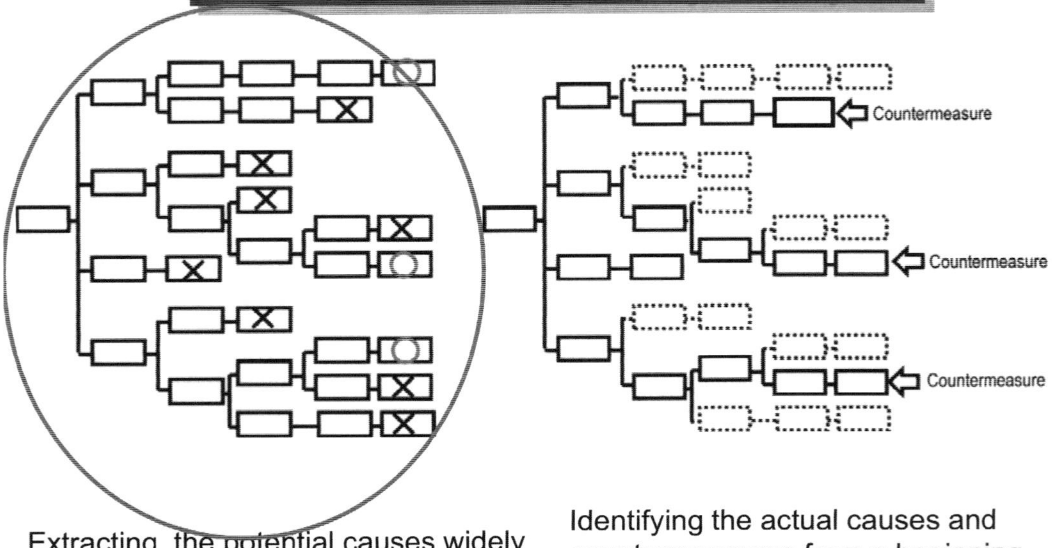


Extracting the potential causes widely

Identifying the actual causes and countermeasures from a beginning of analysis

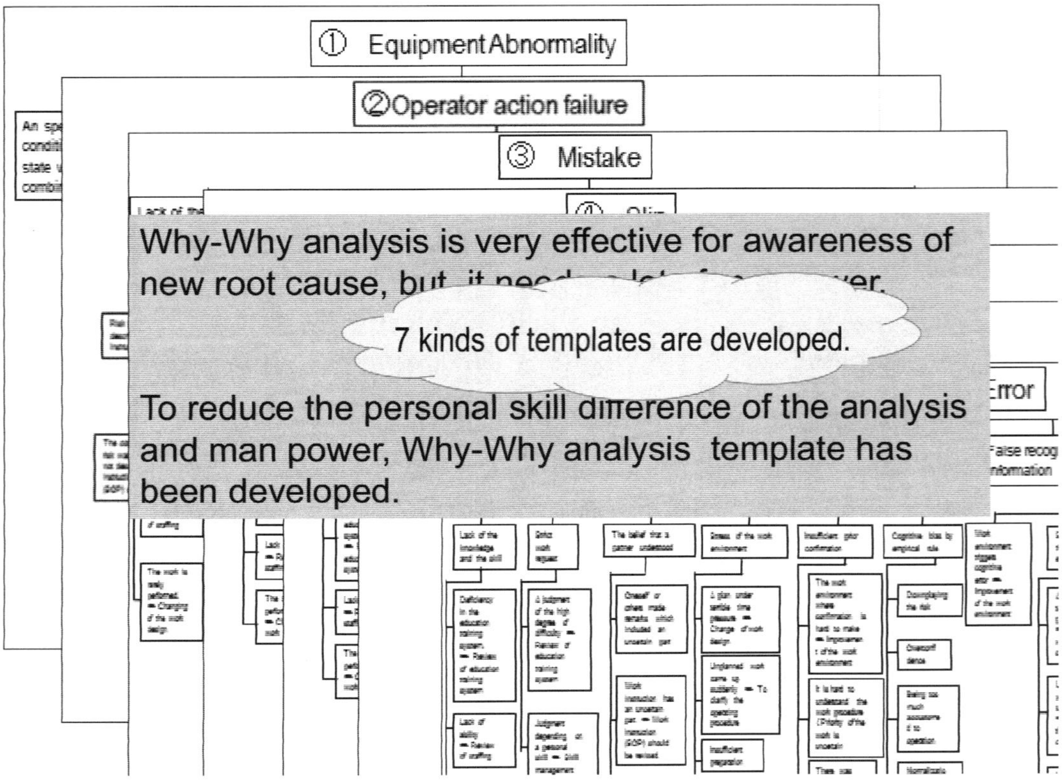

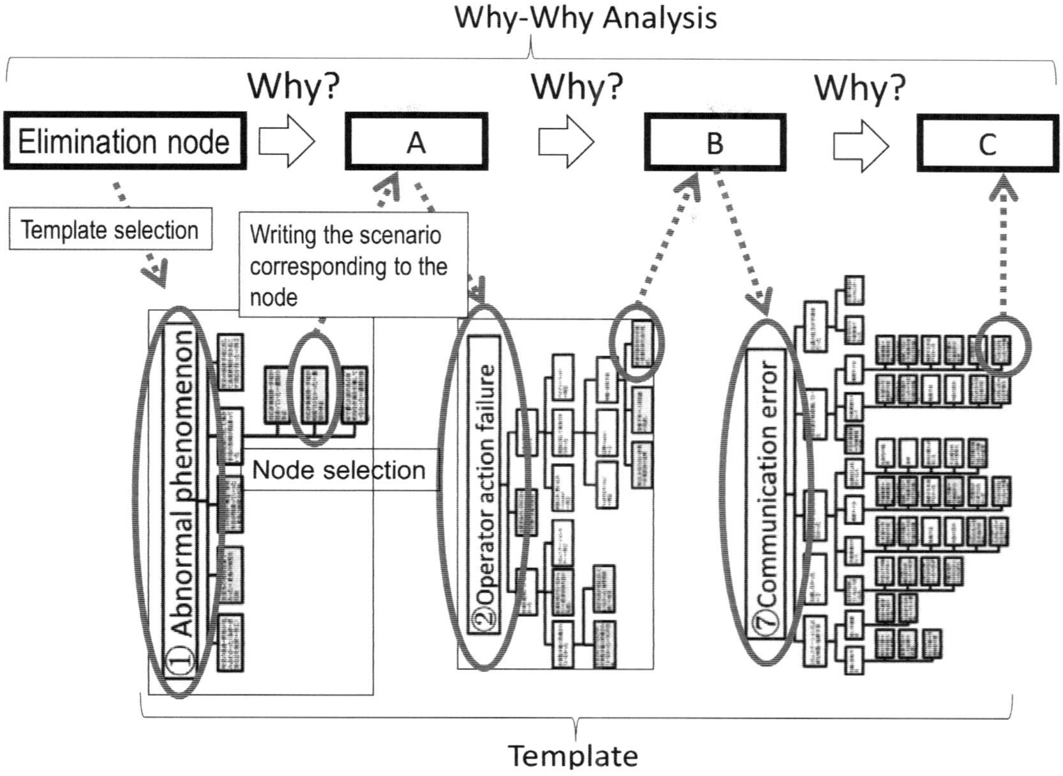


# Outline

THE KAITEKI COMPANY
Mitsubishi Chemical Holdings Group

1. Introduction
2. Objective
3. Variation Tree Analysis (VTA) and Why-Why analysis
   - 3-1. What is VTA and Why-Why analysis?
   - 3-2. Investigation Result
   - ➡ 3-3. Case Study
4. Education program
5. Conclusion

MITSUBISHI CHEMICAL

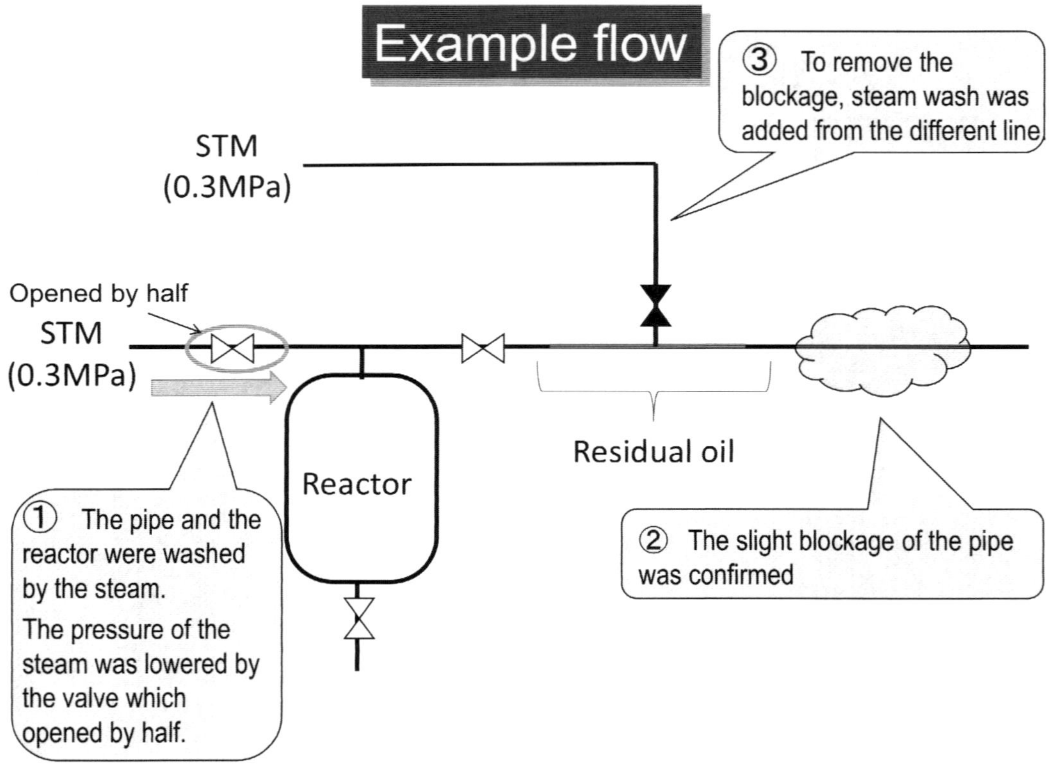
Example flow
③ To remove the blockage, steam wash was added from the different line
STM
(0.3MPa)
Opened by half
STM
(0.3MPa)
Reactor
Residual oil
① The pipe and the reactor were washed by the steam.
The pressure of the steam was lowered by the valve which opened by half.
② The slight blockage of the pipe was confirmed

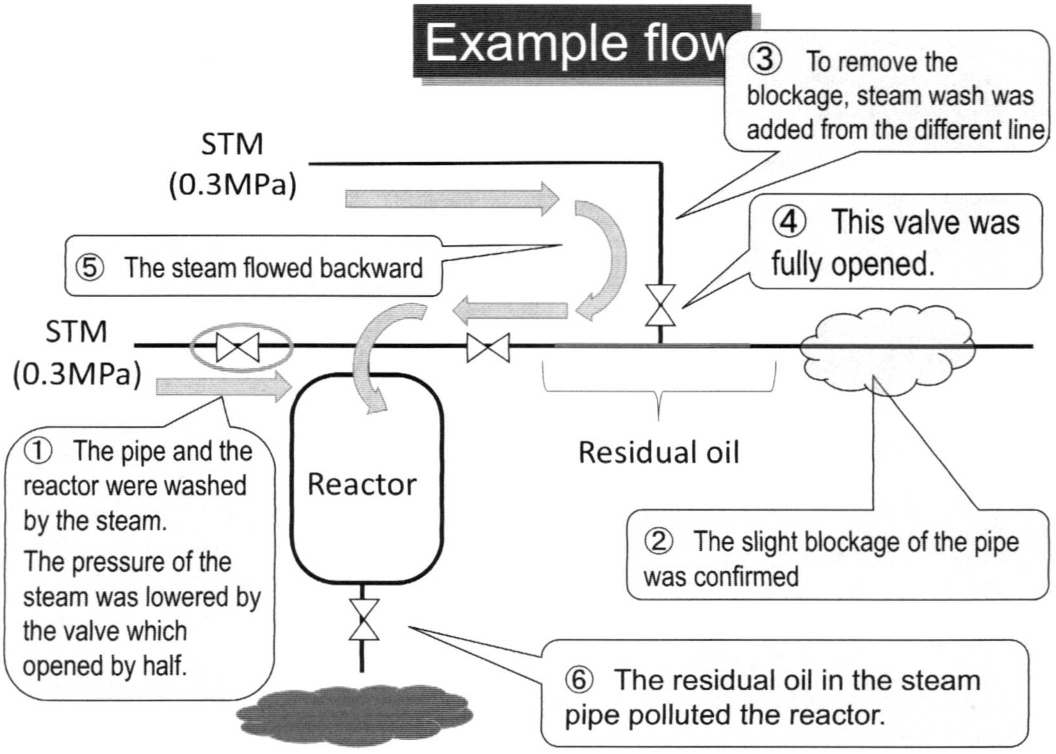
Example flow
③ To remove the blockage, steam wash was added from the different line
STM
(0.3MPa)
④ This valve was fully opened.
⑤ The steam flowed backward
STM
(0.3MPa)
Reactor
Residual oil
① The pipe and the reactor were washed by the steam.
The pressure of the steam was lowered by the valve which opened by half.
② The slight blockage of the pipe was confirmed
⑥ The residual oil in the steam pipe polluted the reactor.

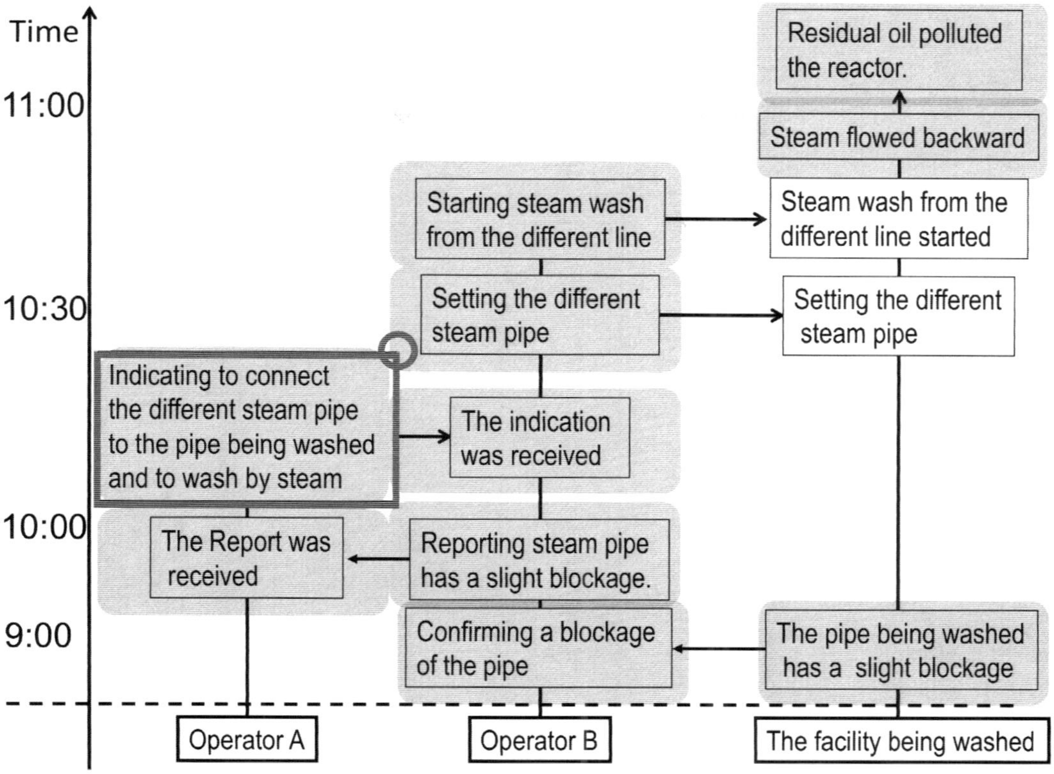
Time
11:00
10:30
10:00
9:00
Residual oil polluted the reactor.
Steam flowed backward
Starting steam wash from the different line
Steam wash from the different line started
Setting the different steam pipe
Setting the different steam pipe
Indicating to connect the different steam pipe to the pipe being washed and to wash by steam
The indication was received
The Report was received
Reporting steam pipe has a slight blockage.
Confirming a blockage of the pipe
The pipe being washed has a slight blockage
Operator A
Operator B
The facility being washed

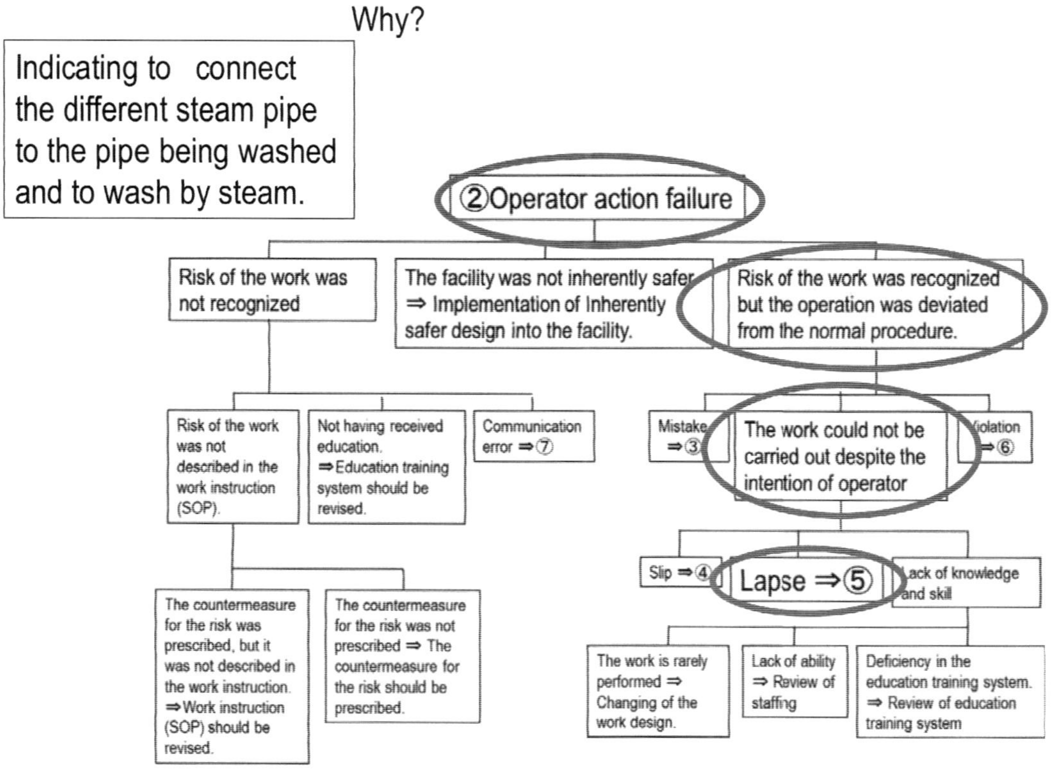
Why?
Indicating to connect the different steam pipe to the pipe being washed and to wash by steam.
②Operator action failure
Risk of the work was not recognized
The facility was not inherently safer ⇒ Implementation of Inherently safer design into the facility.
Risk of the work was recognized but the operation was deviated from the normal procedure.
Risk of the work was not described in the work instruction (SOP).
Not having received education. ⇒Education training system should be revised.
Communication error ⇒⑦
Mistake ⇒③
The work could not be carried out despite the intention of operator
Violation ⇒⑥
The countermeasure for the risk was prescribed, but it was not described in the work instruction. ⇒Work instruction (SOP) should be revised.
The countermeasure for the risk was not prescribed ⇒ The countermeasure for the risk should be prescribed.
Slip ⇒④
Lapse ⇒⑤
Lack of knowledge and skill
The work is rarely performed ⇒ Changing of the work design.
Lack of ability ⇒ Review of staffing
Deficiency in the education training system. ⇒ Review of education training system

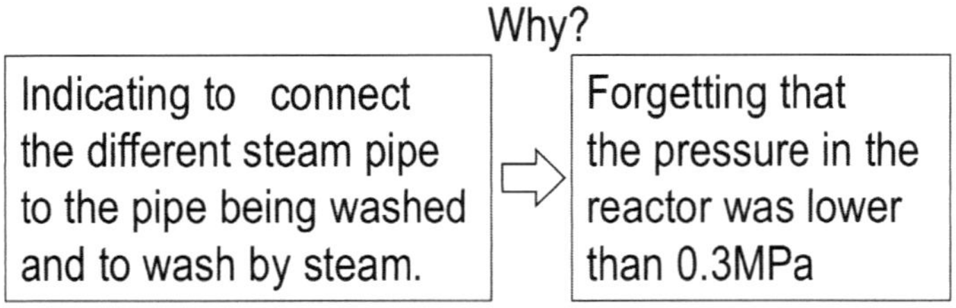
Why?
Indicating to connect the different steam pipe to the pipe being washed and to wash by steam.
Forgetting that the pressure in the reactor was lower than 0.3MPa

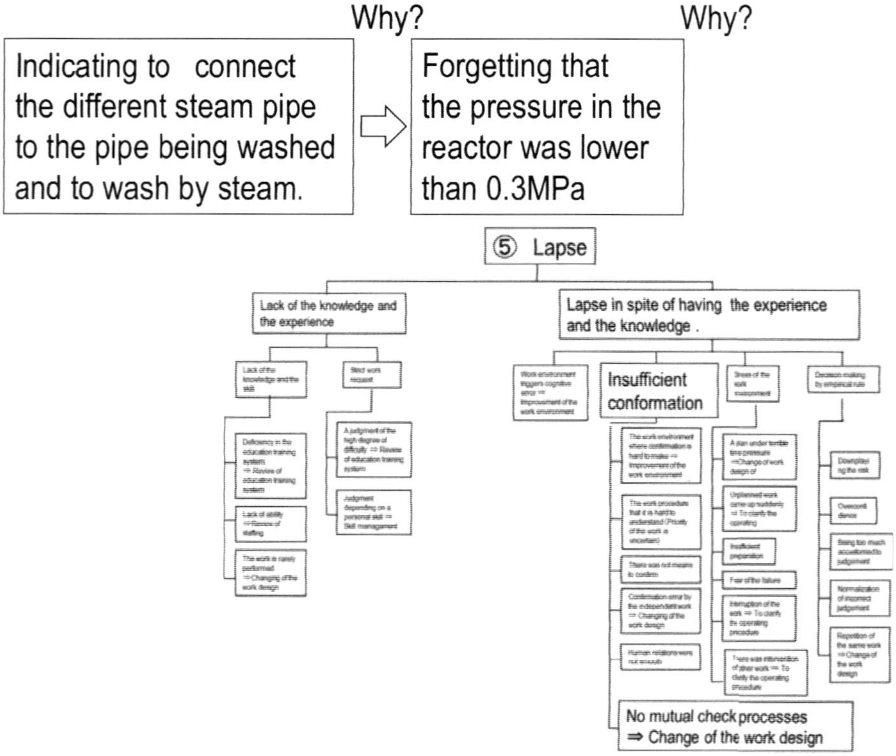
Why?
Why?
Indicating to connect the different steam pipe to the pipe being washed and to wash by steam.
Forgetting that the pressure in the reactor was lower than 0.3MPa
⑤ Lapse
Lack of the knowledge and the experience
Lapse in spite of having the experience and the knowledge .
Insufficient conformation
No mutual check processes
⇒ Change of the work design

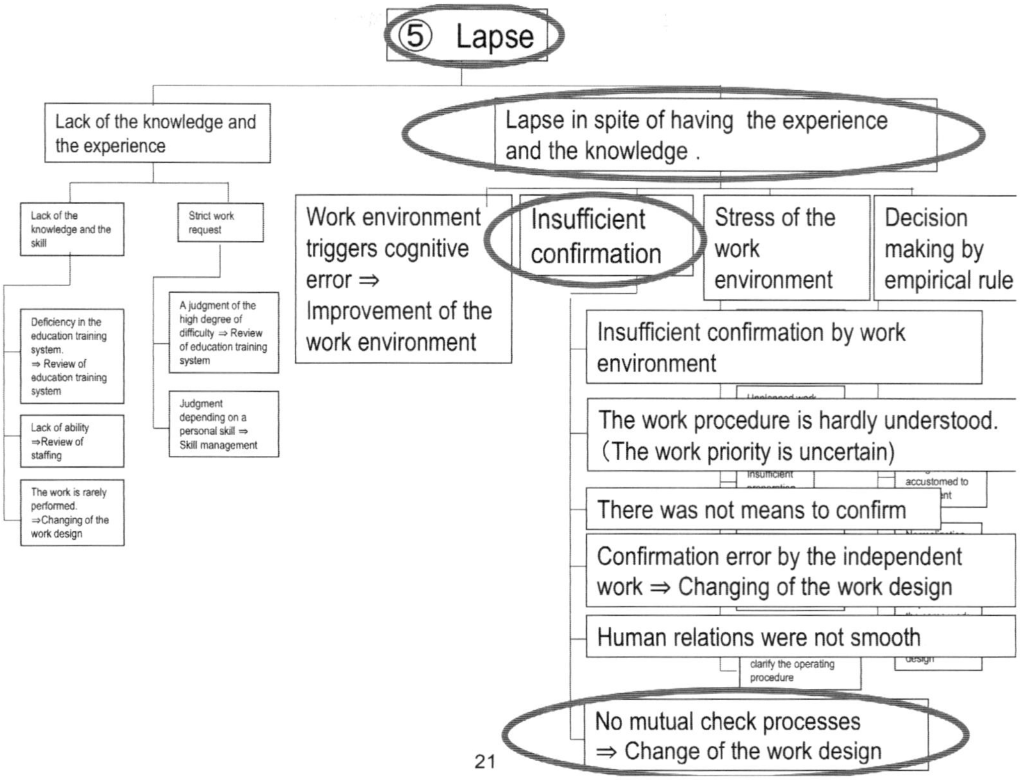


Why? Why?

| | | | | |
|---|---|---|---|---|
| Indicating to connect the different steam pipe to the pipe being washed and to wash by steam. | ⇨ | Forgetting that the pressure in the reactor was lower than 0.3MPa | ⇨ | Not holding a meeting for safe operation before work |

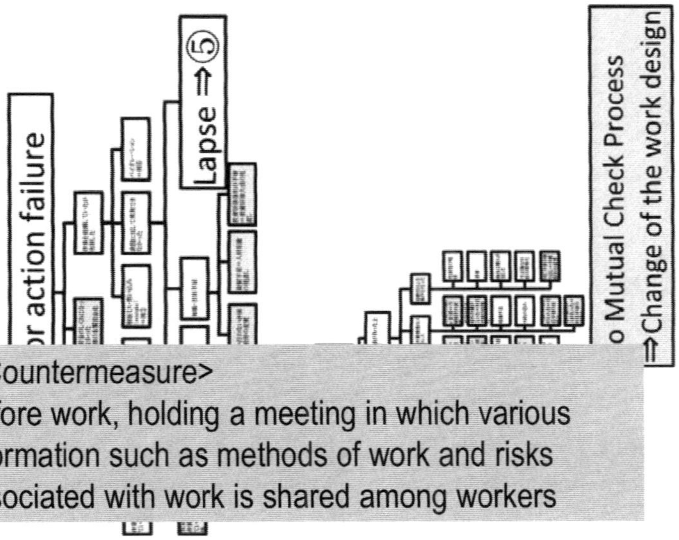


< Countermeasure>
Before work, holding a meeting in which various information such as methods of work and risks associated with work is shared among workers

# Outline

THE KAITEKI COMPANY
Mitsubishi Chemical Holdings Group

1. Introduction
2. Objective
3. Variation Tree Analysis (VTA) and Why-Why analysis
4. Education program
   - 4-1. Process Safety Education Program
   - 4-2. CPSE Education Program
   - 4-3. Practical Safety Training
   - 4-4. Knowledge and Experience Transfer
5. Conclusion

MITSUBISHI CHEMICAL

# Investigation of Accident reports

| < Report > | < Accident factors > | < Countermeasures > |
|---|---|---|
| 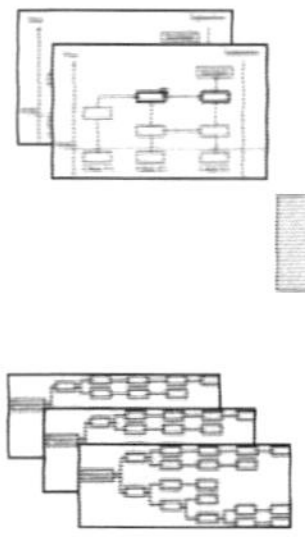 | Plant & management design | Improvement of the design of manufacturing process and maintenance management |
| | Operational design | |
| | Understanding work | Improvement of the management of the SOP and the skill of the operating staff |
| | Collaborative work: | |
| | Personal work: | Improvement of reliability of the individual work execution by operating staff |

It is necessary to take countermeasure by assuming that human error is inevitable. Therefore, to reduce occurrence of accident, it is necessary to change the management style from depending on human factor to depending on inherently safer design or functional safety.

## Investigation of Accident reports

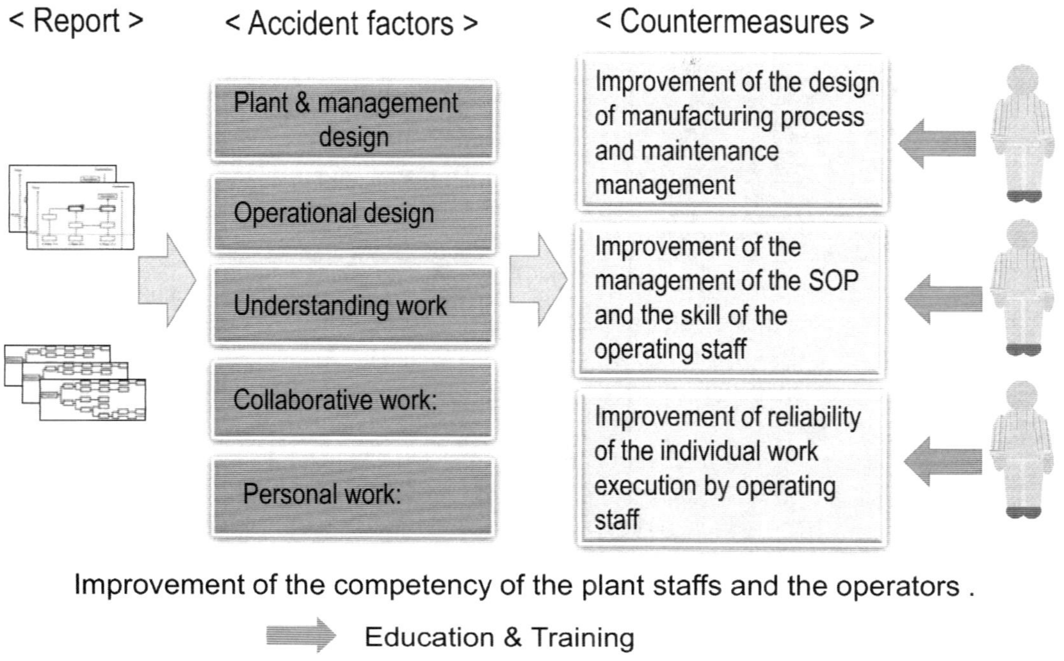


## Process Safety Education Program

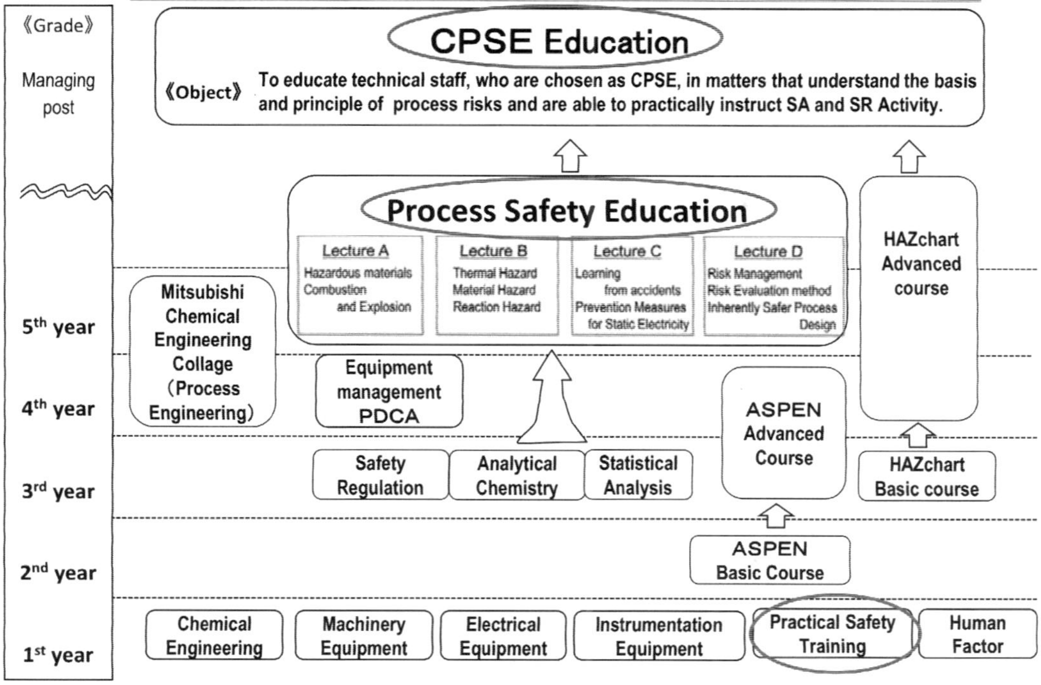

# Process Safety Education

THE KAITEKI COMPANY
Mitsubishi Chemical Holdings Group

<Process Safety Education program>

The safety education class in session

Basic course
Classroom lecture
20-30 attendee (per time, at a plant site)
Middle-tire engineers
12 hours

Objective: To systematically educate mid-career technical staff, who are the core of SA and SR, in matters that form the basis of process safety

| Lecture A | Lecture B | Lecture C | Lecture D |
| --- | --- | --- | --- |
| Hazardous materials<br>Combustion and Explosion | Thermal Hazard<br>Material Hazard<br>Reaction Hazard | Learning from accidents<br>Prevention Measures for Static Electricity | Risk Management<br>Risk Evaluation method<br>Inherently Safer Process Design |

MITSUBISHI CHEMICAL

# CPSE Education

< Chemical Process Safety Engineer (CPSE) program >

Training for CPSE
conducted by visiting instructor

Advanced course
Small group seminars
5-6 attendees (per time, at a plant site)
Chosen engineers

Objective: To educate technical staff, who are chosen as CPSE, in matters that understand the basis and principle of process risks and are able to practically instruct SA and SR.

Feature: Dealing with the theme of process safety in actual workplaces
Lectures centered on rules and principles
Providing solutions through discussions with instructors

Example of Challenging theme:
Safety assessment of the reactor
Evaluation of dust explosion in the pipeline
An investigation of the cause of past accidents and problems

# CPSE Education

THE KAITEKI COMPANY
Mitsubishi Chemical Holdings Group

A report briefing
at Safety Technical Review meeting

A CPSE receives a certificate
of authorization from
the CPSE training instructor

Participant's theme is deepened further through discussions with attendees at the meeting.

53 employees have received CPSE education thus far.

MITSUBISHI CHEMICAL

# Outline

THE KAITEKI COMPANY
Mitsubishi Chemical Holdings Group

1. Introduction
2. Objective
3. Variation Tree Analysis (VTA) and Why-Why analysis
4. Education program
   4-1. Process Safety Education Program
   4-2. CPSE Education Program
   ➡ 4-3. Practical Safety Training
   4-4. Knowledge and Experience Transfer
5. Conclusion

MITSUBISHI CHEMICAL

# Education center
## -Mizushima Plant site-

THE KAITEKI COMPANY
Mitsubishi Chemical Holdings Group

MITSUBISHI CHEMICAL

# Practical safety training

When entering Mitsubishi Chemical, all plant staffs and operators have to receive Practical Safety training.
All plant operators have to receive Practical Safety training every 5 or 6 years to keep safety awareness.

Liquid removal and cleaning experience equipment

Electrical equipment

Instrumentation facility

Spouting and splashes experience equipment

Cut-view models

Hot water circulation facility (small scale plant)

## Practical Safety Training using Miniature Hot Water Recycling System

This training aims to learn the basics of starting up, normal operation and suspension of the system and also to secure stable operation and safety through various problematic phenomena.

**1. Cavitation**

**2. Increasing and decreasing pressure of an atmospheric tank**

**3. Water hammer**

**4. Liquid seal in the piping**

## Fire and explosion experience training

Experiment with the Electrically charged body

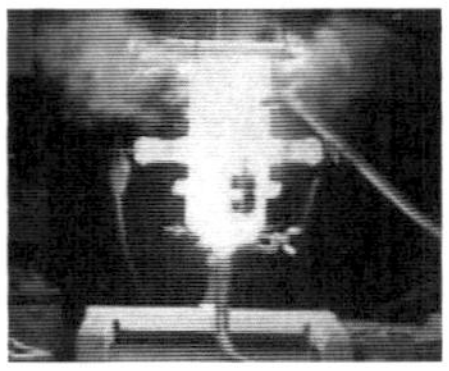

Dust Explosion

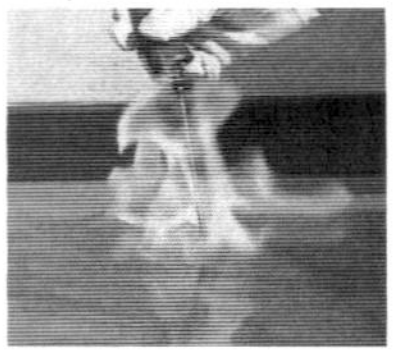

Ignition of a flammable liquid by static electricity

Objective :

To experience various forms of combustion and explosion phenomena
To increase awareness regarding fire and explosions
To utilize this knowledge to prevent accidents.

# Outline

THE KAITEKI COMPANY
Mitsubishi Chemical Holdings Group

1. Introduction
2. Objective
3. Variation Tree Analysis (VTA) and Why-Why analysis
4. Education program
   - 4-1. Process Safety Education Program
   - 4-2. CPSE Education Program
   - 4-3. Practical Safety Training
   - ➡ 4-4. Knowledge and Experience Transfer
5. Conclusion

MITSUBISHI CHEMICAL

# Knowledge and Experience Transfer
## Safety Review (SR) Activity

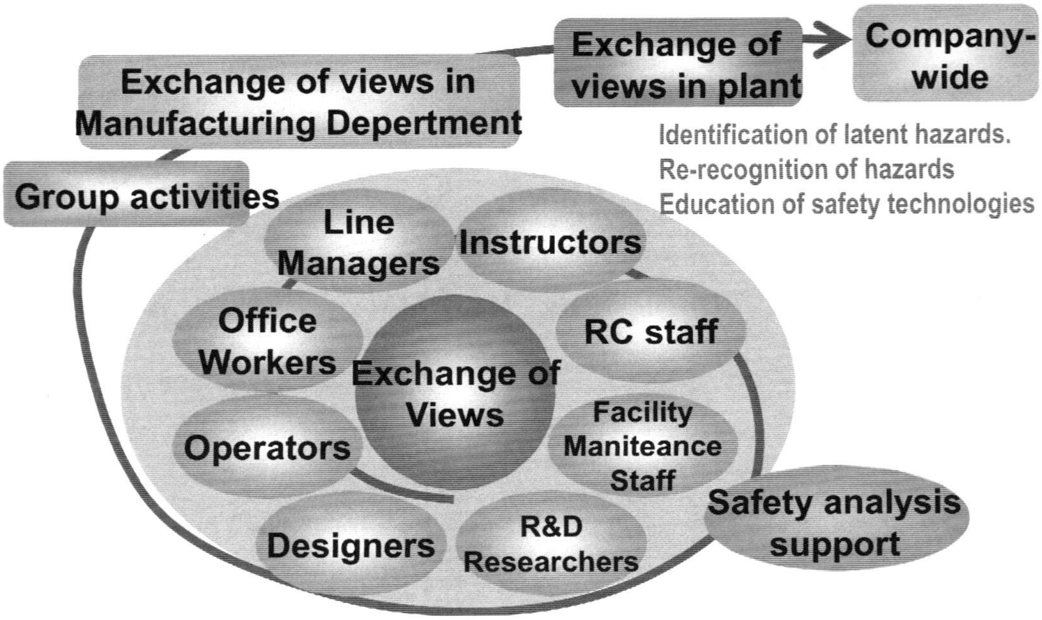

## Knowledge and Experience Transfer
SR instructor

## Knowledge and Experience Transfer
SR instructor


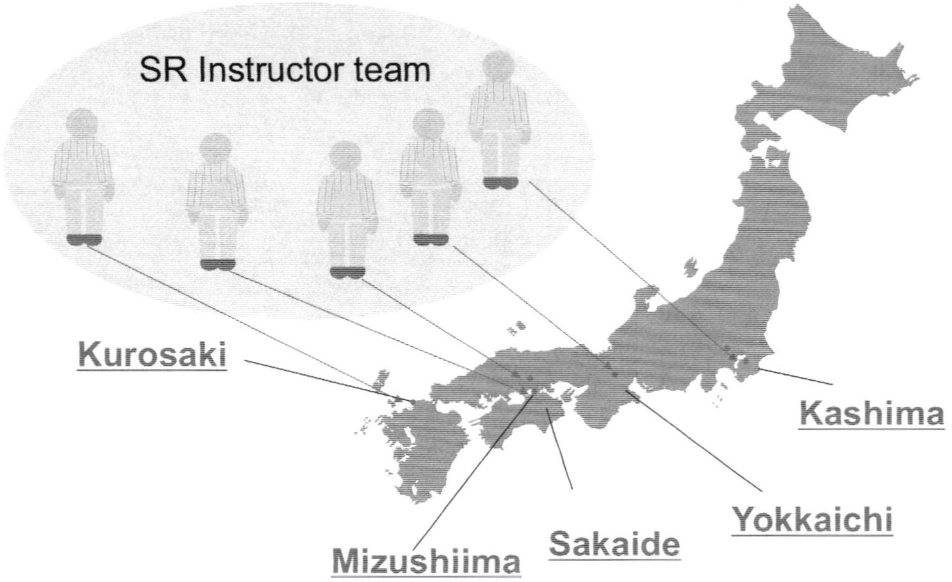

# Outline

THE KAITEKI COMPANY
Mitsubishi Chemical Holdings Group

1. Introduction
2. Objective
3. Variation Tree Analysis (VTA) and Why-Why analysis
4. Education program
5. Conclusion

MITSUBISHI CHEMICAL

# Conclusion

THE KAITEKI COMPANY
Mitsubishi Chemical Holdings Group

Variation tree analysis and Why- Why analysis procedures are effective for analysis of human factors. To reduce the personal skill difference of the analysis and man power, Why-Why analysis template has been developed.

To improve competency of the plant staffs and operators,

- Process Safety Education
- CPSE Education
- Practical Safety Education
- SR activity

are provided to prevent accident.

MITSUBISHI CHEMICAL

THE KAITEKI COMPANY
Mitsubishi Chemical Holdings Group

# Thank you for your attention!

## Questions?

MITSUBISHI CHEMICAL

saudi aramco
research & innovation

# Fire at a Refinery
## (a Process Safety Fatality)?

Mark Fields - Loss Prevention Engineer
4-5/12/2016

where energy is opportunity™

## Presentation Outline

- WHY ?
- Immediate Cause and Outcome
- The Response and Initial (at the scene) recovery
- Lead up to the Fire
- Process safety – it's about people
- The Investigation Findings & Root Causes
- Lessons Learned
- Summary

## WHY?

- My role in this fatality
- Process safety is as much about people as it is about tools and methodologies
- Having a process safety vision, the right process safety tools and systems also requires the full support of management and the buy in of all stakeholders. Without this, process safety related incidents will continue to occur
- This presentation will look at one process safety incident, and consider the process safety related / human interactions that were pre-cursors to this tragic and costly incident




## Immediate Cause and Outcome

### Global Energy Firm fined £1.4m after worker killed in refinery fire

**Date:**
29 July 2015

[redacted] have been fined £1.4 million after a major fire led to the death of a worker at an oil refinery in North Lincolnshire.

Twenty-four year old Robert Greenacre from Grimsby was working near a crude oil distillation unit just before the fire broke out [redacted] in Immingham on 29th June 2010.

The Health and Safety Executive prosecuted the oil giant after an investigation found a fire was caused by an uncontrolled release of crude oil.

Hull Crown Court heard today (29 July) that Mr Greenacre who was a contracted fitter was working with a colleague beneath a distillation column containing hot crude oil. The job required them to open equipment beneath the column. When an item of equipment was opened, the crude oil was released. A short time later it ignited. Mr Greenacre's colleague was able to escape the scene suffering minor burns, Robert did not escape and died beneath the column.

The court also heard operators of major accident hazard establishments must have in place a functioning system of risk assessment for all tasks where hazardous substances could be released.

# Immediate Cause and Outcome

- Double Block and Bleed – re-inserting blind in steam out line beneath crude distillation column (23C-1).

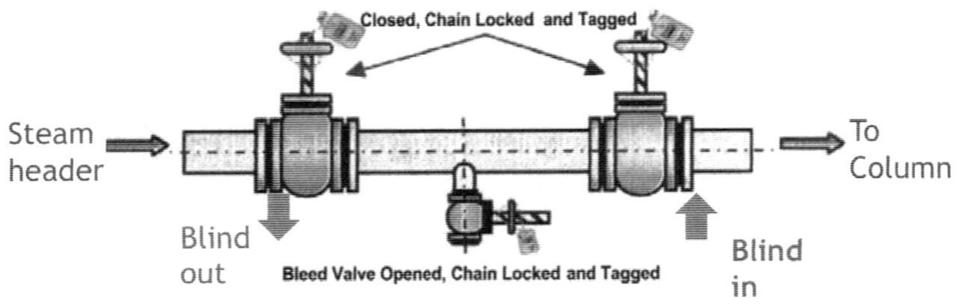




# The Response and Initial (at the scene) recovery

# The Response and Initial (at the scene) recovery

- Emergency call and quick response by the site fire department
- Fire containment and spread prevention + **(asbestos)**
- Accounting for all personnel (mustering/exiting the site)
- Realization of 1 missing person
- Casualty recovery – specialist police forensic recovery team



Lead up to the Fire – not as simple as it seems

- State of the plant and the refinery

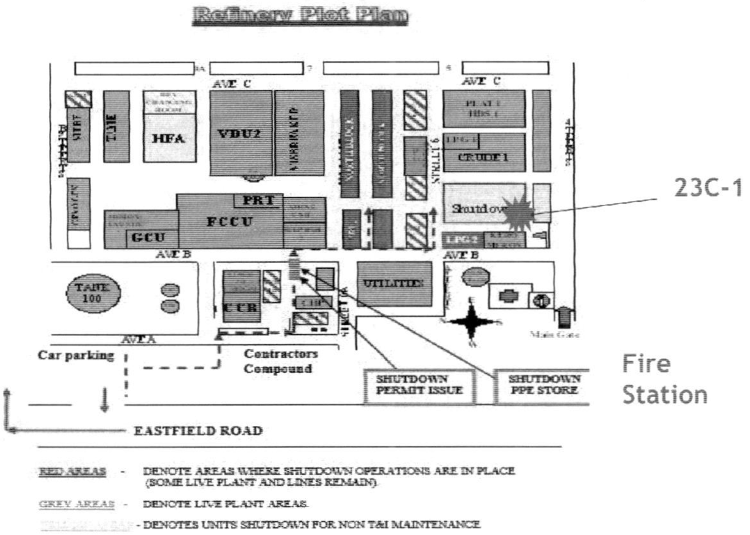

## Lead up to the Fire

- Graphic of re-establishing levels in a crude column, and heating the crude on recirculation.

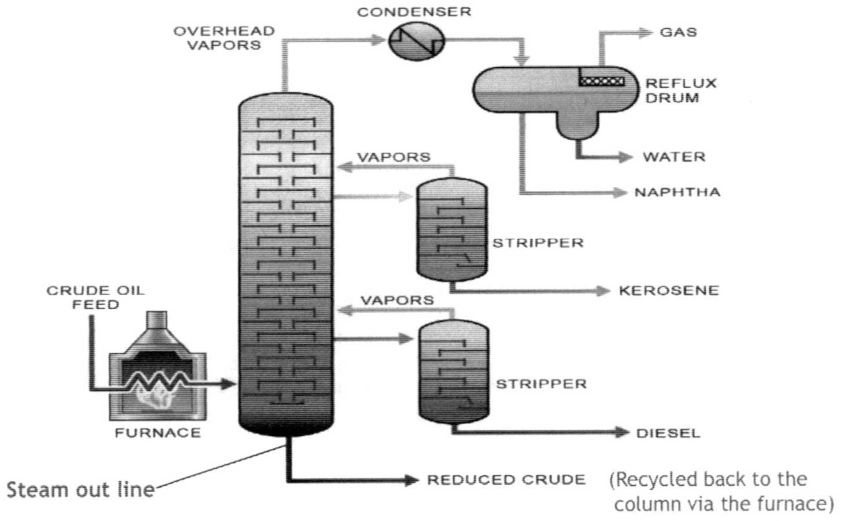





## Lead up to the Fire

- Work permit receivers and supervision (task held until the day shift)
- Procedure (Process) for establishing levels in the column
- Plant isolation / blinding - recent site history (HSG-253) - SIOPE
- Operations coverage - manning study (outside operators)
- Safety coverage
- **FIRE** - engulfs the base of the crude column at around 11:00am

## Process Safety – it's about people

- Plant management (shutdown coinciding with the T&I)
- Shift superintendent (decision making on night shift)
- Shift supervisor (permit issuer)
- Craft supervisor (permit receiver)
- Shift outside operator (retiree)
- Work crew (the deceased and his colleague)






## The Investigation Findings & Root Causes

- HSE lead investigation - area secured - statements taken under oath – prohibition and improvement notices served
- Procedure for establishing levels not followed
- Manual intervention (blinding) error
- SIOPE procedure not audited/reviewed
- Slip Blind found in the debris – Crude 1 (adjoining unit) had a spectacle blind

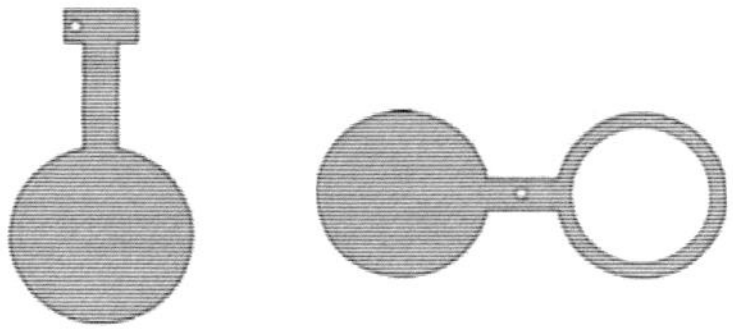

## The Investigation Findings & Root Causes

- Staffing levels - shutdown coincided with a T&I
- A different company were installing the blind
- Procedure for establishing levels in the column not followed - stop point for the blind installation
- Line broken on the wrong side of the block and bleed - bolts were removed by a hacksaw and wedge used to spread the line
- Why did this not seem strange for re-inserting a blind that had been removed just a couple of weeks earlier?
- Court ruling took a long time!



## Lessons Learned

- Startup procedures for the column - not followed
- SIOPE procedures - if not followed diligently can lead to major process safety type incidents
- Keeping products inside lines and vessels is really the most important thing (Lord Cullen _ Piper Alpha 25 years on)
- Process safety tools, - shutdown / startup procedures, work permits / risk assessments / JSA / pre hazard checklists ***do not*** make a job safe
- They support PEOPLE in completing tasks safely

## Summary (what can we learn from this tragic and costly event)?

- Processes and procedures are developed to enable safe operations and safe intervention activities
- Simple and easy to understand processes/procedures are more likely to be followed
- Bad habits and "Custom and Practice," can spread like wildfire
- You don't always get what you expect - you get what you inspect
- ***The Aim*** - Competent, well trained and motivated staff, following well written and established procedures and supported by good supervision, make the job safe
- **What is Competence?**

***'Competence is the continuing ability of individuals and teams to perform reliably the MAH elements of their roles, responsibilities & tasks, and for this to be demonstrable against standards'*** **- Taken from COMAH.**



## Summary

- Process Safety - Its all about people!
- To reiterate - In our Oil and Gas Industry

**Keeping products inside their intended lines and vessels is really the most important thing !!**

# Thank you

## Any Questions

# Case Study - Thermal Expansion on Non-Blocked Pipelines

Nipun Soni and Anup Elias
Process Engineering

6th December 2016

Creating Value Through Chemistry

## Outline

- Objectives
- Background
- Scope
- Study Approach
- Conclusions

## Objectives

- Case study: Investigated the reasons for thermal expansion in non-blocked 1-Butene pipeline
- Verified design & made simple design changes to eliminate overpressure by thermal expansion

## Background

- Process Safety Analysis is a critical step at conceptual design
  - Flammable fluids with high thermal expansion coefficient
- Thermal (Hydraulic) expansion as per API 521
  - Blocked-in piping, vessel, cold side of exchanger
  - No pressure relief device required as long as piping with fluid in it is not blocked
- Non-blocked pipelines (continuous use)
  - What if volumetric rate of thermal expansion during a hot summer day is greater than outflow of fluid, such as during a startup?

# Schematic

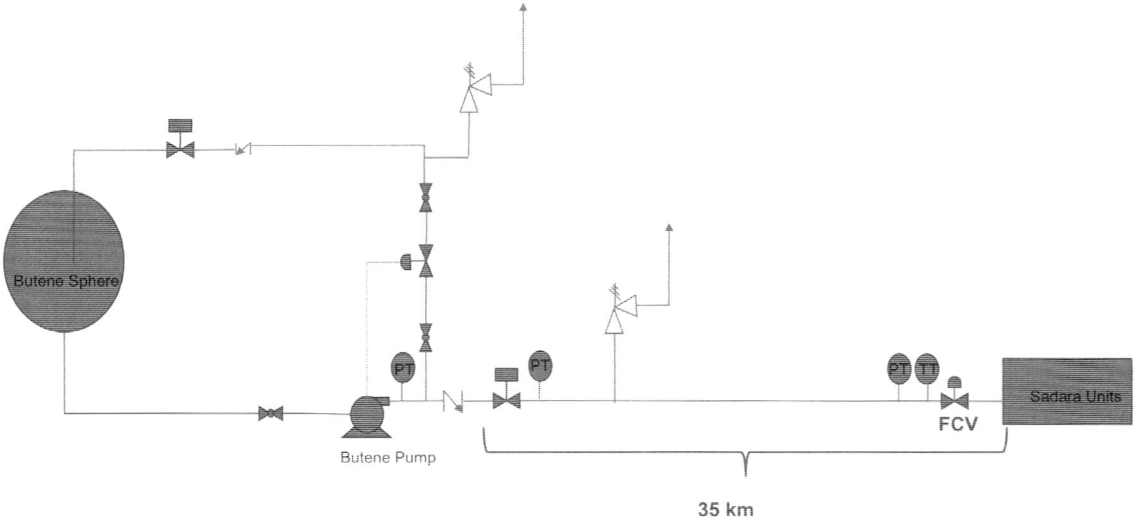


# Methodology

- Average rate of thermal expansion is estimated as a function of temperature rise, pipe volume, fluid densities and time of exposure to sun
  - For this case study we used an actual temperature profile to estimate average rate of thermal expansion
  - For a new project, temperature rise needs to be estimated by performing a heat balance, considering maximum solar radiation heat flux, wind velocity, thermal conductivity, heat transfer co-efficient of fluid and insulation resistance

# Pipeline temperature profile

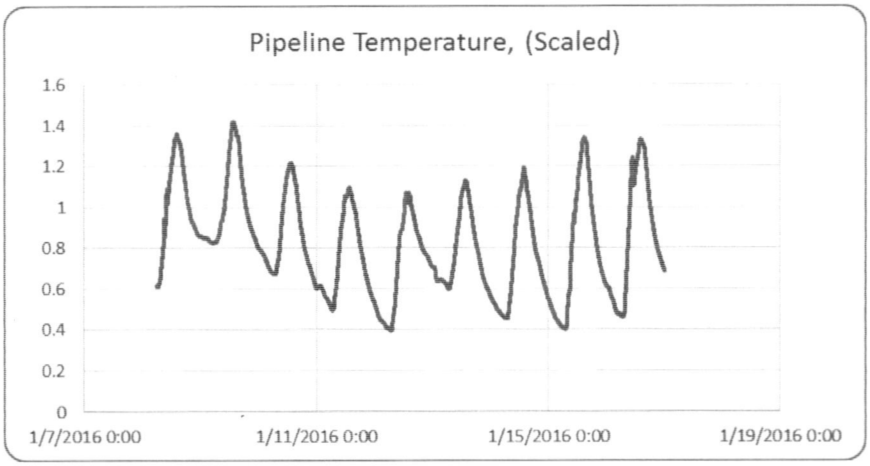


- Temperature rise: Maximum: 8 °C/hr, Average: 3 °C/hr

# Effect of liquid temperature on pipeline pressure

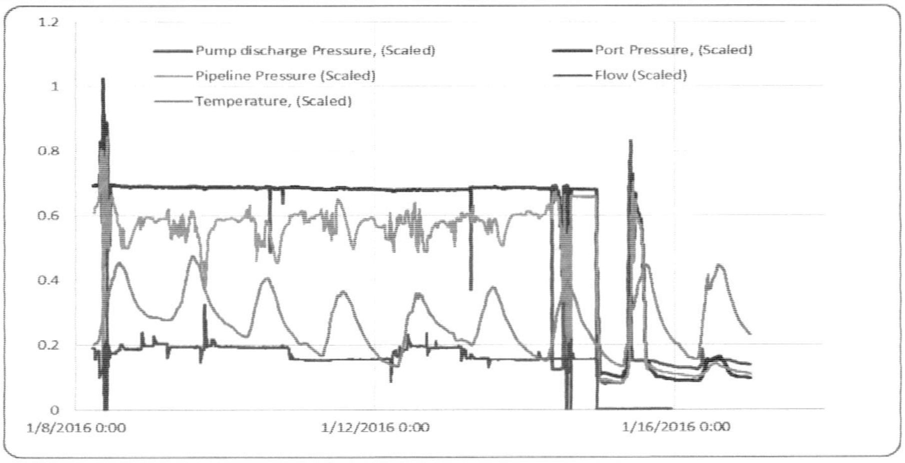


- Flow was at 20% of rated flow. Pump discharge pressure remained constant. Pipeline pressure increased with liquid temperature.

# Average thermal relief rate

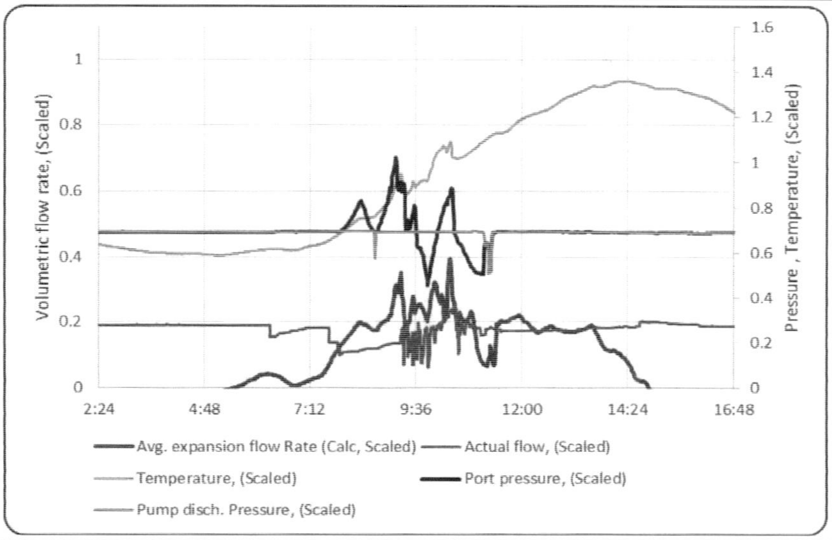


Thermal expansion rate correlates with observed pressure

# Case 2: Thermal expansion in depressurized liquid full blocked pipe

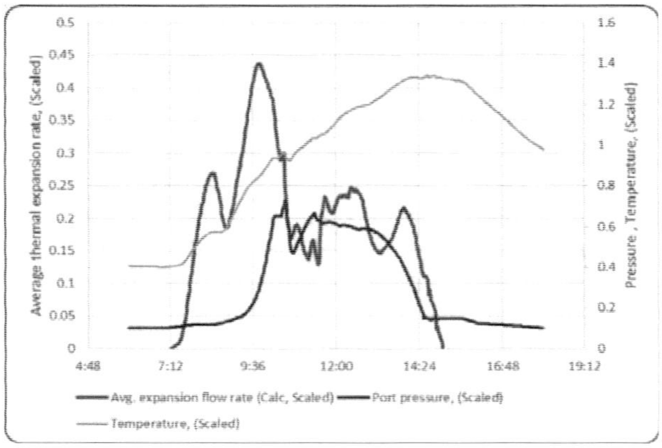


Average thermal expansion rate can be used to analyze thermal relief requirement.

# Upgraded System

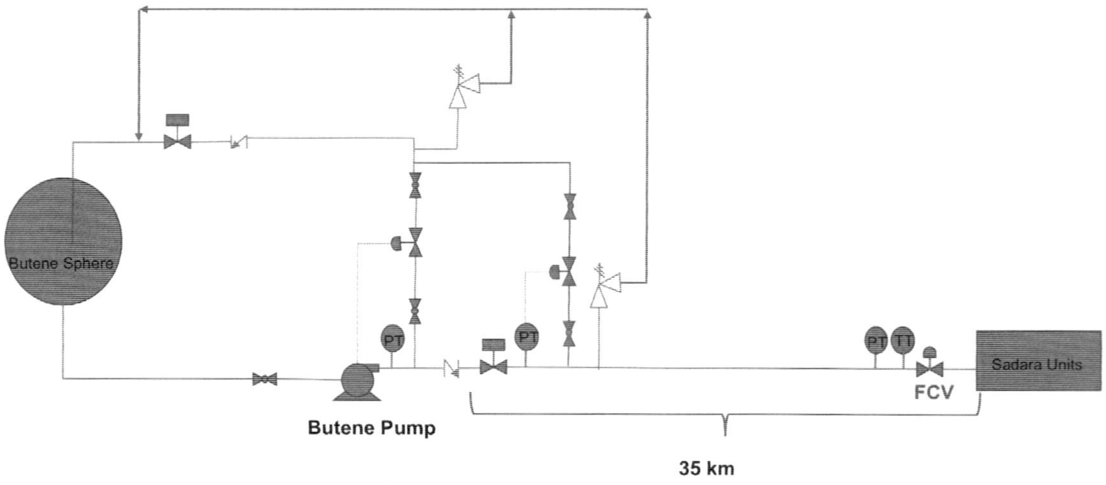


# Upgraded System

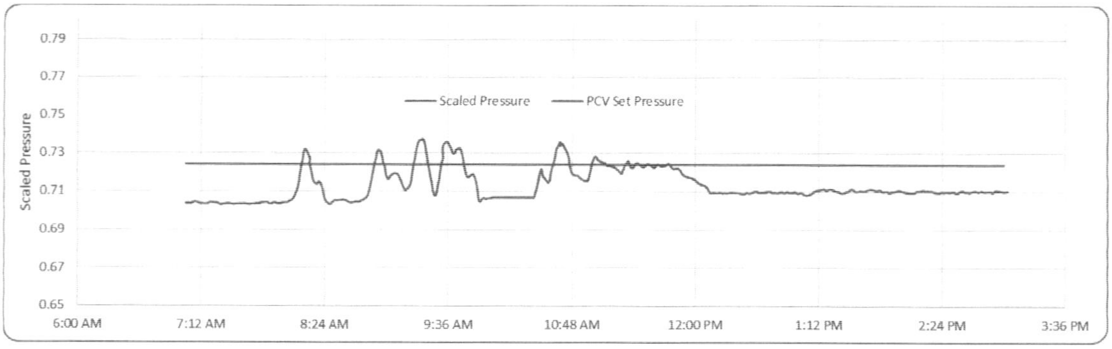


- At low flow, fluid expands due to thermal expansion, thermal relief goes back to the tank through a pressure control valve (PCV).

## Other options

- Pump minimum circulation line downstream of check valve
- Expansion bellows
- Thermal relief feed forward in a cascade to the destination tank

## Conclusions

- Thermal relief requirements for flowing fluid may not be obvious during conceptual design. It should be analyzed for other operating scenarios such as low flow startup condition, especially in case of cross-country pipelines and fluids with high thermal expansion coefficients.
- A safer and more economic design was achieved with a new recycle loop and thermal relief valve discharge routed to the source or destination tank.

Thank you

**Q & A**

صدارة
Sadara

# HIDDEN THREATS IN O2 SYSTEMS & UNITED APPROACH FOR MITIGATING POTENTIAL PROCESS SAFETY INCIDENTS

YOUSEF AL-HAWSAWI

CHEMISTRY THAT MATTERS™

## CONTENT

- INTRODUCTION & BACKGROUND.
- KNOW O2 RISK & TAKING NECESSARY ACTIONS.
- FINDINGS.
- PREVENTIVE ACTIONS.
- LEARNING SUMMARY & RECOMMENDATIONS

No. 1

HIDDEN THREATS IN O2 SYSTEM

# 1. INTRODUCTION & BACKGROUND

## RELIBILITY ENHANCMENT ANALYSIS

**2008** SABIC lunched Its reliability program (STRP) covering **all plants**.

**2010** EOEG Plants evaluation cycle completed.

O2 Feed system classified as **High Risk** & failure likelihood **1/5** years

Business loss more than **M$ 100 ,** calling for high & immediate attention !

No. 3

## RELIABILITY ANALYSIS OVERVIEW

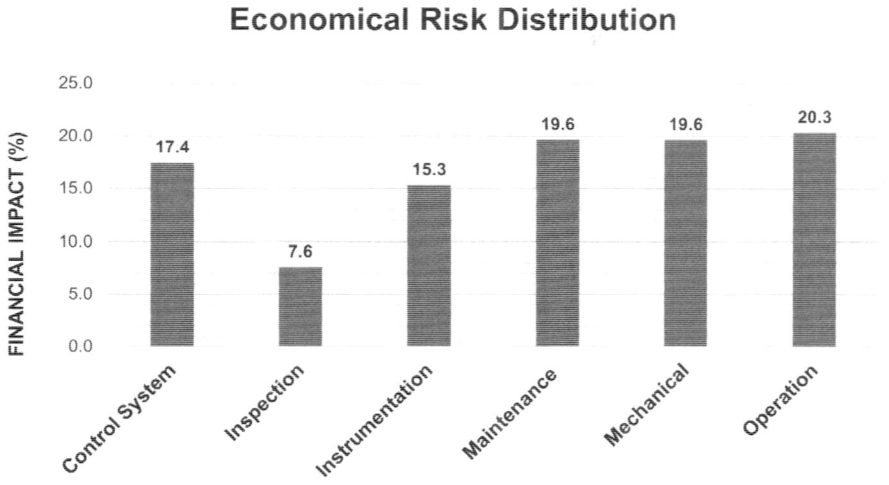


No. 4

## IMMEDIATE ACTIONS

Detail review of leading/Lagging indicators **PMs**, **Failure frequencies, etc.**

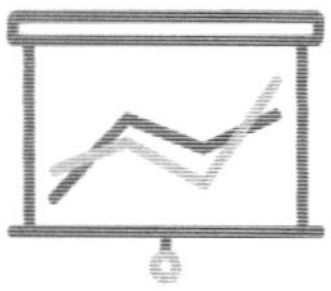

Establishing monthly performance & reliability meetings and KPI's.

Establishing Learning From Incident & Best Practices System (**HVL**)

Ensure engagement in **knowledge Sharing meetings.**

No. 5

HIDDEN THREATS IN O2 SYSTEM

# 2. KNOW O2 RISK & TAKING NECESSARY ACTIONS.

## WHAT IS O2 ?

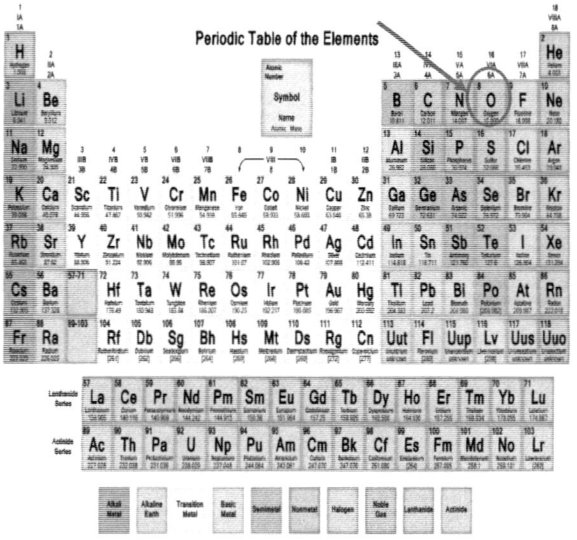


- Discovered by Swedish pharmacist Carl Wilhelm in 1777.
- Most oxidizing gas & supports combustion.
- Highly Reactive & major contributor in 100% of fire prorogation.
- Oxygen enriched atmosphere is ≥ 25 mol% per ASTM G63
- Most industrial applications require 90-93% pure or 99+ % pure
- In 2014, over 4 Trillion Cu. Ft globally were produced.
- ARCO channelview & Terria Industries Explosion caused by O2.

## LEARNING PLATFORMS

**SABIC** EHSS safety reports

**Plant licensor** Safety papers, TEM & Licenses Meetings

SABIC SES & International Standards Updates

**SABIC MCE** Domain & **SABIC Affiliates** High Value Learning

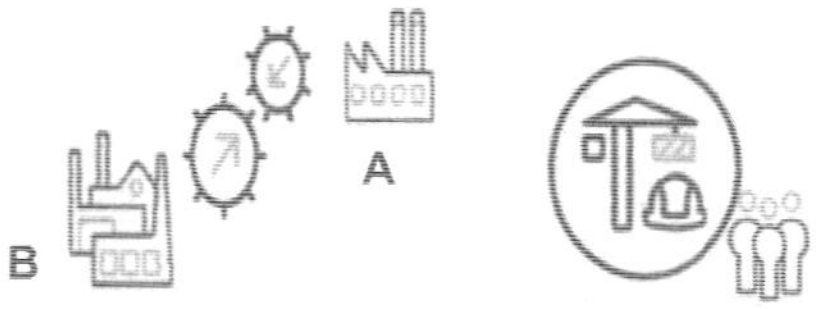




## UNITED ACTIONS VS. PSM ELEMENTS

The implemented actions were aligned with following SHEM's (Corresponding to PSM Elements in OSHA) :

- EHSS Documentation & Control of Records
- SAFER & Process Safety Risk Assessment
- Operating Procedures
- Training & Competence
- Third Party EHSS Management
- Pre-Startup SHE Review
- Mechanical Integrity
- Work Permits
- Management of Change
- EHSS Incident Reporting, Classification, Investigation & Analysis
- Emergency Planning and Response
- Compliance Audits

## UNITED MANDATED ACTIONS ON O2

**Group all PS Information on O2** & its handling risk.

**Perform Risk Assessment** for un-comfort operation on O2 system.

**Trained & certified personnel's** allowed for maintenance activities on O2 system.

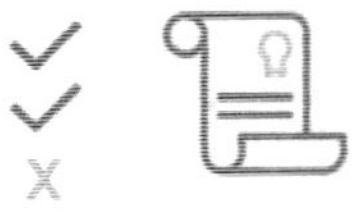

**Preventing the use of Non- Metallic material** on except approved ones.

PTFE

PCTFE

FEP

Teflon

Graphite



## UNITED MANDATED ACTIONS ON O2

**Cleaning system by DMW & approved CGA chemical.**

**Reducing critical & impingement velocities** to prevent sparks.

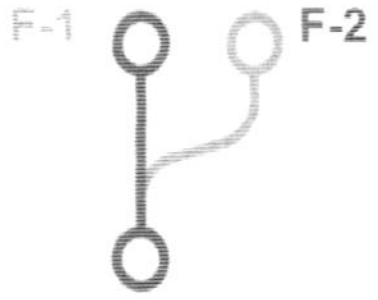


**Test O2 valves** to max design pressure & review response timing.

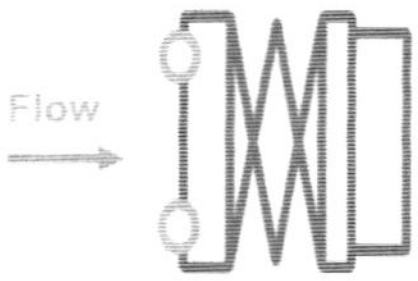


Complete **review of all instruments** & wiring at field.

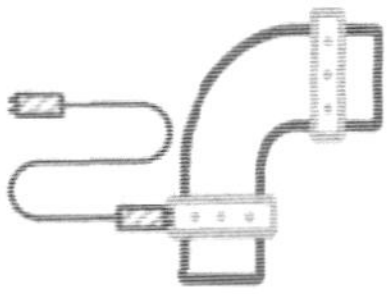

## UNITED MANDATED ACTIONS ON O2

**Stop the use** oily or adhesive maintenance and tonguing equipment's.

**Developed PM notification** capturing all actions to be done ever TA

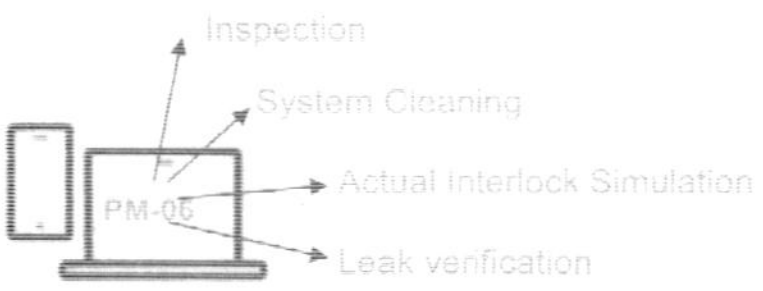


**Audit the activities by rigorous Checklist covering** critical elements

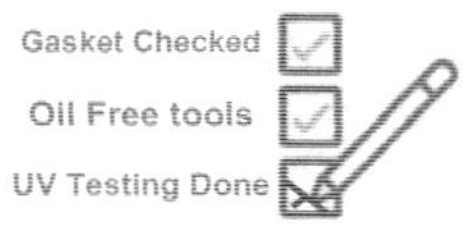


**Review S/D findings and benchmark** with previous years

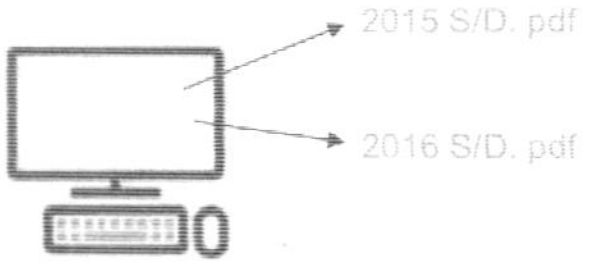


No. 12

## UNITED MANDATED ACTIONS ON O2

**Inspect O2 System** every T/A & perform baroscopic for piping

**Analyze process findings** available on system including any black soot.

**Test O2 system interlocks** randomly

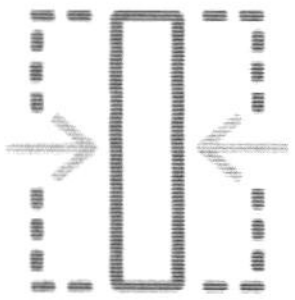

**Communicate findings** to Licensor for investigation.

No. 13

HIDDEN THREATS IN O2 SYSTEM

# 3. FINDINGS

## O2 MIXER

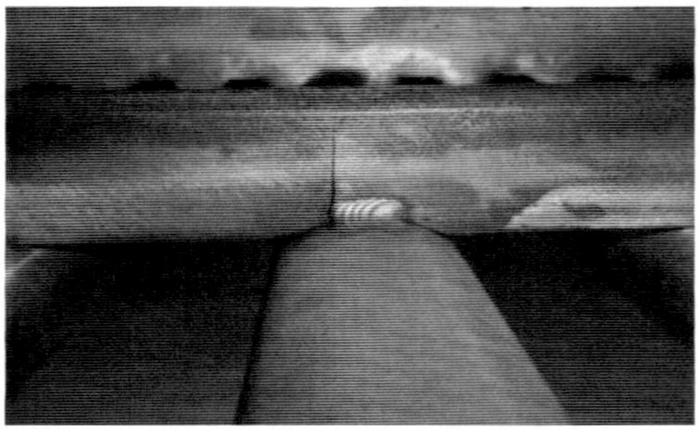

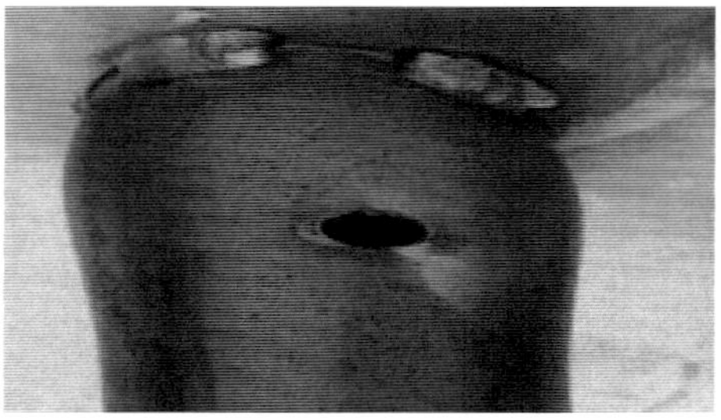

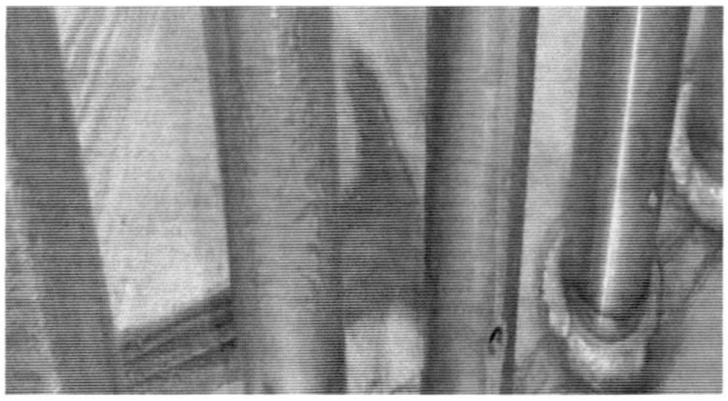

سابك
sabic

## O2 FILTERS

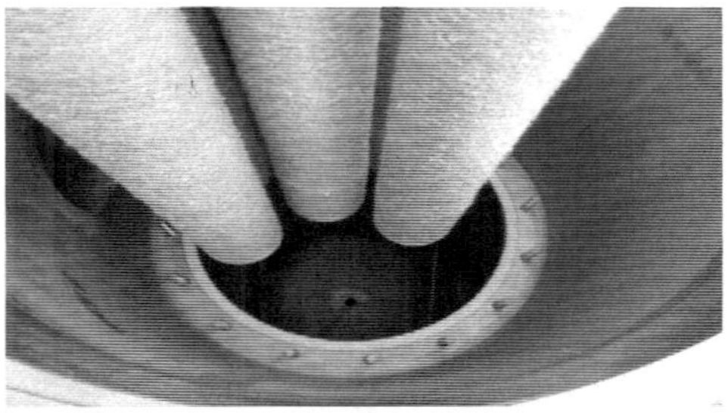

No. 16

سابك
sabic

## HP N2 SYSTEM

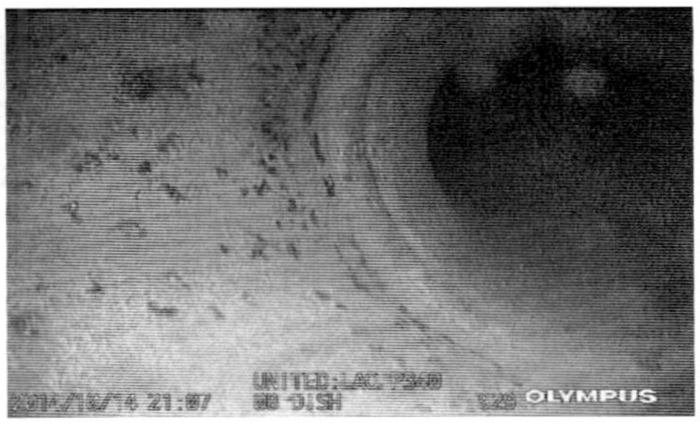

No. 17

سابك
sabic

## O2 VALVES

No. 18

سابك
sabic

## O2 FEEDLINE

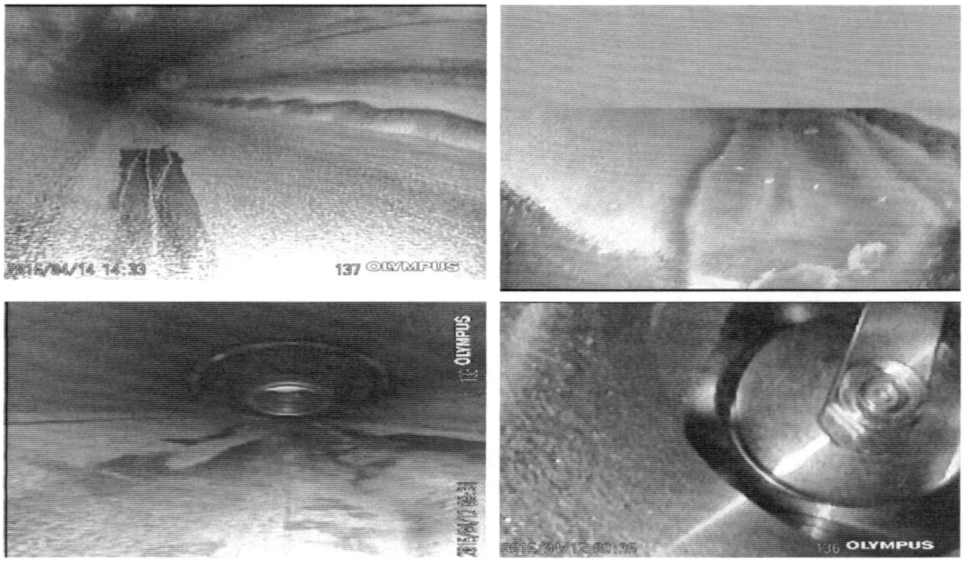


No. 19

## O2 FEEDLINE

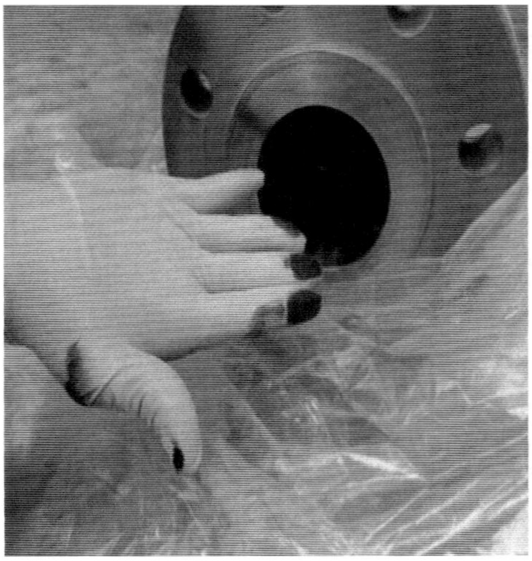

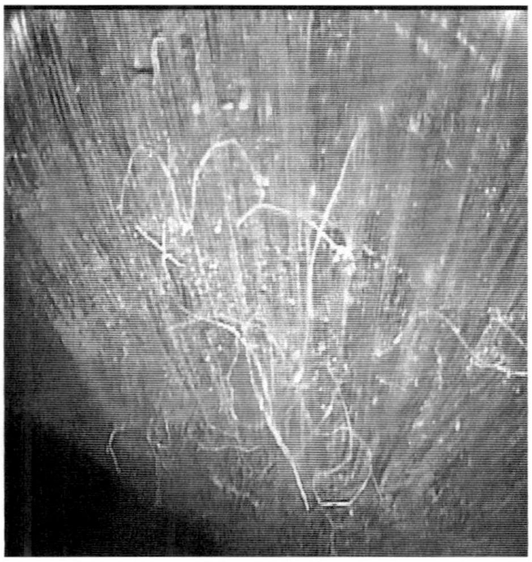

No. 20

## CG PIPING

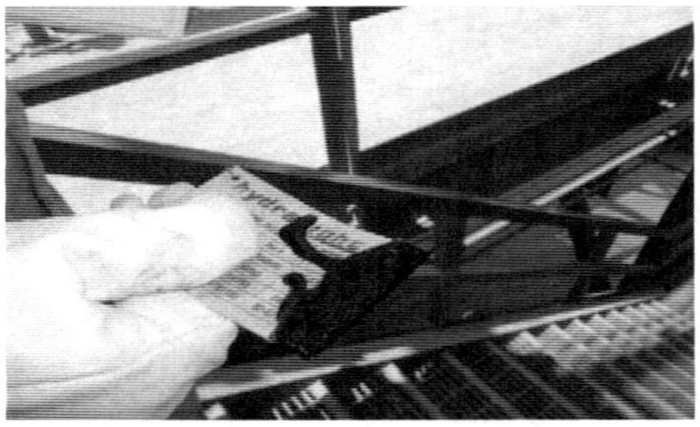

No. 21

## INSTRUMENTATION

- 3 Valves out of 26 didn't respond per design time.
- 1 critical instrument failure due to loss of wiring.
- Differences in interlock design vs. actual set point

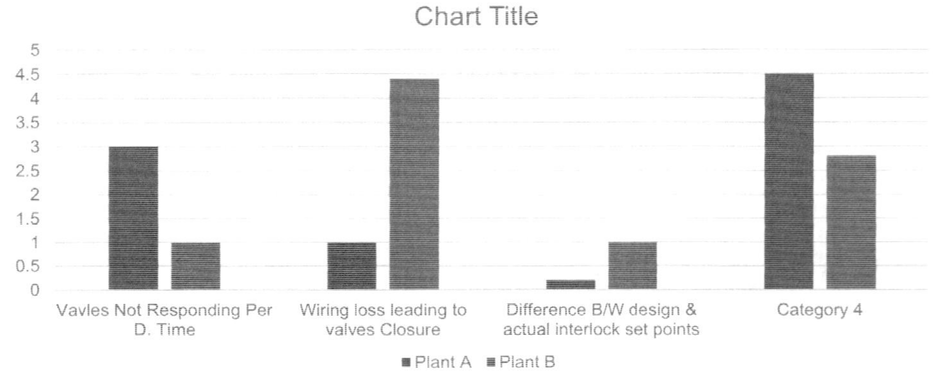


No. 22

HIDDEN THREATS IN O2 SYSTEM

# 4. PREVENTIVE ACTIONS

سابك
sabic

## OMS INSPECTION & LAB ANALYSIS

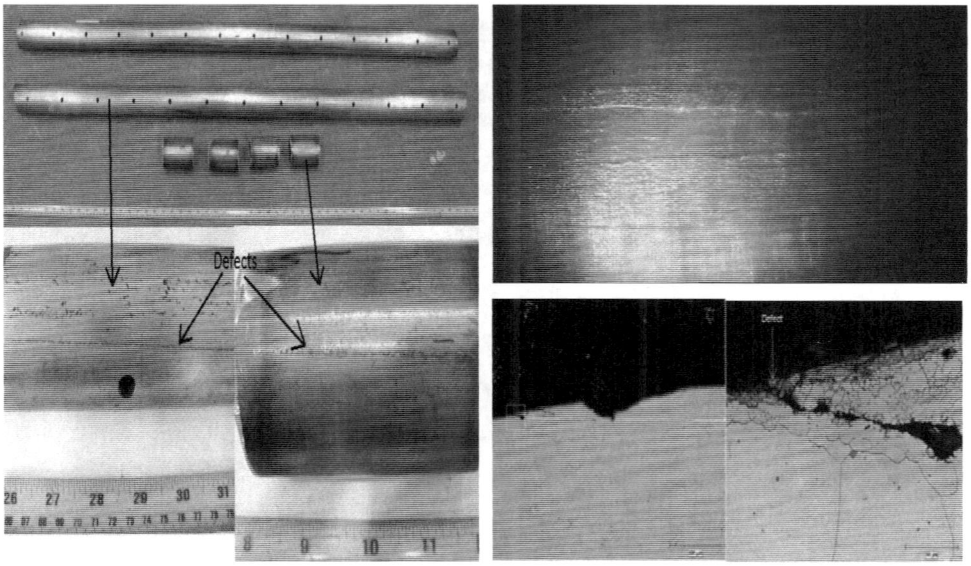


No. 24

سابك
sabic

## O2 MIXING UNIT CLEANING & INSPECTION

No. 25

## O2 FILTER REPLACEMENT

No. 26

## HP N2 FILTERS CLEANING

No. 27

سابك
sabic

## O2 VALVES CLEANING

No. 28

سابك
sabic

## O2 VALVES TESTING

No. 29

سابك
sabic

## O2 PIPING INSPECTION

No. 30

سابك
sabic

## CG PIPING CLEANING

No. 31

HIDDEN THREATS IN O2 SYSTEM

# 5. LEARNING SUMMARY & RECOMMENDATIONS

## OPERABILITY

Assuring Effective Use of SOP can reduces failures up to **40%.**

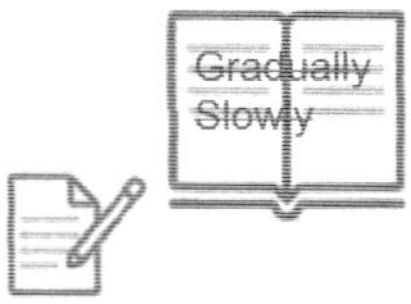


Leaks, trips, and operation challenges review team shall be from different divisions.

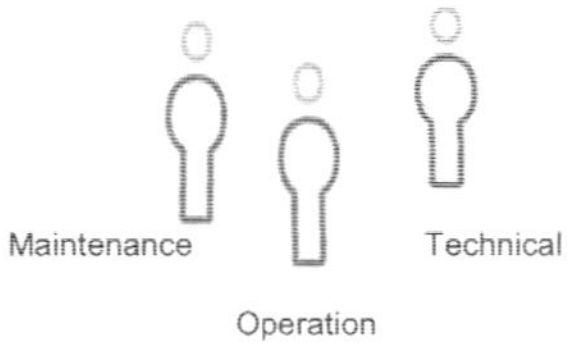


Recognizing hidden failures start with **operation alertness**

Sense of "**Un-Ease**" is a skill that can prevent catastrophic failures.

***Operability is*** *ability to keep an equipment, a system or a whole industrial installation in a safe and reliable functioning condition, according to pre-defined operational requirements*

## RELIABILITY

Best maintenance strategies are defined based on **failure consequence**.

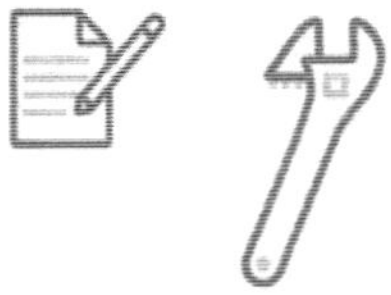

Deviation in instrument reading is **reliability deficiency signs**.

Performance management of **process instrumentation & SIF** to be done together.

Functional failures are **leading indicators** of reliability issues.

***Reliability is*** *An attribute of any system that consistently produces the same results, preferably meeting or exceeding its specifications.*





## INTEGRITY

Integrity programs shall cover **both sides** of unit.

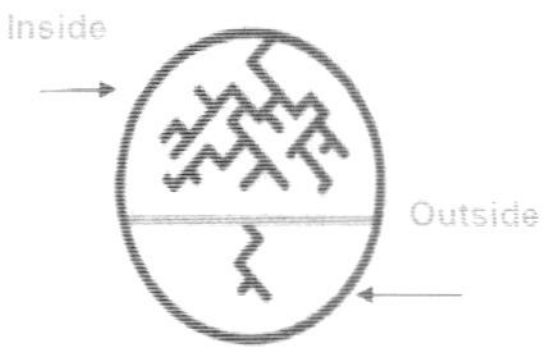


Mechanical integrity management systems must **link probability & consequence** of failure.

**No form** of maintenance can increase what asset failed to do.

Only **10%** of failures were related to age, **90%** were not !

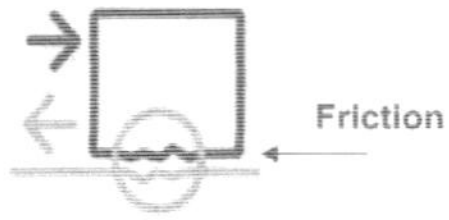


***Integrity is*** *State of a system where it is performing its intended functions without being degraded or impaired by changes or disruption in its internal or external environment.*

## RECOMMENDATIONS

**Perform rigorous inspection** of O2 system every TA.

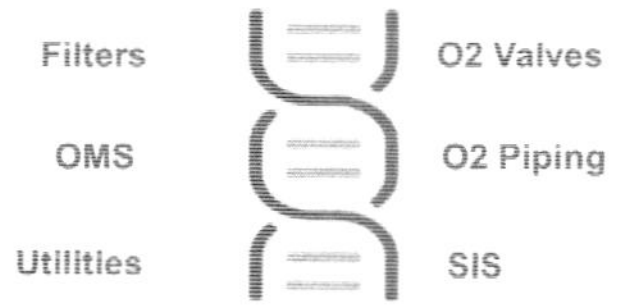


**Analyze** utilities & feedstocks, and process findings always.

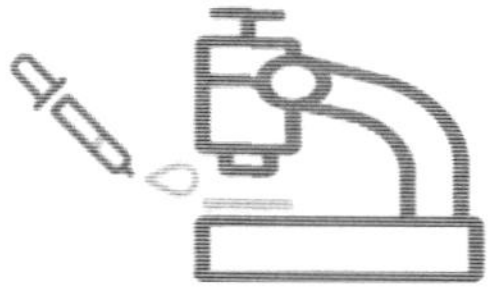

**Clean critical** equipment's in O2 system using approved chemicals.

**Do baroscopic** inspections for unaccusable points/lines

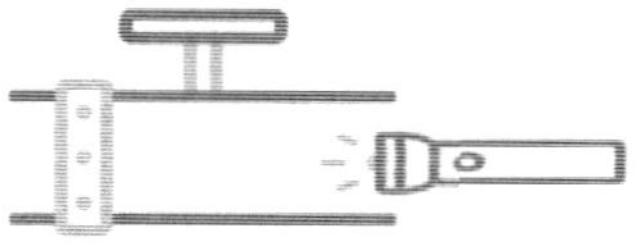



## RECOMMENDATIONS

**Review SOPs** when catalyst is changed

Consider **Failure consequence** not **failure history** during PM frequency optimization.

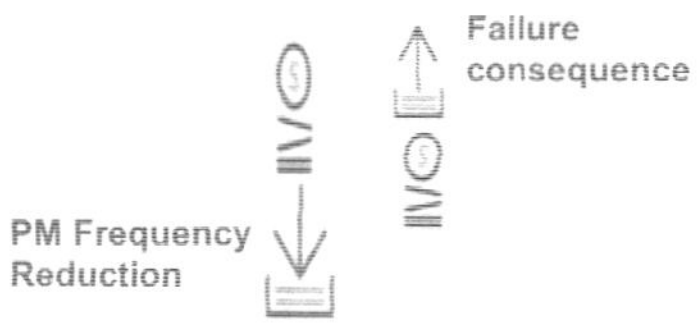


**Convert** best practices to procedures.

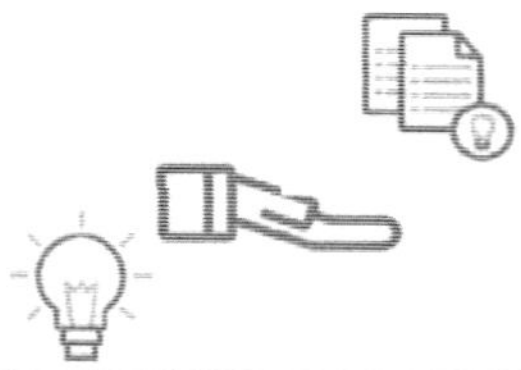

**Maintain a findings record** and seek for vendor consultation.

سابك
sabic

Thank you

شكراً

No. 38

سابك
sabic

# PROCESS SAFETY FIELD ENGAGEMENT PROGRAM AT PETROCHEMICAL COMPANY **(P-SAFE)**

December 2016

CHEMISTRY THAT MATTERS™

صدف
sadaf

## TABLE OF CONTENTS

## PROCESS SAFETY LEADERSHIP AND COMMITMENT-*BACKGROUND*

- Asset integrity, maintenance and inspection become important concerning Process Safety as plant ages.

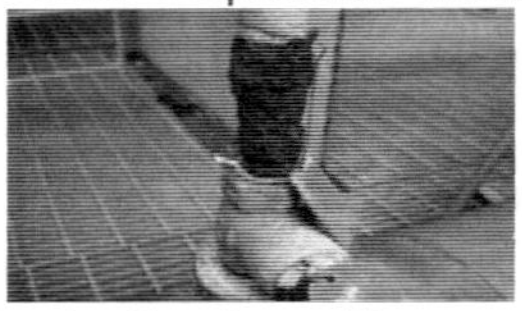

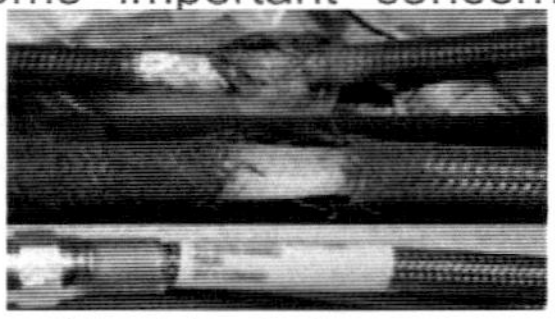

- Process safety requires constant attention. Absence of incidents is not a good indicator that process safety is being managed.
- Aiming to manage Process Safety incidents by controlling of Process Safety leading indicators.
- It is a crucial role and process in term of process safety excellence where P-SaFE is part of it.

*"P-SaFE is a periodic formal verification rounds(Go and See/Walk the Talk) of the health & effectiveness of existing Process Safety barriers using predefined checklists to identify the potential for the design barriers' missing and/or defect, which may result in a significant Process Safety Incidents".*

No. 2

## PURPOSE AND OBJECTIVE

*Purpose:*

- Real Process Safety is out in the plant and not behind our desks.
- Demonstrate Visible Leadership Commitment.
- Good interaction via conversations with front liners.
- Listen to the concerns of those who are closest to the process.
- Employees' motivation by looking at actions of business leaders.
- Resolving chronic unease and promote process safety culture.

*Objective:*

To prevent incidents resulting from unintentional release of energy or hazardous substances. It is about "keeping the materials and product inside the designated containments such as the pipes and vessels". This is accomplished by Process Safety barriers.

No. 3

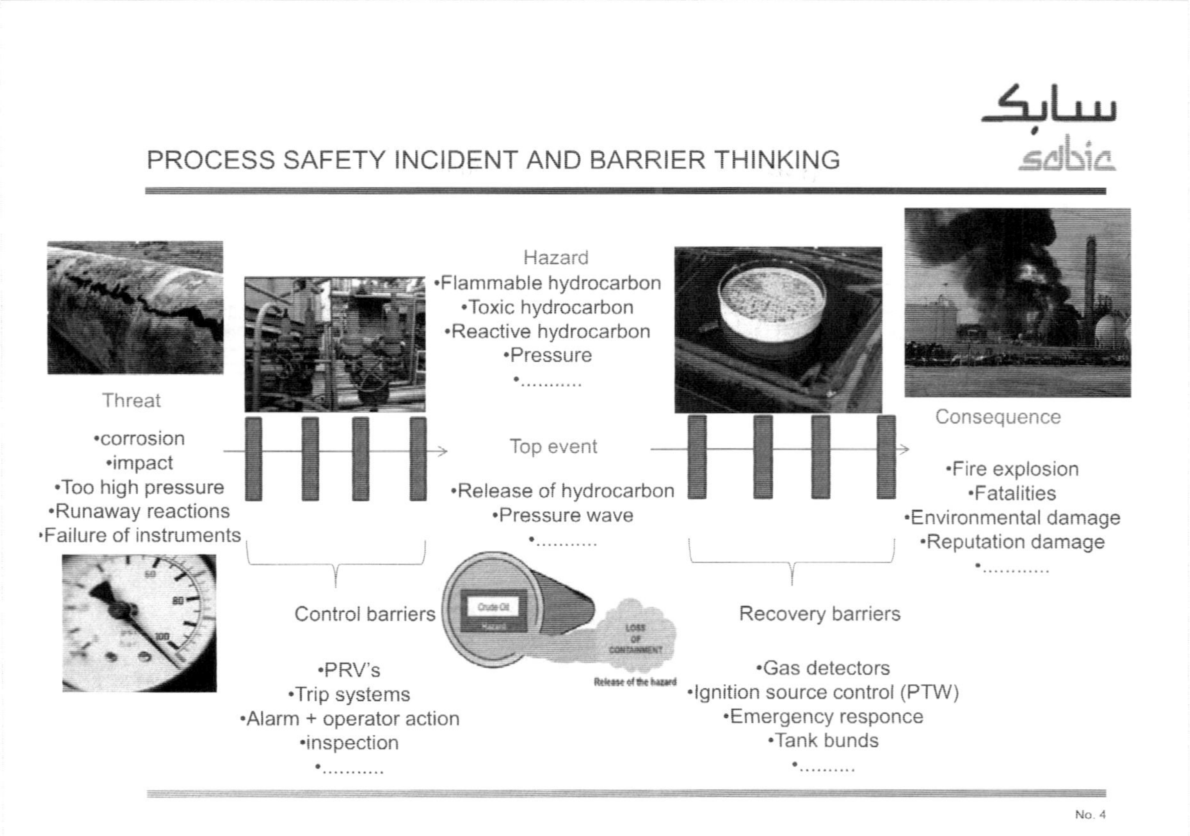
سابك
sabic
PROCESS SAFETY INCIDENT AND BARRIER THINKING
Threat
•corrosion
•impact
•Too high pressure
•Runaway reactions
•Failure of instruments
Hazard
•Flammable hydrocarbon
•Toxic hydrocarbon
•Reactive hydrocarbon
•Pressure
•..........
Top event
•Release of hydrocarbon
•Pressure wave
•..........
Consequence
•Fire explosion
•Fatalities
•Environmental damage
•Reputation damage
•...........
Control barriers
•PRV's
•Trip systems
•Alarm + operator action
•inspection
•..........
Release of the hazard
Recovery barriers
•Gas detectors
•Ignition source control (PTW)
•Emergency responce
•Tank bunds
•.........
No. 4

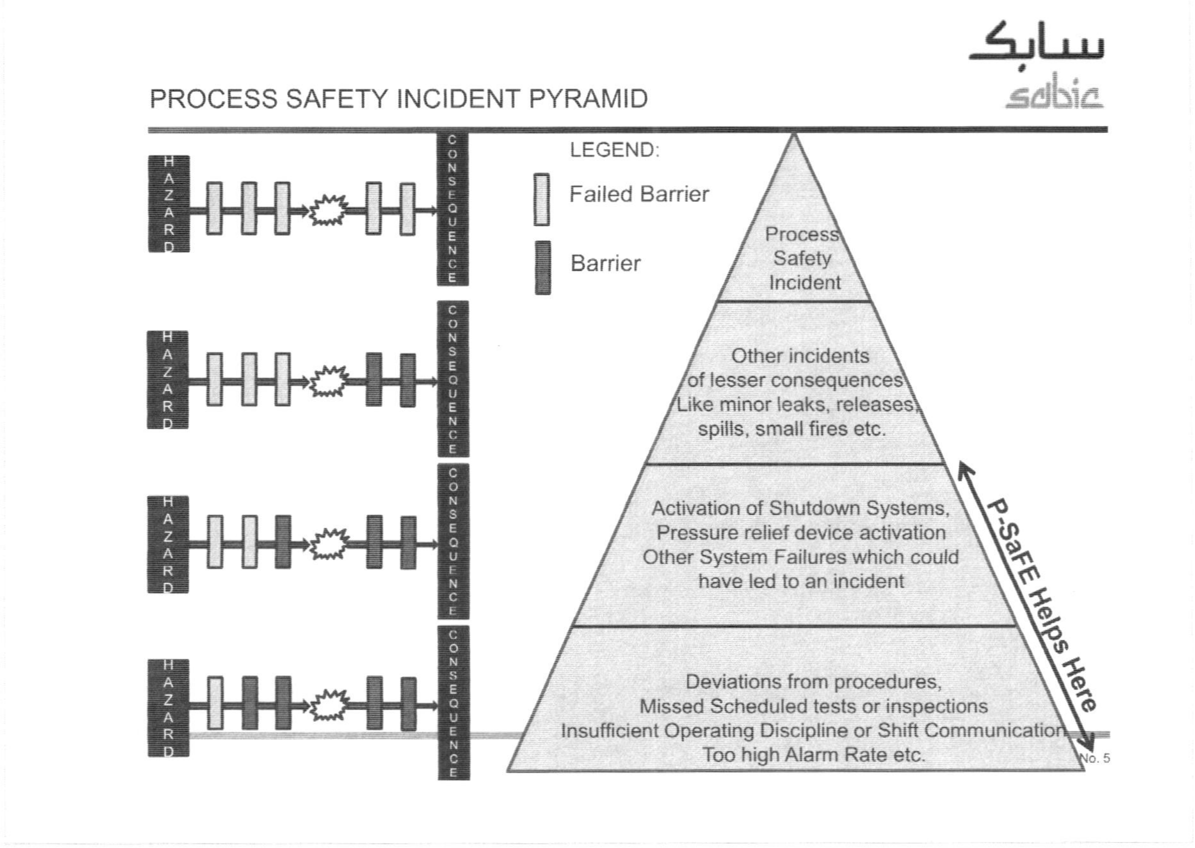
سابك
sabic
PROCESS SAFETY INCIDENT PYRAMID
HAZARD
CONSEQUENCE
LEGEND:
Failed Barrier
Barrier
Process Safety Incident
Other incidents of lesser consequences Like minor leaks, releases, spills, small fires etc.
Activation of Shutdown Systems, Pressure relief device activation Other System Failures which could have led to an incident
Deviations from procedures, Missed Scheduled tests or inspections Insufficient Operating Discipline or Shift Communication Too high Alarm Rate etc.
P-SaFE Helps Here
No. 5

## P-SAFE PROGRAM AT SADAF

- P-SaFE program started at SADAF in 2011 as a pilot.
- 5 checklists developed and were refined to suit site specific conditions.

- *Equipment & Process Condition:* This category mainly includes the designed barriers for piping and equipment systems.

- *Relief system:* This category mainly consists of relief valve, rupture disc, and other type overpressure (including vacuum scenario) protected facilities.

- *ESD & MOC:* This category mainly consists of Emergency Shut-down systems (ESD) and Management of Change (MOC) procedure.

- *Operations:* This category mainly consists of Operation related activities such as Shift Communication, Alarm handling and plant monitoring.

- *Recovery Measures & Procedures:* This category mainly consists of Recovery Barriers such as system isolation, fire-fighting & deluge equipment.

No. 6

## P-SAFE PROGRAM AT SADAF

- P-SaFE program rolled out site wide in 2012.
- A management system had been initiated and approved in 2014.
- As part of Process Safety continuous improvement, a newly checklist had been introduced this year and that is about *Tanks Safety* : This category mainly focuses on Tanks Safety in terms of protections, construction & integrity and operation.

*Apart of that:*

- The schedule is closely tracked and all the findings are shared with respective production teams and driven to logical conclusion.
- Systemic issues are addressed by PS group whereas plant specific issues by respective PT.
- NO. of P-SaFE rounds conducted is tracked and reported as a monthly PS leading KPI.

No. 7

## HOW THE CHECKLISTS WERE DEVELOPED

A review of a collection of 130 existing Bow Ties had identified 25 major barriers
give 80% of the safeguards. The P-SaFE checklists are made around these safeguards.

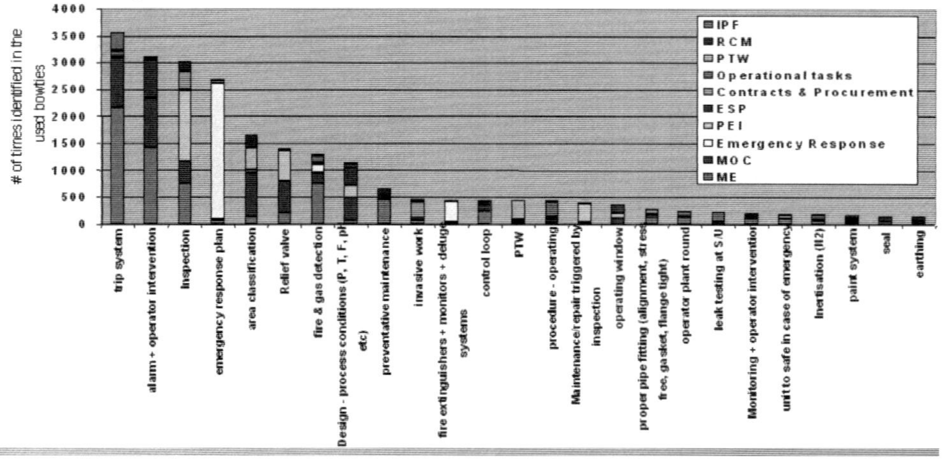


No. 8

## SOME REMEDIAL ACTIONS TAKEN AFTER P-SAFE OBSERVATIONS

*Examples:*

- PMs and functional testing
- Checking of Balanced Bellow PSVs bonnet vents for pluggage or fouling
- Vents/Drains caps and end blinds
- Process Safety knowledge promoted
- Roads close to process were closed as a part of facility siting recommendations.
- Compliance increased for ESD bypass authorization
- Operations Emergency procedures availability.
- No. of controllers in manual mode/bypassed of operation is being addressed.

*Many more findings still remain outstanding and need attention.*

No. 9

## SOME REMEDIAL ACTIONS TAKEN AFTER P-SAFE OBSERVATIONS

*Examples*

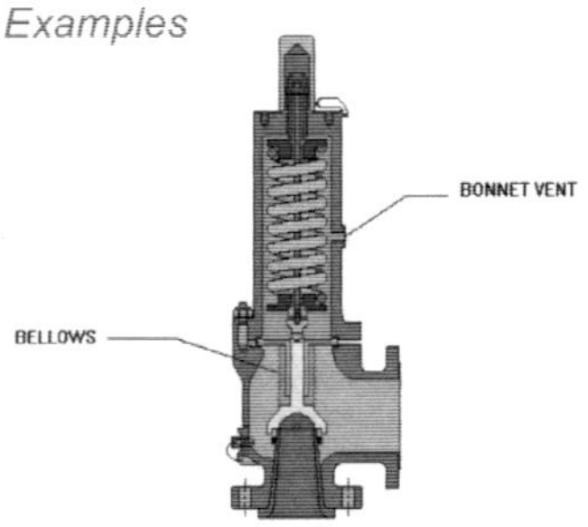


- Balanced Bellow PSVs bonnet vents

- PMs and functional testing

- Vents/Drains caps and end blinds

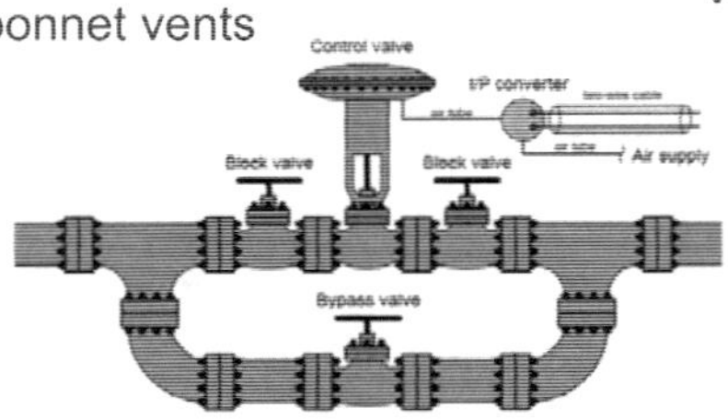


- controllers in manual mode/bypassed

No. 10

## THINGS TO KEEP IN MIND WHILE DOING P-SAFE

This is not a structured audit. This is more about management visibility and demonstrating priorities.

How :

- ➢ Restricted to a specific production unit.
- ➢ The prompts should be used as a starting point for the conversation (not a checklist).
- ➢ Do not try to cover too much or complete the questionnaire.
- ➢ Don't try to cover all areas. When uncertain about the response received, follow up with a subject matter expert.
- ➢ Time should be spent both in the control room and the plant.

No. 11

## EXPECTED OUTCOMES

- Effective Process Safety Barriers
- Maintaining the integrity of our assets
- Zero Loss Of Primary Containment (LOPC)
- Effective Process Safety performance
- Zero Process Safety and EHSS Incident.

No. 12

# THANKS FOR YOUR KIND ATTENTION

3RD CCPS GLOBAL SUMMIT ON PROCESS SAFETY
Excellence is now operational
PETROTECHNICS
Achieving Operational Excellence: Connecting the Impact of PSM to Daily Operations
Andrew Bartlett – Operational Excellence Consultant

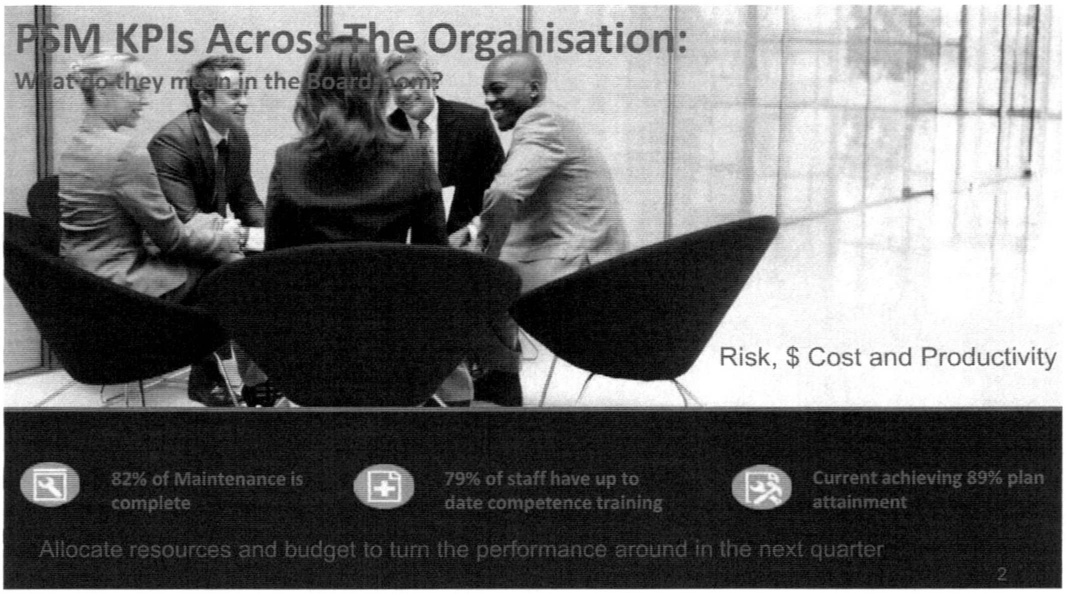
PSM KPIs Across The Organisation:
Risk, $ Cost and Productivity
82% of Maintenance is complete
79% of staff have up to date competence training
Current achieving 89% plan attainment
Allocate resources and budget to turn the performance around in the next quarter
2

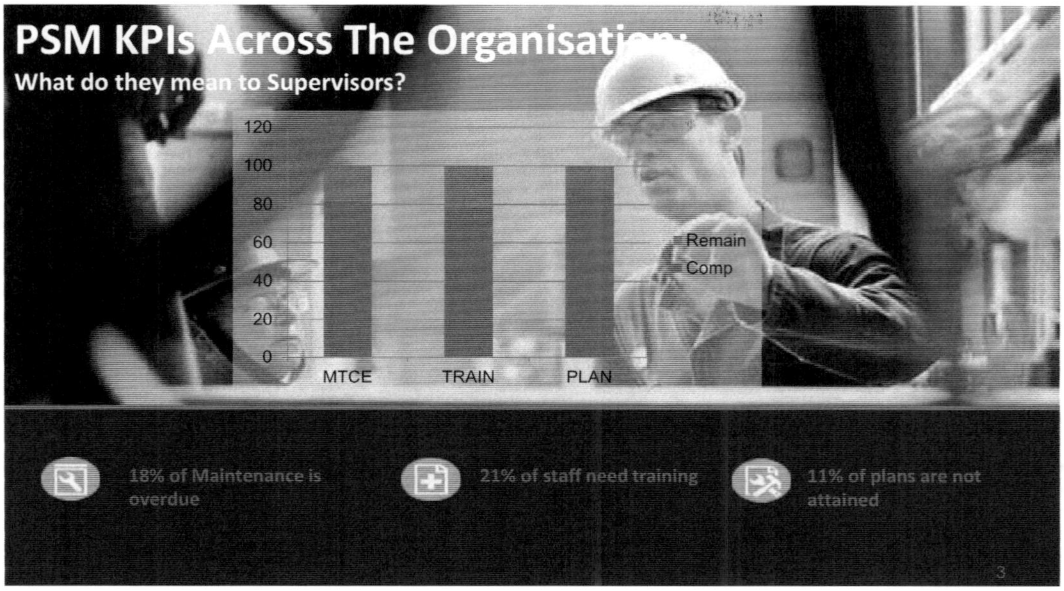
PSM KPIs Across The Organisat
What do they mean to Supervisors?
120
100
80
60
40
20
0
MTCE
TRAIN
PLAN
Remain
Comp
18% of Maintenance is overdue
21% of staff need training
11% of plans are not attained
3

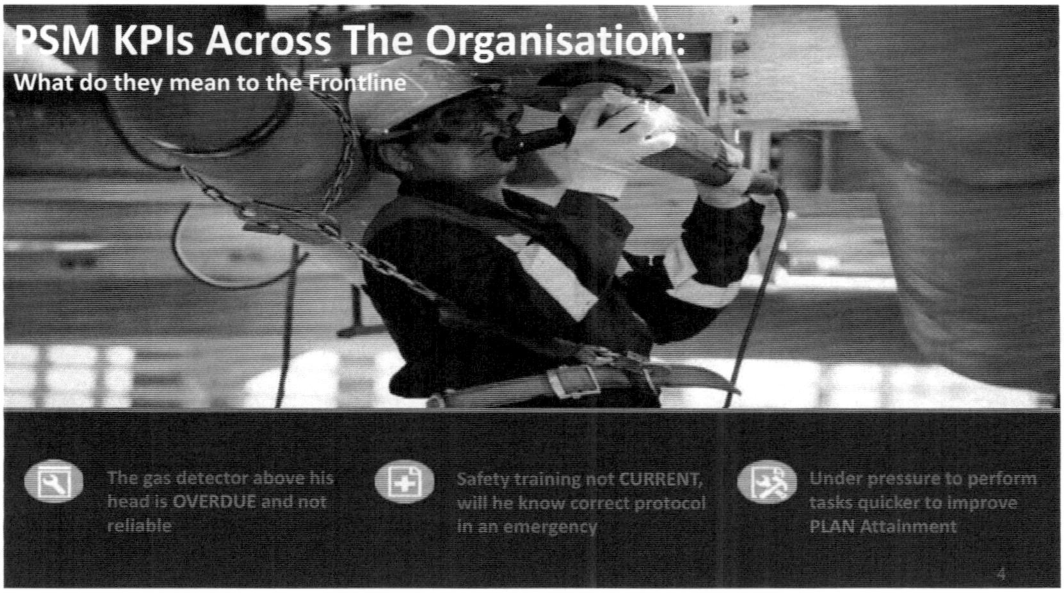
PSM KPIs Across The Organisation:
What do they mean to the Frontline
The gas detector above his head is OVERDUE and not reliable
Safety training not CURRENT, will he know correct protocol in an emergency
Under pressure to perform tasks quicker to improve PLAN Attainment
4

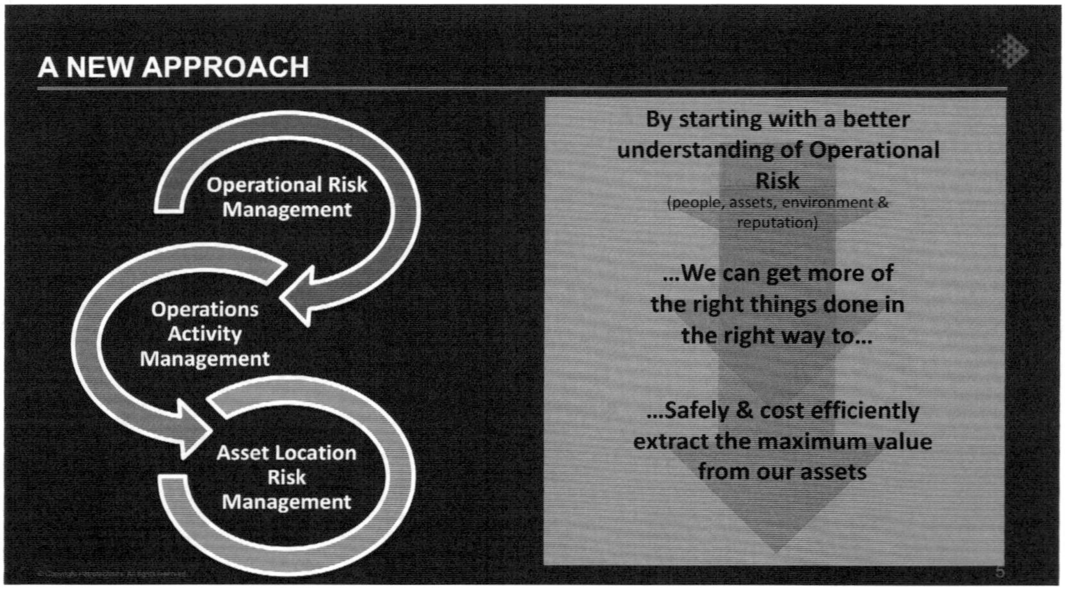
A NEW APPROACH
Operational Risk Management
Operations Activity Management
Asset Location Risk Management
By starting with a better understanding of Operational Risk
(people, assets, environment & reputation)
...We can get more of the right things done in the right way to...
...Safely & cost efficiently extract the maximum value from our assets
5

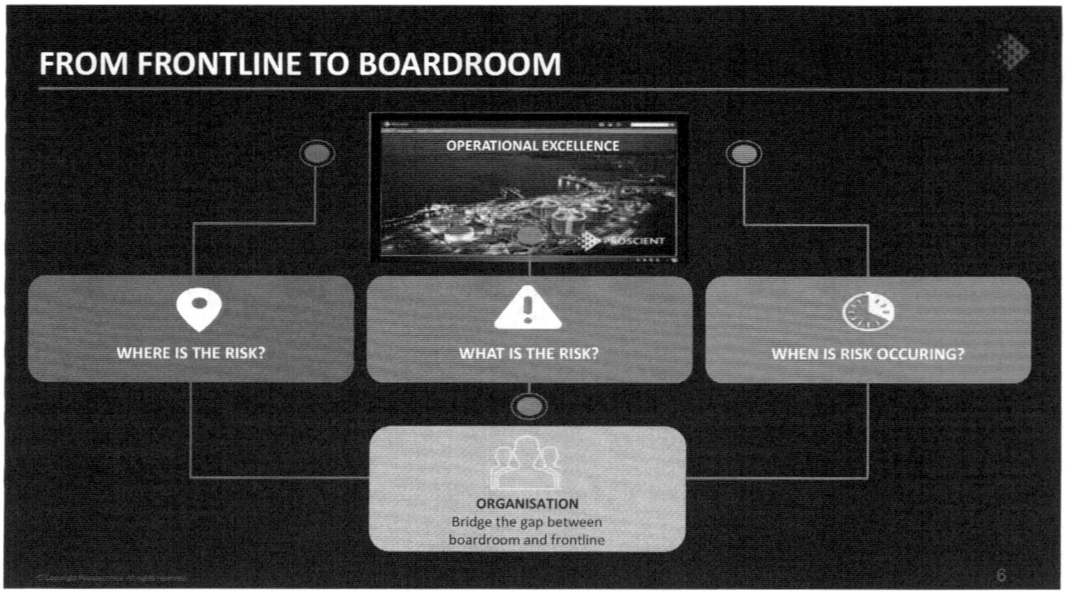
FROM FRONTLINE TO BOARDROOM
OPERATIONAL EXCELLENCE
PROSCIENT
WHERE IS THE RISK?
WHAT IS THE RISK?
WHEN IS RISK OCCURING?
ORGANISATION
Bridge the gap between boardroom and frontline
6

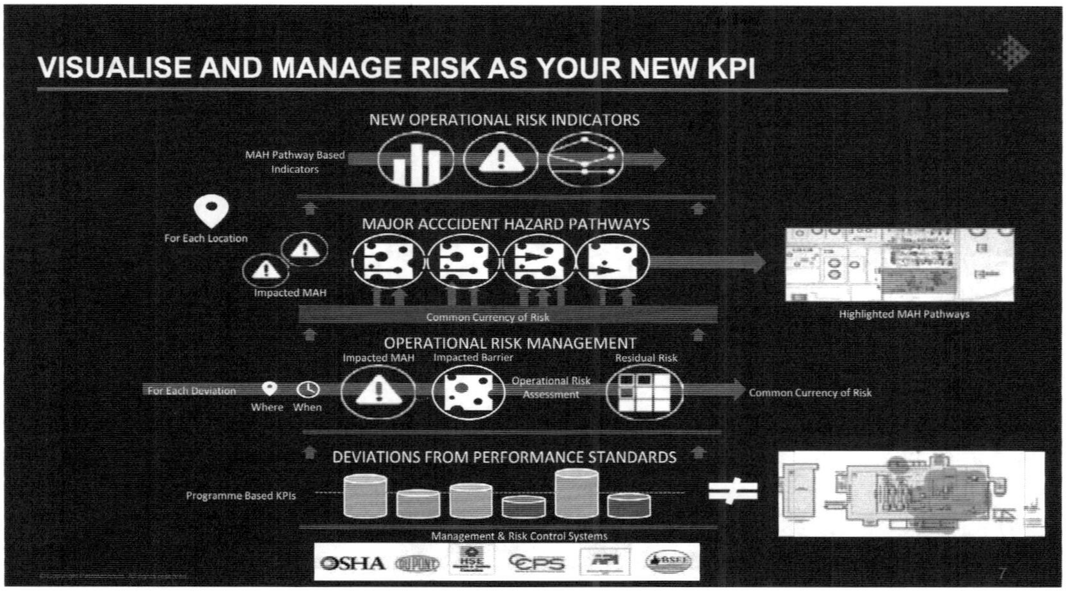
VISUALISE AND MANAGE RISK AS YOUR NEW KPI
NEW OPERATIONAL RISK INDICATORS
MAH Pathway Based Indicators
For Each Location
MAJOR ACCCIDENT HAZARD PATHWAYS
Impacted MAH
Common Currency of Risk
Highlighted MAH Pathways
OPERATIONAL RISK MANAGEMENT
Impacted MAH
Impacted Barrier
Residual Risk
For Each Deviation
Where
When
Operational Risk Assessment
Common Currency of Risk
DEVIATIONS FROM PERFORMANCE STANDARDS
Programme Based KPIs
Management & Risk Control Systems
OSHA
CCPS
API
7

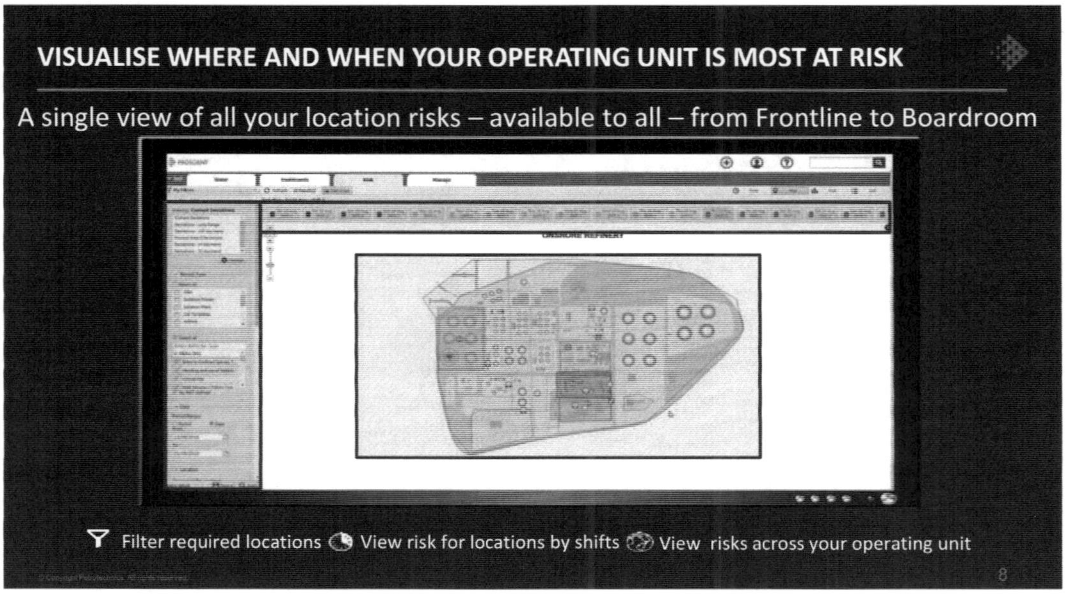
VISUALISE WHERE AND WHEN YOUR OPERATING UNIT IS MOST AT RISK
A single view of all your location risks – available to all – from Frontline to Boardroom
Filter required locations
View risk for locations by shifts
View risks across your operating unit
8

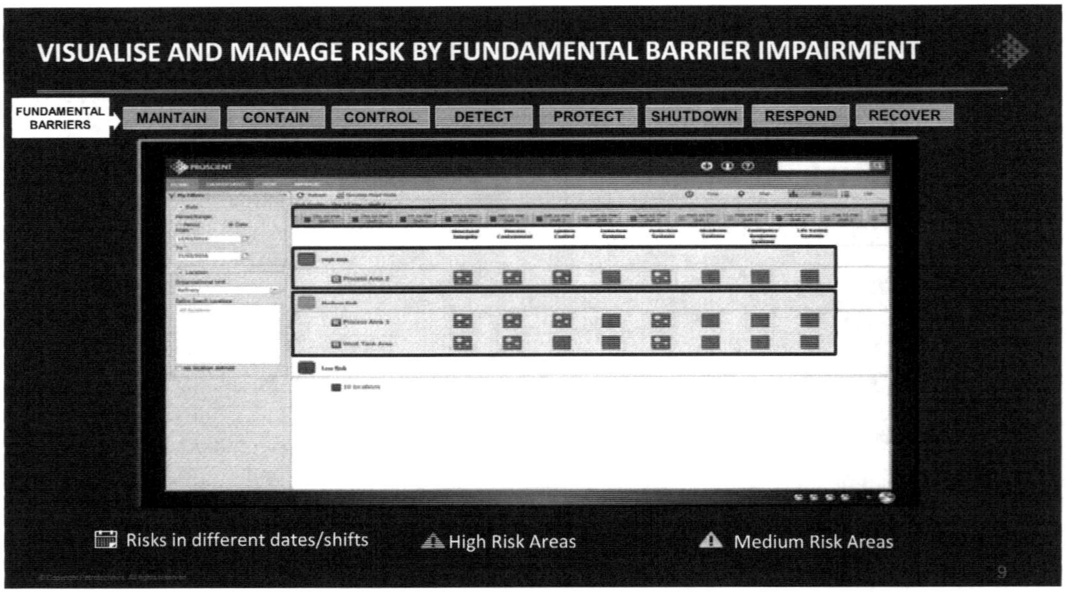
VISUALISE AND MANAGE RISK BY FUNDAMENTAL BARRIER IMPAIRMENT
FUNDAMENTAL BARRIERS
MAINTAIN
CONTAIN
CONTROL
DETECT
PROTECT
SHUTDOWN
RESPOND
RECOVER
Risks in different dates/shifts
High Risk Areas
Medium Risk Areas
9

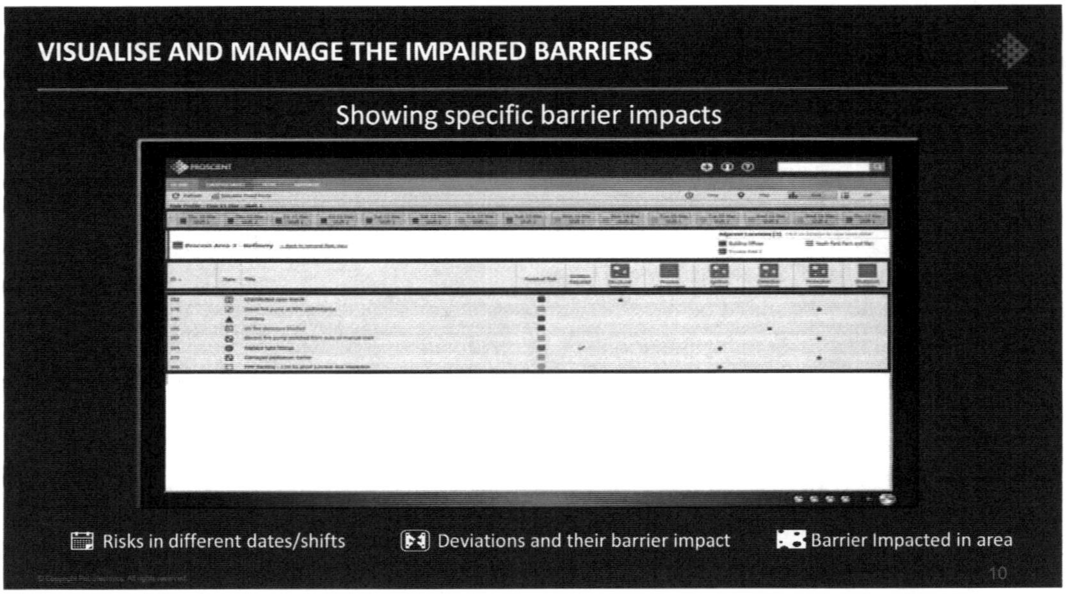
VISUALISE AND MANAGE THE IMPAIRED BARRIERS
Showing specific barrier impacts
Risks in different dates/shifts
Deviations and their barrier impact
Barrier Impacted in area
10

## VISUALISE THE MAJOR ACCIDENT PATHWAYS

Showing major accident scenarios

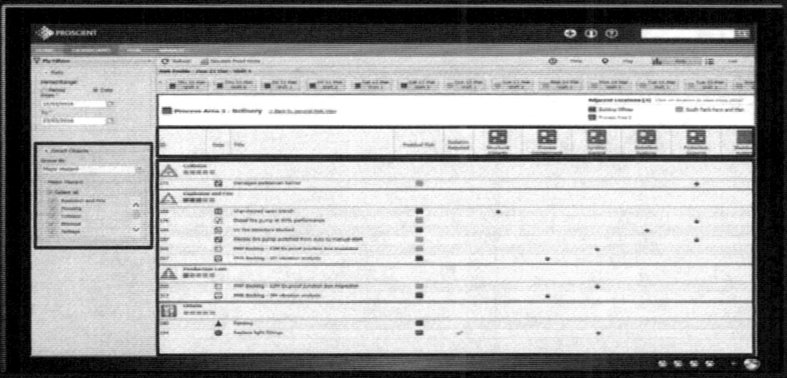

Filter Major Hazards | Accident Pathways | Barrier Management in area

11

## DYNAMICALLY MANAGE FRONTLINE ACTIVITIES

Optimise the operations plan on the basis of workload, risk and resources

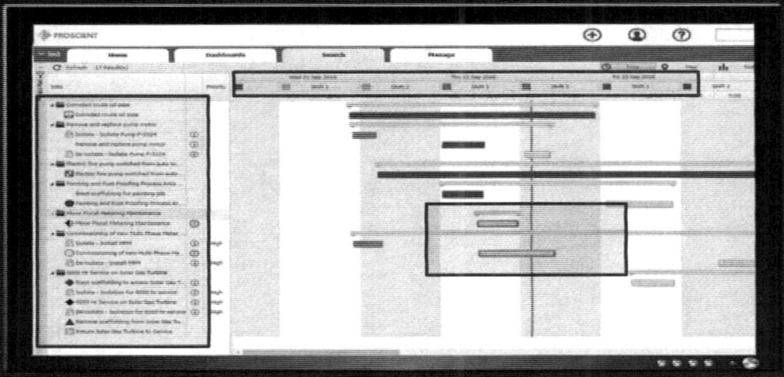

View all the operational jobs | Job conflict based on Safety Policy | View risk across all shifts

12

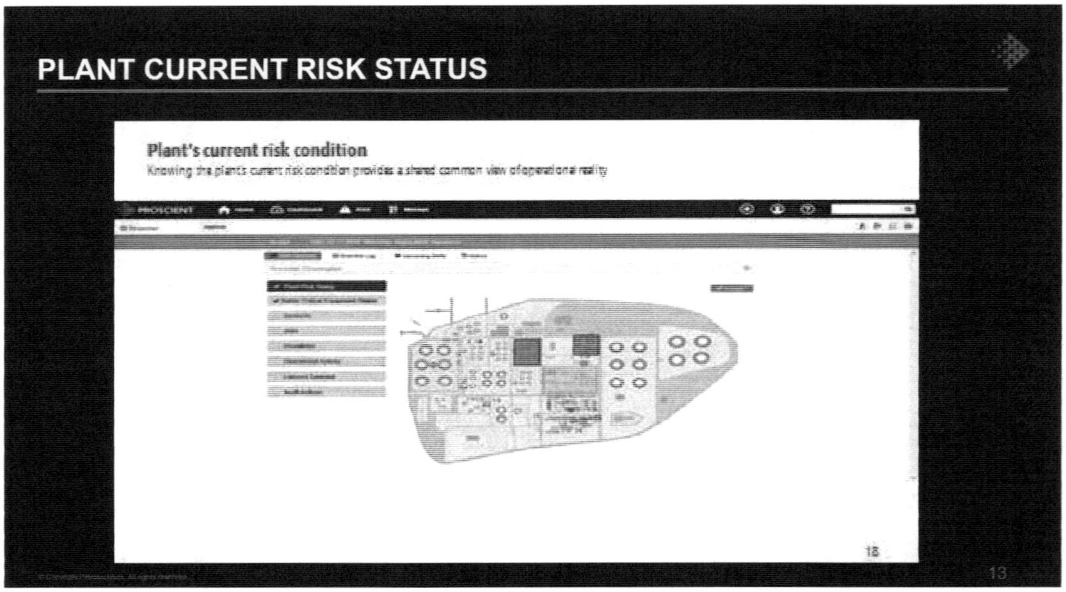
PLANT CURRENT RISK STATUS
Plant's current risk condition
Knowing the plant's current risk condition provides a shared common view of operational reality
PROSCIENT
18
13

VISUALISE AND MANAGE ACTIVITY
Enterprise transparency
"Operational decisions are now connected to Process Safety Management and I can see all risk across the operation"
14

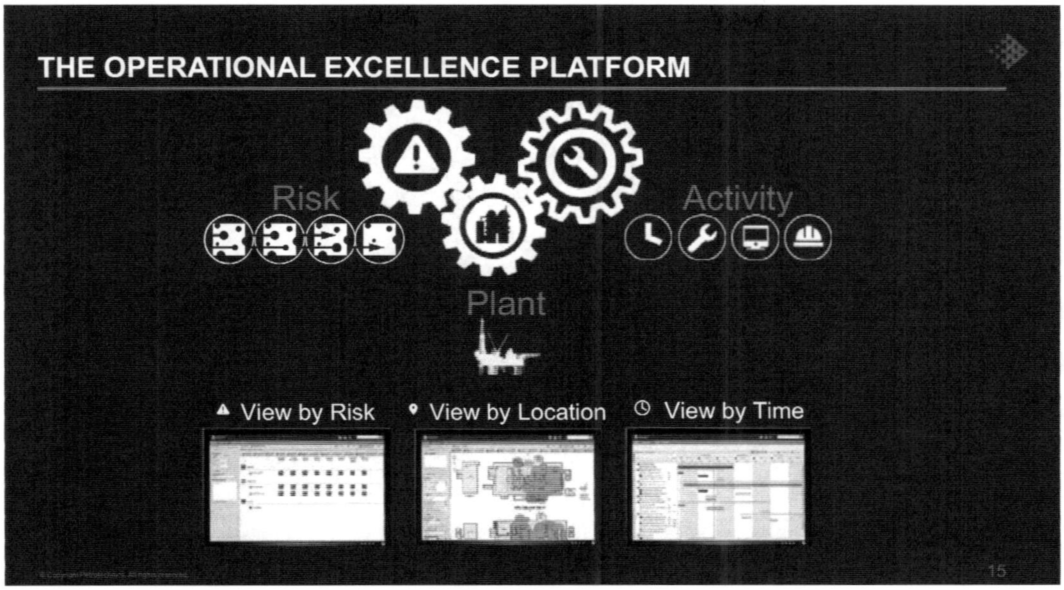
THE OPERATIONAL EXCELLENCE PLATFORM
Risk
Activity
Plant
View by Risk
View by Location
View by Time
15

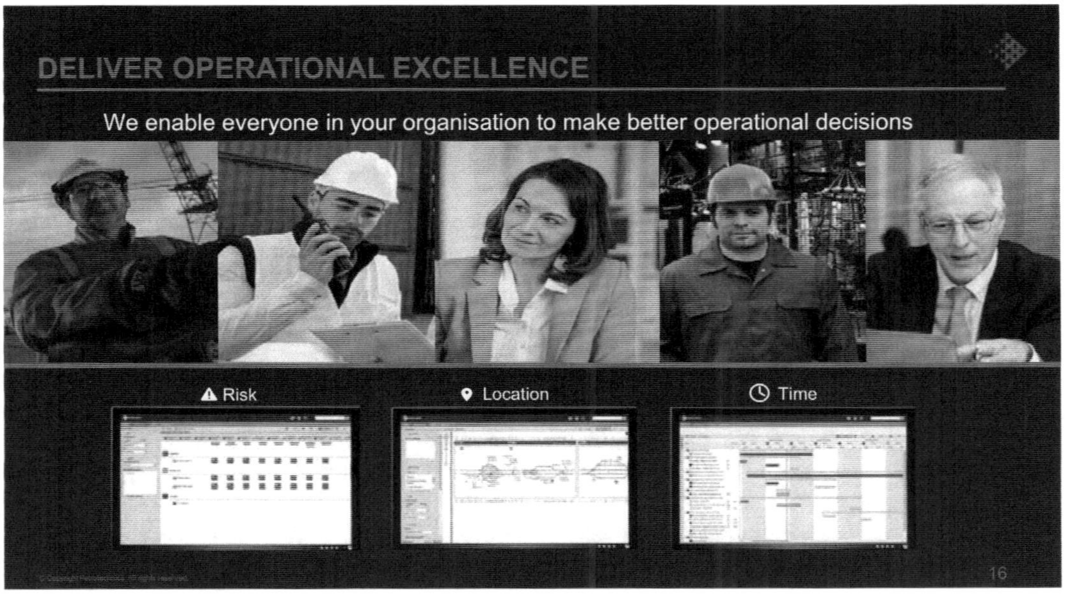
DELIVER OPERATIONAL EXCELLENCE
We enable everyone in your organisation to make better operational decisions
Risk
Location
Time
16

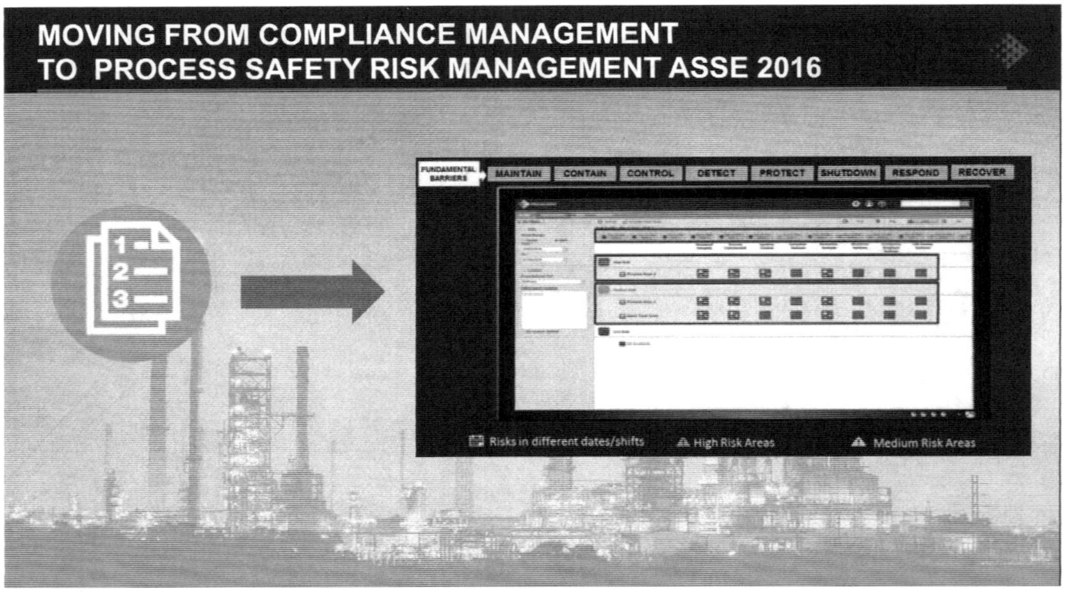
MOVING FROM COMPLIANCE MANAGEMENT
TO PROCESS SAFETY RISK MANAGEMENT ASSE 2016
FUNDAMENTAL BARRIERS
MAINTAIN
CONTAIN
CONTROL
DETECT
PROTECT
SHUTDOWN
RESPOND
RECOVER
Risks in different dates/shifts
High Risk Areas
Medium Risk Areas

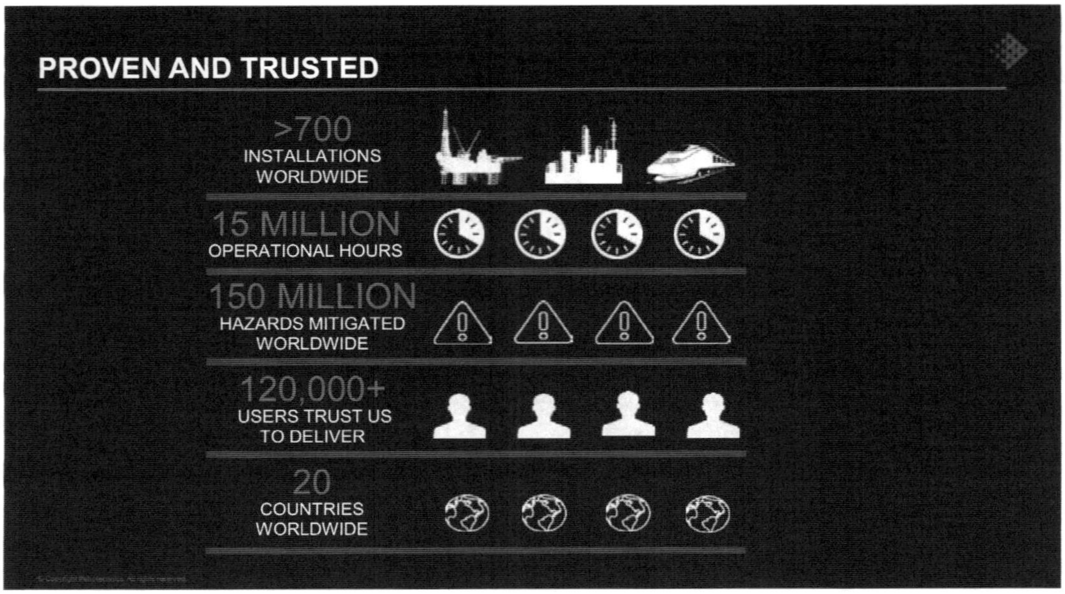
PROVEN AND TRUSTED
>700
INSTALLATIONS WORLDWIDE
15 MILLION
OPERATIONAL HOURS
150 MILLION
HAZARDS MITIGATED WORLDWIDE
120,000+
USERS TRUST US TO DELIVER
20
COUNTRIES WORLDWIDE

## OUR CLIENTS

Network Rail
Saudi Aramco
Crossrail
amec
OXY
Chevron
EnQuest
Quadrant Energy
ExxonMobil
TEEKAY PETROJARL
Cameron LNG
Golden Pass LNG
GDF SUEZ
nexen
TOTAL
Apache
ENGIE
Enriching life | Petro Rabigh
nationalgrid
Canadian Natural
eni
bp
KPOC
BG GROUP
Petrofac
TAQA
centrica storage
TALISMAN ENERGY
WOOD GROUP PSN

19

## THANK-YOU FOR LISTENING

## ANY QUESTIONS?

*Excellence is now operational* PETROTECHNICS

Find Out More www.Petrotechnics.com/Solutions | #Petrotechnics | #Proscient

Contact Me: Andrew.Bartlett@petrotec .com

TASNEE التصنيع

# Safety Performance vs. Behavioral


**by: Mr. Hamad A. Al-Shbrain,**
TASNEE – Petrochemicals BU ,
TQM – Manager


**5th of December 2016**

TASNEE التصنيع

- **Safety Performance**
- **Organization Model**
- **Measuring the Culture**
- **Define Want and Need**
- **Define the benefits of Behavioral Based Safety**
- **Success Factors**
- **Conclusions**

# Safety Performance

TASNEE التصنيع

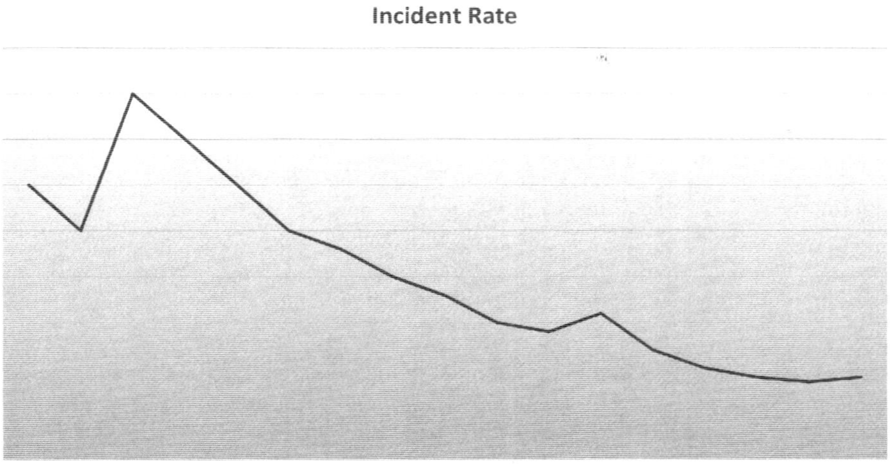


*Most of the Organization focus in reducing the Incident Rate IR and sustain low IR.*

3

# Safety Performance

TASNEE التصنيع

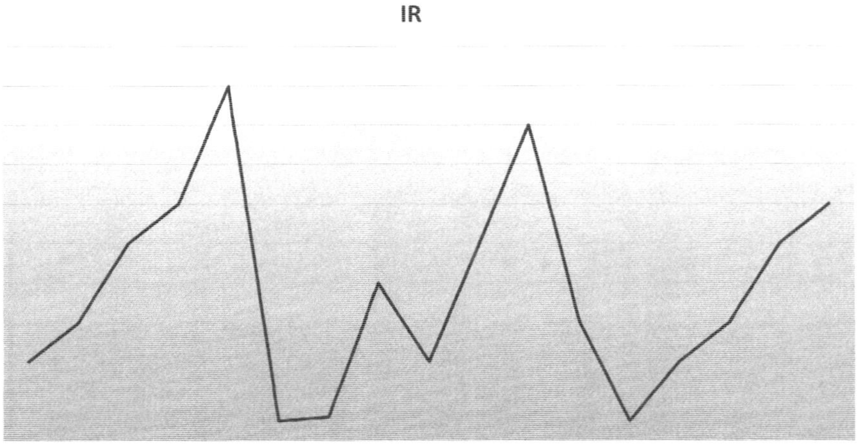


*BUT THE REALITY IS DIFFERENT*

4

## Organization Model

TASNEE التصنيع

*Culture is the glue that integrates people, environment & their behavior into effective management systems.*

5

## Can we measure the Culture ?

TASNEE التصنيع

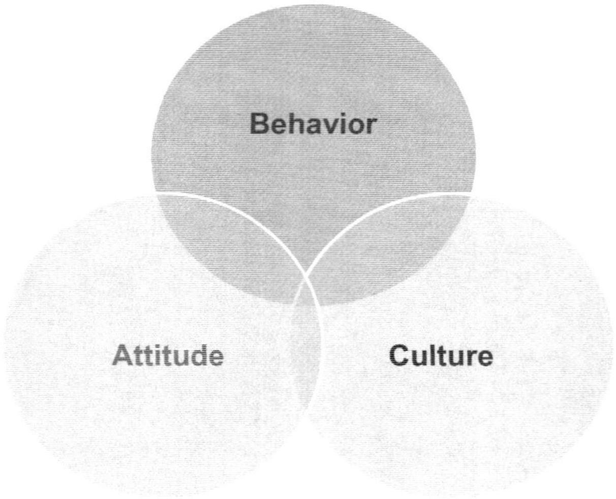


*To change the Culture and Attitude of our work force is not easy and that is why we will focus on critical Behaviors by means of observations (Behavior is an observable act).*

6

## Process safety pyramid(API-754)

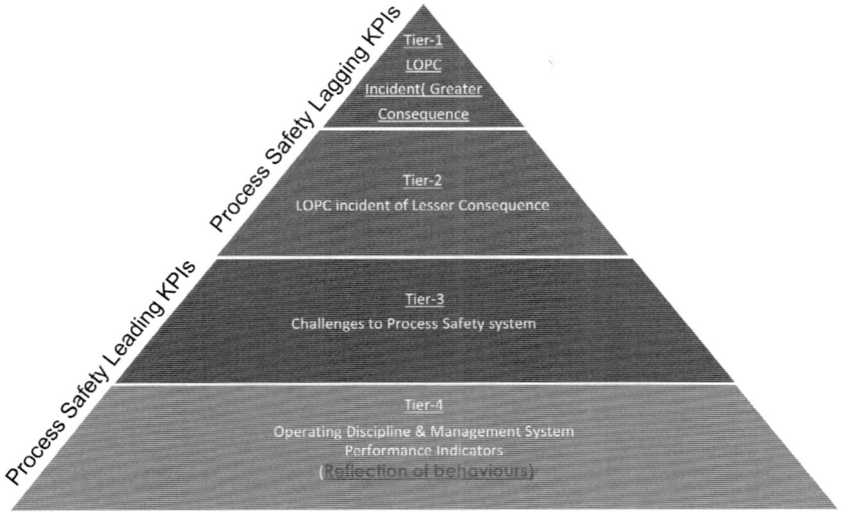


## Incident Triangle

TASNEE التصنيع

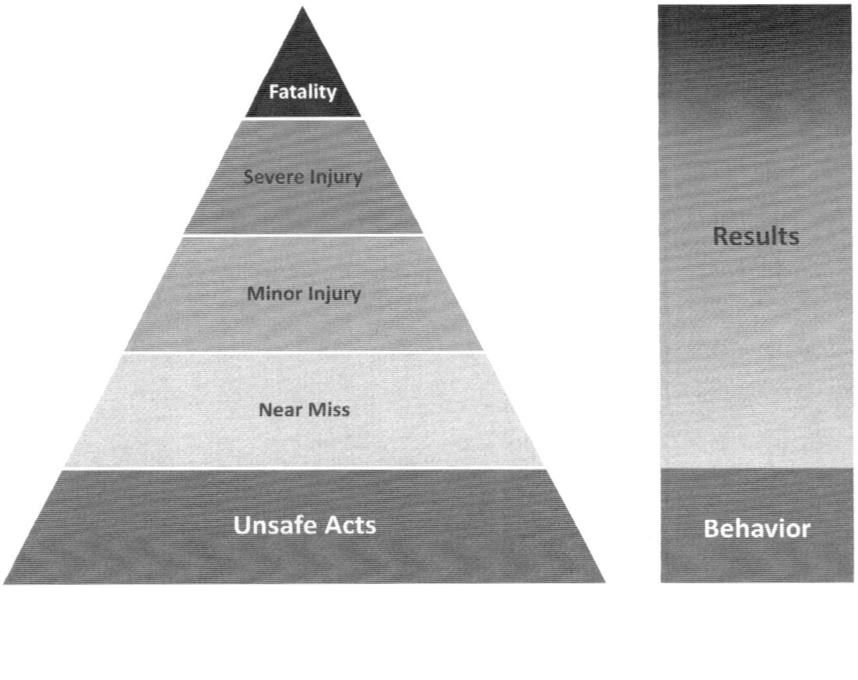


8

## Iceberg

TASNEE التصنيع

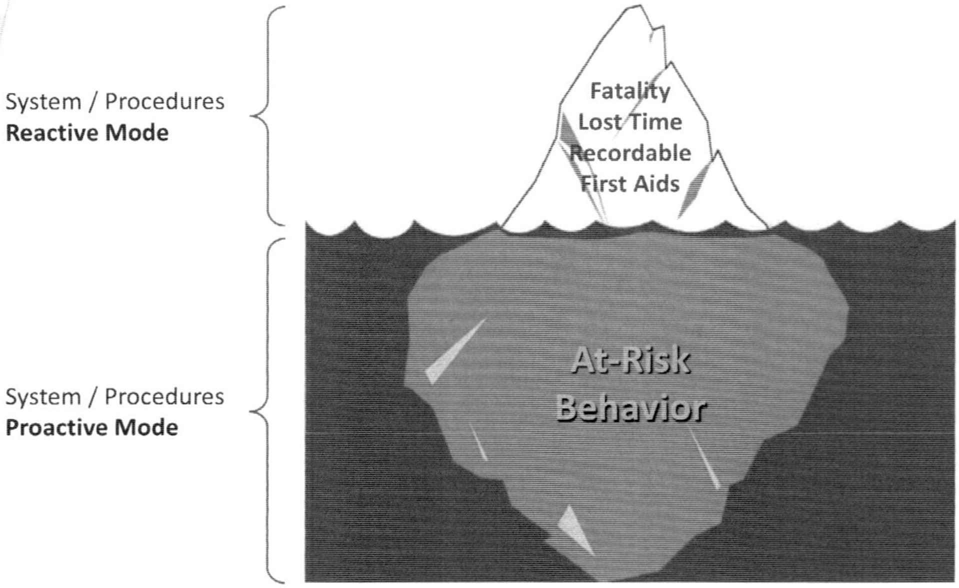


9

## What We Want ?

TASNEE التصنيع

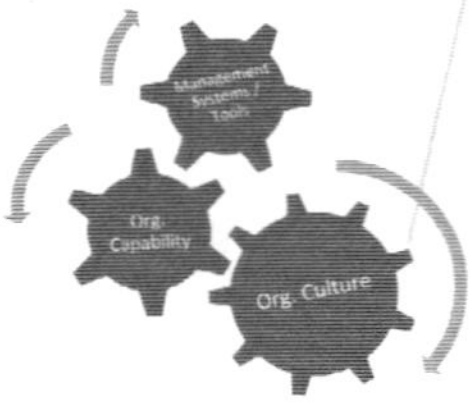


- A process to deal with the behaviors that are observable and can be seen as a supporting initiative to all other existing safety activities (Proactive approach).
- A safety culture forms part of the values, attitudes, goals and proficiency of an organization’s health and safety program.
- A focus on the three contributing factors to any incident ; people, the environment and their behavior.

10

## What We Need ?

TASNEE التصنيع

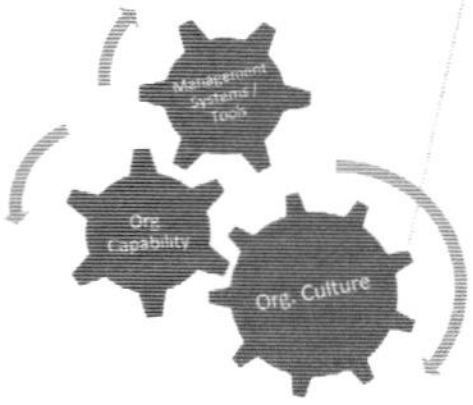


***We need process to;***

- Identify and correct unsafe worker behaviours.
- Recognize good worker behaviours.
- Create a pro-active safety behavior , and
- Educate the organization that loss prevention is a core value and not just a priority.

11

## What are the benefits of Behavioral Based Safety?

TASNEE التصنيع

***Creating a Culture and Attitude in our work force*** *of doing the right things, in the right way, at the right time, every time!*

***This will then contribute in;***

- Reducing the number of Incidents in our workplace
- Thinking more proactively
- Empowering our employees
- Maintaining a good reputation

*That will lead us direct or indirect to the reduction of cost within the Company.*

12

## *Proposed - Process Flow Chart*

TASNEE التصنيع

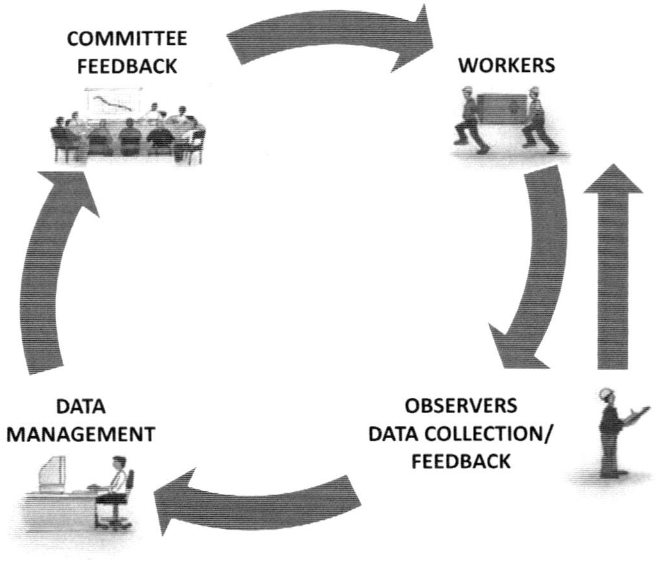


13

## *Deployment Project Map*

TASNEE التصنيع

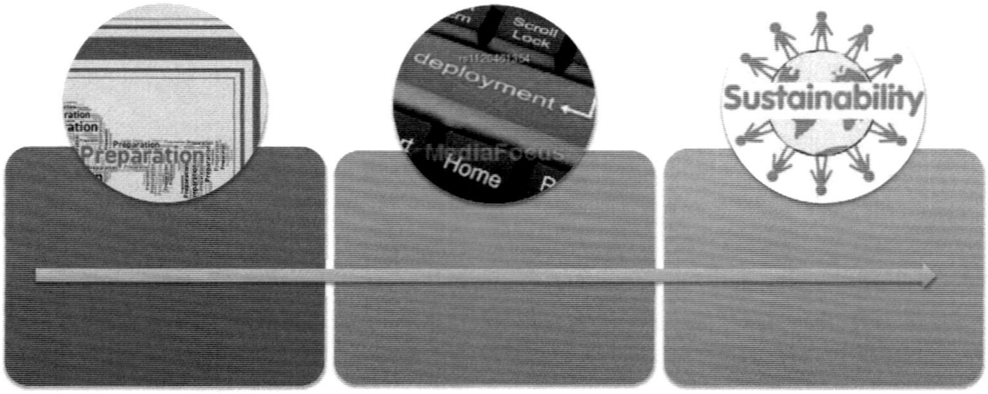


14

## Success Factors

TASNEE التصنيع

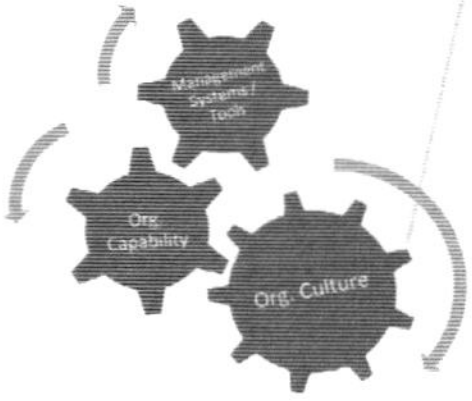


15

## Management Commitment

TASNEE التصنيع

- No names of those being observed will be recorded during BBS observations.
- Information recorded on the BBS observation forms will not be used as a basis for disciplinary action. The information will be only used to determine areas of safety concerns and/or improvement.
- Managers, Superintendents and Supervisors, will not be evaluated on the basis of observation data from their areas (i.e., percentage of employees doing observations). BBS Participation will not be included as a leadership KPI.
- Employees will not be evaluated on their annual performance review on the basis of their participation in BBS.
- Supervisors, Managers and Superintendents will be evaluated on their support of BBS as part of leadership support behavior.
- All employees are encouraged to volunteer in the process.

16

## Conclusions

TASNEE التصنيع

The major component s to consistently strive and achieve an excellent performance are :

**Engagement**

**Engagement** in process at **all levels** from senior executives through front-line workers are required.

**Anticipation**

The organization must **encourage the reporting** of any deviations and also report the good behavior(share good practice).

**Focus**

Focusing in the unsafe act **identification** , analysis and implementation process.

**Review**

**Creating a culture** in which people are aware of, and act in ways to minimize the potentially problematic areas and to **influence** those.

**Performance**

Consistently performing , improving , controlling and **monitoring** as intended.

17

***Message , defiantly we will improve what we measure***

## Thank You ..

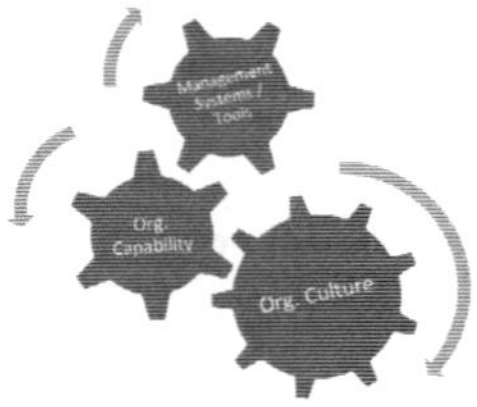

# Concluding Remarks

## Shakeel Kadri
## Executive Director, CCPS

"The Global Community Committed to
Process Safety"

MAIN SPONSOR

PLATINUM SPONSOR

GOLD SPONSORS

SILVER SPONSOR

BRONZE SPONSORS

## 3rd Global Summit Metrics

- Four Keynotes delivered
- Two panel – Leadership and Academia] held
- Leadership Forum held – 17 Leaders participated
- 36 Technical papers delivered in various sessions
  - Process Safety Leadership, Committed Culture, Asset Integrity Management, Metrics and Responsible Collaboration, Managing Risk, Human Factors, Enhanced Lessons Learned
- Total participants -- 487

*The Global Community Committed to Process Safety*

3RD CCPS GLOBAL SUMMIT ON PROCESS SAFETY

## SABIC CEO's Inspiring Message

- Key points
  - Process Safety importance
  - Organizational Culture,
  - Leadership and
  - Human Resources.
  - He stressed the participants to commit themselves to the principles of Process Safety and it is fundamental to survival of our businesses.

*The Global Community Committed to Process Safety*

3RD CCPS GLOBAL SUMMIT ON PROCESS SAFETY

## Leadership Panel Summary

- Leaders must drive process safety and lead by example
  - But leadership itself is not enough
- Strong Process Safety Management program
  - Always look to simplify to aid understanding by non-engineers
- Engagement across all levels of the organization
  - Not just compliance, but true understanding
  - More proactive, less reactive
- Developing new talent, the process safety leaders of the future
- Ageing assets – a recurring theme

*The Global Community Committed to Process Safety*

3RD CCPS GLOBAL SUMMIT ON PROCESS SAFETY

## Academia Panel Summary

- Education of Process Safety in Engineering is a critical need
- Region has some effort started and 1st batch will come out in 2017
- AICHE is working on the global solution
- SABIC has initiated a program to educate engineers in process safety
- Education of process safety in Japan is critical
- Process safety should be taught in all engineering disciplines

*The Global Community Committed to Process Safety*

3RD CCPS GLOBAL SUMMIT ON PROCESS SAFETY

4th

ON

**Venue: Okayama Convention Center, Japan**
**12-13 September 2017**

## Sponsored and supported by

More than 30 companies in Japan

## Organized by

**CENTER FOR CHEMICAL PROCESS SAFETY**

CCPS An AIChE Technology Alliance
Center for Chemical Process Safety

**JAPAN SOCIETY OF SAFETY ENGINEERS (JSSE)**

**CENTER FOR SAFE AND DISASTER-RESISTANT SOCIETY, OKAYAMA UNIVERSITY**

# Middle East Process Safety Conference (MEPSC) 2017 co-located with MEPEC

**9-10 OCTOBER 2017 IN BAHRAIN**

# MEPSC – Review of 2015

CCPS
MIDDLE EAST
PROCESS SAFETY
CONFERENCE

**MEPSC 2015 IN NUMBERS**

- **746** Participants
- **60** Hours of Knowledge-sharing over 2 Days
- **54** renowned names in the Technical Program
- **22** Sponsors and Exhibitors

**MEPSC 2015 HIGHLIGHTS**

- Opening Session with CEOs of ADCO, GASCO & Borouge
- Student Program
- Pre-conference workshops

**Participants: 746 unique participants**

Participants' Profile

- Process Safety Engineers and Leaders
- Manufacturing and Operations Leaders
- Senior Leaders whose role involves operations
- Compliance Officers & Inspectors
- Process Industry Professionals
- Managers, Scientists, and Educators

## MEPSC 2017

Days 1 & 2 –
Single programming track that covers key process safety subjects

Keynotes and panel discussions at the start of each day.

Possible programming topics

- **Process safety metrics**
- Inherently safer design
- Incident management and business continuity
- **Process safety leadership & culture**
- **Lessons learned and best practices**
- Management of change
- Mechanical integrity
- Hazard identification & risk analysis
- Pre-startup safety review

## MEPSC 2017

Day 3 – 2-3 workshops from which the attendees can choose from to get a deeper understanding of select process safety topics.

Possible workshop topics

- Process safety leadership
- Risk-based process safety
- Recognizing catastrophic incident warning signs

Other topics of interest?

# 3rd Global Summit

THANKS!

"The Global Community Committed to Process Safety"

# AUTHOR INDEX